MECHANICAL PROPERTIES OF CERAMICS

MECHANICAL PROPERTIES OF CERAMICS

JOHN B. WACHTMAN

Rutgers University

A Wiley-Interscience Publication

JOHN WILEY & SONS, INC.

New York / Chichester / Brisbane / Toronto / Singapore

This text is printed on acid-free paper.

Library of Congress Cataloging in Publication Data:

Wachtman, John B., 1928–
Mechanical properties of ceramics / by John B. Wachtman.
p. cm.
"A Wiley-Interscience publication."
Includes bibliographic references and index.
ISBN 0-471-13316-7 (cloth : acid-free paper)
1. Ceramic materials—Mechanical properties. I. Title.
TA455.C43W38 1996
620.1′40423—dc20 95-36612
CIP

Printed in the United States of America

10 9 8 7 6 5 4 3 2 1

CONTENTS

PREFACE

The mechanical properties of ceramics are currently an active field of research and development. Many property improvements and much fundamental understanding have been achieved quite recently. Developments in theory in the last decade or so now permit a reasonably unified treatment of the role of microstructure in determining the effect of mechanical properties of ceramics even though some open questions remain. This book attempts to present a systematic introductory treatment of the current state of fundamental understanding supplemented by the empirical facts.

This book is intended for use as a text. On one level, it can be used by undergraduate students (juniors and seniors) in ceramics, materials science, or related fields and assumes no prior study of mechanics. It assumes only a general background including typical undergraduate courses in physics and calculus. It is also intended for this book to be useful to graduate students and it thus presents more advanced material as well. It is hoped that by selective use the book will be useful to both the beginning student of mechanical properties and to the somewhat more advanced student. The first three chapters offer an introduction to readers with minimal background in mechanical properties. The more advanced reader can begin with Chapter 4 or read selectively in the later chapters on topics to suit his or her interest. Chapters 4 through 20 present the basic science of the mechanical properties of ceramics. These chapters contain some data on specific ceramics to illustrate the general treatment. However, the major portion of the data on specific families of ceramics is given later in Chapter 21 through 27.

This book is also intended to be useful as a reference for materials scientists and engineers who must deal with the mechanical behavior of ceramics but whose primary interests are in other aspects such as processing or electrical properties. Good mechanical properties of ceramics depend strongly on details of microstructure. The treatment of microstructure–mechanical property relationships given here is intended to be useful to developers of ceramic materials and to people involved in quality control during production. Summaries of test procedures including equations for stress, strain, and fracture mechanics parameters are provided.

Another purpose of this book is to provide information for users of ceramics such as engineers in disciplines outside materials who are designing devices or structures. Chapters on each of the major types of structural ceramics provide a guide to achievable mechanical properties. The reader interested in a specific material (e.g., alumina) may go directly to the chapter on that material for a brief summary of its primary mechanical properties. Design for specified levels of safety and for minimal life under service conditions is treated.

ACKNOWLEDGMENTS

The writer appreciates many helpful conversations with his colleagues: W.R. Cannon, M.J. Matthewson, and D.E. Niesz.

Helpful conversations and correspondance with the following people are gratefully acknowledged, but no endorsement of the treatments given here is implied: O.L. Anderson (University of California at Los Angeles); P.F. Becher (Oak Ridge National Laboratory); R.C. Bradt (University of Alabama at Tuscaloosa); B. Budiansky (Harvard University); R.M. Cannon (University of California at Berkeley); D.R. Clarke (University of California at Santa Barbara); R.F. Cook (IBM—Yorktown Heights); S.F. Duffy (NASA—Lewis Research Center); A.G. Evans (Harvard University); K.T. Faber (Northwestern University); S.W. Freiman (National Institutes of Standards and Technology); D.J. Green (Pennsylvania State University); S.T. Gulati (Corning Inc.); R. Hannink (CSIRO Australia); D.P.H. Hasselman (Virginia Polytechnic Institute and State University); A.H. Heuer (Case-Western Reserve University); K. Jakus (University of Massachusetts at Amherst); B.L.P. Keyes (Oak Ridge National Laboratory); C.R. Kurkjian (AT&T Bell Laboratories); T.G. Langdon (University of Southern California); D.C. Larsen (Corning Inc.); B.R. Lawn (National Institutes of Standards and Technology); D.B. Marshall (Rockwell International Science Center); J.J. Mecholsky, Jr. (University of Florida); K.M. Prewo (United Technologies Research Center); G.D. Quinn (National Institutes of Standards and Technology); R.W. Rice (Consultant, Alexandria, Virginia); J.S. Ritter (University of Massachusetts at Amherst); S.J. Schneider (National Institutes of Standards and Technology); D.K. Shetty (University of Utah); R.J. Stokes (University of Minnesota and University of Warwick); M.V. Swain (Sydney University); S.M. Wiederhorn (National Institutes of Standards and Technology).

I also wish to thank the authors and publishers who have graciously permitted the reproduction of tables and figures from their publications, including: The American Ceramic Society, The American Institute of Physics, Annual Reviews, Inc., Ceramics International, Chapman and Hall, Plenum Press, and Roy W. Rice. A detailed reference is given with each table or figure. Special thanks are extended to the American Ceramic Society from whose publications many of the figures were taken.

JOHN B. WACHTMAN

New Brunswick, New Jersey

NOTATION

The principal symbols used are listed here. Some symbols with subscripts are listed for clarity, other subscripts are thought to be evident from context and are not listed. Generally subscripts *m*, *f*, and *c* refer to matrix, fiber, and composite, but subscript *c* also denotes a critical stress value in inelastic behavior. A few symbols that are used only in one place and are defined there are also omitted from this list.

A	area, coefficient in power law for crack growth rate
A, B	constants in interatomic potential; factors in modified indentation technique; functions in torsional resonance frequency equation for rectangular bar
A, B, C, A', B', C'	functions in Voight and Reuss bounds
B	bulk modulus
b	constant in temperature dependence of elastic modulus; constant in porosity dependence
b, c	semiaxes in elliptical crack
c	elastic stiffness constant; crack radius; crack half-length for internal or through crack, full length for edge crack
c_m	maximum stable crack length in Lawn's formulation of internal stress combined with indent stress
D	distance from support to load in four-point bending, diameter of cylinder, diameter of grain
d	depth; spacing of obstacles in Lange's crack bowing formulation

E	Young's modulus
F	force
$\boldsymbol{F}$	force vector
f_f, f_t	flexural and torsional resonance frequencies
G	mechanical energy release rate
I	moment of inertia
J	joule; stress invariant; J-integral for mechanical energy release rate
K	stress intensity factor for mode I loading
K_C^∞	critical stress intensity factor for a long crack
$K_{\rm I}, K_{\rm II}, K_{\rm III}$	stress intensity factors for opening, shearing, and tearing modes
$K_{\rm I\mathit{C}}$	critical stress intensity factor for mode I failure
k	dimensionless integral in Weibull calculations with varying stress in body
L	length, distance between supports in bending
M	bending moment
m	Weibull m parameter in general
m_f	exponent in Paris fatigue law; not related to Weibull parameter m
$m_{t=0}$	Weibull m parameter determined from strength without subcritical crack growth
m_σ	Weibull m parameter determined from time to failure under constant stress with subcritical crack growth
$m_{\dot{\varepsilon}}$	Weibull m parameter determined from strength measured at constant strain rate with subcritical crack growth
m_R	Weibull m parameter with no subcritical crack growth but with R-curve behavior
m_s	exponent of true strain rate in empirical expression for true stress during steady-state creep in superplasticity; not related to Weibull parameter m
N	number of specimens
n	exponent of stress intensity factor in power law for crack growth rate
$n_{\rm cr}$	exponent of stress in power law for creep rate
n_K	exponent in R-curve equation
n'	exponent in power law for crack growth rate measured from indentation cracks
n_i	components of unit normal vector
n_s	exponent of true strain in empirical expression for true stress during plastic behavior
P	load
p	pressure; equivalent stress on crack surface from bridging, transformation, etc.; stress intensity factor expo-

p	nent in velocity growth law for crack growth under creep conditions
$p(\sigma)$	probability of failure in interval σ to $\sigma + d\sigma$
P	load in general; indent load
P^*	internal stress parameter in Lawn formulation expressed as an equivalent indentation load
$P_f(\sigma)$	cumulative probability of failure for all stresses less than or equal to σ
$P_s(\sigma) = 1 - P_f(\sigma)$	cumulative probability of survival
Q	charge; heat
Q_1, Q_2	constants in interatomic potential function
R	radius of curvature, resistance to crack growth (= critical strain energy release rate)
r	radius of crack, radius of fiber
S	stress
s	elastic compliance constant
s^2	variance
sf	safety factor
T	absolute temperature; toughness (= critical stress intensity factor)
T_1, T_1'	correction factor in flexural resonance equation for rectangular bar, prime for cylindrical rod
t	thickness, time
t_f	time to failure
$t_{f,t=0}$	time to failure without subcritical crack growth
$t_{f,\sigma}$	time to failure under constant stress with subcritical crack growth
$t_{f,\dot{\varepsilon}}$	time to failure under constant strain rate with subcritical crack growth
t_s	estimated minimum failure time
$\boldsymbol{u}$	displacement vector
U	strain energy per unit area in J-integral
U	potential energy
U_e	elastic stored energy
U_s	surface energy of crack
u	crack opening displacement (total opening is $2u$)
u, v, w	displacements
V_f, V_m	volume fraction of fibers and matrix in composite
Vol	volume
v	velocity of crack growth under stress
v_t, v_L	velocity of transverse and longitudinal elastic waves
w	width of test specimen, half-width of process zone around crack
W	work done on specimen by system
x	interatomic separation

x, y, z	coordinates
Y	dimensionless factor in stress intensity equation
α	numerical factor in hardness equation
β	angle of shear
χ	numerical factor in stress intensity as function of indent crack length
ε	strain, deviation from mean in statistical calculations
ε^T	strain accompanying transformation
δ	deflection of center of bar in bending
γ	surface energy per unit area
γ_{gb}	grain boundary surface energy per unit area
η	numerical factor in stress intensity factor as function of inert strength
κ	factor in stress intensity equations
λ	Lame elastic constant
μ	shear modulus; dimensionless constant in Lawn's expression for effective local stress intensity factor from internal stress
ν	Poisson's ratio
θ	angle
ρ	radius of curvature at crack tip
σ_0	constant in Orowan force law; Weibull parameter
σ	stress; sometimes strength
$\bar{\sigma}$	mean (average) strength
σ_c	critical value of stress for inelastic behavior as in transformation toughening
σ_f	fracture stress (i.e., strength)
σ_{max}	maximum stress in specimen
σ_u	minimum stress for failure in three-parameter Weibull distribution
σ_y	yield stress
τ	shear stress; frictional shear stress in fiber pullout
ψ	numerical factor in stress intensity factor in Lawn's formalism (equivalent to Y)
ζ	numerical factor in stress intensity factor as function of indentation load

INTRODUCTION

The mechanical behavior of a ceramic part is clearly important when the part is used for the primary purpose of carrying a load. Mechanical behavior of ceramics is also of interest in many cases when the primary purpose of the ceramic is to perform some other function, such as an electrical, magnetic, optical, or chemical function. This is the case because the incidental forces acting on the ceramic in such other applications may be sufficient to cause failure unless the ceramic is designed to resist their effect. Ceramics are very sensitive to the presence of flaws so that failure may occur unexpectedly. The characteristics of crack propagation in most ceramics are such that a failure is likely to be instantaneous and total. Thus the failure is often not only unexpected but catastrophic. Understanding the nature of the mechanical behavior of ceramics and designing for it in both the design of the microstructure of the ceramic and the design of the device using the ceramic can do much to alleviate this problem. Any user of ceramics thus needs to be aware of the characteristics of their mechanical behavior and to take it into account.

Practical coverage of the mechanical properties of ceramics in a book of reasonable length requires combining judicious summaries of several topics, each of which could be treated at much greater length. The concepts and basic mathematical treatment of continuum mechanics are a necessary foundation. Recognition of the critical role that defects play and quantitative descriptions of their behavior under stress are also essential. With these topics in hand one can turn to the real theme of mechanical properties of ceramics from a materials scientist's point of view—development of a fundamental science of mechanical properties and the use of this theory to understand the effect of microstructure on these properties. Much of the progress in ceramics in recent decades comes from the development of microstructures that enhance their properties. This is especially true of mechanical properties for which important improvements have been achieved through use of special microstructures to achieve transformation toughening, whisker and fiber reinforcement, and in situ reinforcement by elongated grain growth.

Numerous authors have presented theoretical models to explain various aspects of mechanical behavior and there is considerable overlap in some of the treatments. An attempt is made to present a theoretical treatment here with two goals in mind: first, to present a treatment sufficiently complete to contain the significant basic science and with sufficient sophistication to quantitatively describe the main aspects of behavior at least approximately; and second, to give priority to the original workers. The great number of later theoretical treatments and the complexity of some of these make a complete account of all theoretical efforts impractical and, in any event, progress is still continuing. An attempt is made to provide references to typical current papers to lead the reader to the latest work at the time of writing.

The approach taken in the present book is to begin with room-temperature mechanical behavior and go on later to high-temperature behavior. Accordingly, elastic deformation and brittle failure are treated before plastic deformation. Linear elastic fracture mechanics is summarized to provide a basis for the subsequent treatment of brittle failure under various conditions. Simple catastrophic brittle failure is considered and the statistical treatment of the highly variable strength of most ceramics is covered. Next, the phenomenon of subcritical crack propagation associated with stress-assisted environmental effects is taken up. Attention then shifts to stable crack propagation in materials having a toughness that increases with the initial stage of crack propagation (the R-curve effect). With these fundamentals in hand microstructural effects are introduced through a survey of toughening mechanisms in ceramics. Two of these mechanisms—inelastic deformation in the process zone near a crack and partial bridging near the crack tip—are then treated in more detail in subsequent chapters. Fiber-reinforced ceramic-matrix composites are treated both for their own importance and because the treatment casts light on some of the toughening mechanisms of polycrystalline ceramics.

Attention is then shifted to plastic deformation and high-temperature behavior of ceramics. The fundamentals of dislocation behavior in general and particular dislocation behavior in ceramics are reviewed. High-temperature creep and stress rupture including safe-life design under creep conditions are then treated.

This book concludes with chapters giving tables and figures showing representative data on mechanical properties of important types of ceramic materials. Glass is taken up first. The mechanical behavior of polycrystalline ceramics in general and design considerations for their use are considered next. The mechanical behavior of each of four major families of advanced polycrystalline structural ceramics (alumina, silicon nitride, silicon carbide, and zirconia) is then reviewed in the light of the fundamental science previously given.

MECHANICAL
PROPERTIES OF
CERAMICS

CHAPTER 1

STRESS AND STRAIN

1.1 INTRODUCTION

Quantitative treatment of the mechanical behavior of ceramics (or any solid) requires the mathematical description of stress and strain. Each of these quantities is a second rank tensor. The full three-dimensional treatment of stress and strain requires a somewhat cumbersome notation that will be taken up later. It is convenient to begin with a simple treatment of tensile stress and tensile strain and discuss types of mechanical behavior in terms of these. Subsequently, shear stress and strain can be introduced using a simple two-dimensional treatment. This approach permits qualitative ideas about the mechanical behavior of ceramics to be explained without obscuring them within the complexity of a full three-dimensional treatment of all the components of stress and strain. The scheme followed here is to introduce stress and strain in terms of an easily visualized picture of deforming a bar.

Deformation of a material under a constant load may occur very quickly and come to a stop or it may continue slowly for a long time. It is not necessary to introduce the complexity of time-dependent deformation in order to consider stress, strain, and many of the basic ideas of deformation behavior. Accordingly, consideration of time-dependent deformation is reserved for later treatment. That is, deformation under load will be taken as instantaneous until otherwise specified.

A simple visualization of stress and strain can be done for a linearly elastic, isotropic material. That is, the material has the same mechanical properties in all directions (i.e., the material is isotropic), deformation under load is fully recovered when the load is removed (i.e., the material is elastic), and the amount of recoverable deformation is proportional to the load (i.e., the material is linearly elastic). The restrictions of isotropic deformation and linear elasticity will be assumed only for portions of the treatment. It provides insight to allow the deformation behavior to be more general in these respects and to introduce these restrictions as they are needed to simplify the treatment.

Consider a bar (as shown in the top portion of Figure 1.1) of length L, width w, and thickness h (the dimension into the paper in Figure 1.1) with a force F acting parallel to the length on each end (i.e., uniaxial loading). The bar will generally deform under the action of the force. Accordingly, the bar is taken to deform instantaneously an amount δL_1 in the direction of the force F and to deform an amount δw_1 in the direction of the width, w, as shown in the middle portion of Figure 1.1. In general, the bar will also deform an amount δh in the third dimension.

Consider the situation if two identical bars are subject to the same total load F as shown in the lower portion of Figure 1.1. A tie is shown at each end of the two bars to indicate sharing of the load equally between them. Each bar must support only $F/2$ so that the deflections will be smaller in magnitude: that is, $\delta L_2 < \delta L_1$, $|\delta w_2| < |\delta w_1|$, and $|\delta h_2| < |\delta h_1|$. For a tensile force F as shown, the deformation δL is an extension, but the deformation at right angles to F is usually a contraction. Accordingly, δw and δh are generally negative for an applied tensile force. Absolute value bars are therefore used here to indicate that it is the magnitude of these deformations at right angles to the applied uniaxial loading that is less than the deformation in the direction of loading.

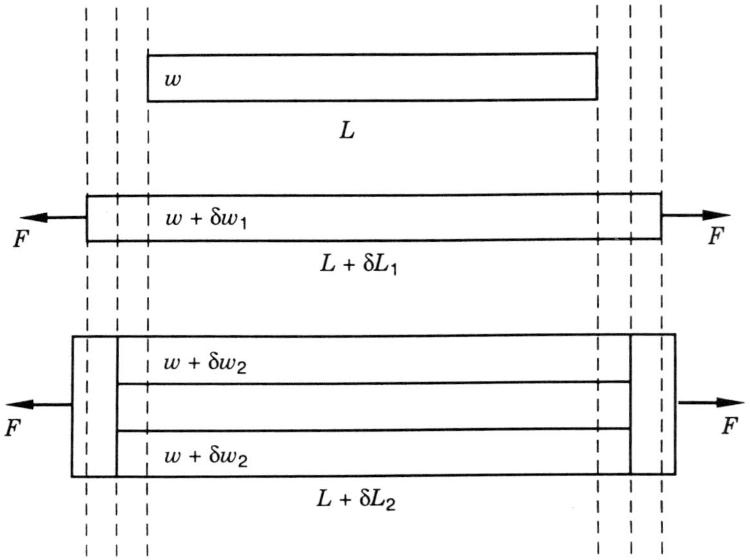

Figure 1.1 Bar under tensile stress.

For certain directions in certain single crystals, the deformation δw or δh can actually be positive for an applied tensile stress. These are exceptional and rather rare cases. It is convenient to think of δw and δh as negative and to define Poisson's ratio (the ratio of transverse to lateral strain) to be a positive number for this situation, as is done later.

The deformation in the lower portion of Figure 1.1 is less than that in the upper portion because the load is carried by a larger area. This suggests that the variable controlling the deformation behavior is not total load, but load per unit area. Accordingly, the tensile stress σ is defined as the force F divided by the area A, so that

$$\sigma = F/A = F/wh \tag{1.1}$$

A similar discussion leads to defining strain as a ratio of change in length to initial length. Figure 1.2 shows change in length of δL_1 and $2\delta L_1$ in two bars of length L and $2L$, respectively. Strain is accordingly defined by

$$\varepsilon = \delta L/L \tag{1.2}$$

This is actually the definition of engineering strain in contrast with the quantity unfortunately termed "true strain," which will be discussed in Chapter 17. The term strain will refer to engineering strain in this book.

1.2 TENSOR NOTATION FOR STRESS

Stress (and strain) must be referred to a set of axes, which we take as the cartesian axes xyz. Stress is generally written (Sines, 1969)

$$\sigma_{\text{on } x \text{ plane in } y \text{ direction}} = \sigma_{xy} = \frac{\textit{force in y direction}}{\textit{area of plane perpendicular to x}} \tag{1.3}$$

Here "x plane" means "plane perpendicular to the x axis." In terms of indices denoting xyz by x_1, x_2, x_3, the above stress component is written

$$\sigma_{\text{on } x_i \text{ plane in } x_j \text{ direction}} = \sigma_{ij} \tag{1.4}$$

There are thus nine possible stress components in three dimensions. However, it can be shown that $\sigma_{ij} = \sigma_{ji}$ for $i \neq j$, so that there are only six independent components. The components with $i = j$ are termed tensile stresses (with compressive stresses considered a negative tensile stress). The components with $i \neq j$ are termed shear stresses. (Discussion of conventions for the signs assigned to shear stresses is deferred until later.) The shear stresses are sometimes written τ_{ij} instead of σ_{ij} to emphasize their nature as shear stresses.

Written as a matrix the stresses in three dimensions are

$$\sigma = \begin{pmatrix} \sigma_{11} & \sigma_{12} & \sigma_{13} \\ \sigma_{21} & \sigma_{22} & \sigma_{23} \\ \sigma_{31} & \sigma_{32} & \sigma_{33} \end{pmatrix} = \begin{pmatrix} \sigma_{xx} & \sigma_{xy} & \sigma_{xz} \\ \sigma_{yx} & \sigma_{yy} & \sigma_{yz} \\ \sigma_{zx} & \sigma_{zy} & \sigma_{zz} \end{pmatrix} \tag{1.5}$$

with

$$\sigma_{12} = \sigma_{21}\,,\ \sigma_{13} = \sigma_{31}\,,\ \sigma_{12} = \sigma_{21} \tag{1.6}$$

It is easier to visualize matters in two dimensions and attention will be restricted accordingly in the next few pages. For two dimensions (see Figure 1.2) there are a total of four stress components of which only three are independent. Written as the matrix, the stresses in two dimensions are

$$\sigma = \begin{pmatrix} \sigma_{11} & \sigma_{12} \\ \sigma_{21} & \sigma_{22} \end{pmatrix} \tag{1.7}$$

with

$$\sigma_{12} = \sigma_{21} \tag{1.8}$$

The stress components in two dimensions are displayed in Figure 1.2, which shows a square under two force systems, both of which are balanced so that the square is under zero net force and zero net torque. In the first force system the forces are normal to the faces and produce tensile stresses. The force F_x on the right face (the $+x$ face) is balanced by the force $-F_x$ on the left face (the $-x$ face). Similar results hold for the upper and lower faces. In the second force

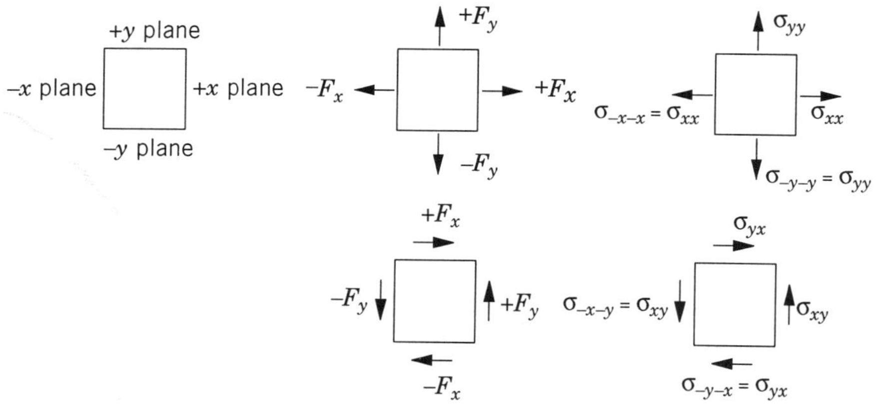

Figure 1.2 Definition of stress in two dimensions.

system the forces are along the faces and produce shear stresses. For the square to be under no net torque the forces must be such that $\sigma_{xy} = \sigma_{yx}$.

Some of the units commonly used for stress and energy are summarized in Table 1.1.

1.3 STRESS IN A ROTATED COORDINATE SYSTEM

We now examine how a chosen stress is expressed in a rotated coordinate system. Considering the bar of Figure 1.1 and taking the x axis along the long axis of the bar, the stress caused by a load in this direction acting on the cross-section area A of the bar is

$$\sigma = \frac{F_x}{A_x} = \sigma_{x_1 \text{ plane}, x_1 \text{ direction}} = \sigma_{11} = \sigma_{xx} \tag{1.9}$$

For this bar and load the other stress components in the xy system are

$$\sigma_{xy} = 0 \text{ and } \sigma_{yy} = 0 \tag{1.10}$$

The question now arises as to how the same force and plane should be described in a rotated coordinate system. To answer this question consider a rotation through θ of the xy axes to become the $x'y'$ axes as shown in Figure 1.3. Now

$$\sigma'_{xx} = \frac{F'_x}{A'_x} \tag{1.11}$$

From the figure

$$F'_x = F_x \cos\theta \tag{1.12}$$

and

$$A'_x = \frac{A_x}{\cos\theta} \tag{1.13}$$

TABLE 1.1 Units of Stress, Energy, and Energy per Unit Area

Quantity	International Unit	Value in Other Commonly Used Units
Stress	N/m^2 = Pa	10 dyn/cm^2 = 1.450×10^4 psi = 1.020×10^{-5} kg/cm^2
Energy	J	10^7 erg = 0.239 Cal (g)
Surface energy per unit area	J/m^2	1000 erg/cm^2

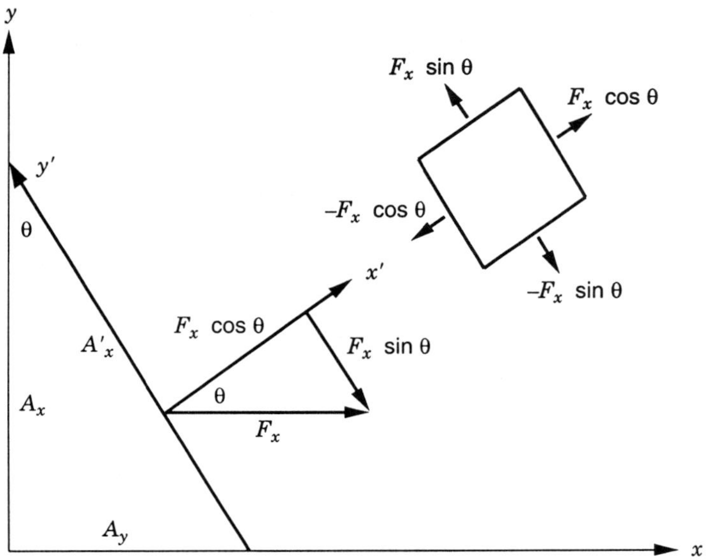

Figure 1.3 Stress in rotated coordinate system.

so that

$$\sigma'_{xx} = \frac{F_x}{A_x} \cos^2 \theta \tag{1.14}$$

Using the trigonometric identity

$$\cos^2 \theta = \frac{1 + \cos 2\theta}{2} \tag{1.15}$$

gives

$$\sigma'_{xx} = \sigma_{xx} \frac{(1 + \cos 2\theta)}{2} \tag{1.16}$$

For shear stresses a sign convention must be introduced. A shear stress is taken as positive if the force acts in a positive direction on a plane whose normal is in a positive direction or equivalently for a force acting in a negative direction on a plane whose normal is in a negative direction.

The stress σ_{xx} also produces components σ'_{xy} and σ'_{yy}

$$\sigma'_{xy} = \frac{F'_y}{A'_x} \tag{1.17}$$

Since

$$F'_y = F_x \sin \theta \tag{1.18}$$

the result is

$$\sigma'_{xy} = \frac{F_x}{A_x} \sin\theta \cos\theta \tag{1.19}$$

Using the trigonometric identity

$$\sin 2\theta = 2 \sin\theta \cos\theta \tag{1.20}$$

gives

$$\sigma'_{xy} = \sigma_{xx} \sin\theta \cos\theta = \frac{\sigma_{xx} \sin 2\theta}{2} \tag{1.21}$$

For the rotated coordinate system one must also consider the forces acting on the y plane. The area of this plane is

$$A'_y = \frac{A_x}{\sin\theta} \tag{1.22}$$

The resulting stress component is

$$\sigma'_{yy} = \frac{-F'_y}{A'_y} = \frac{F_x}{A_x} \sin^2\theta \tag{1.23}$$

Using the trigonometric identity

$$\sin^2\theta = \frac{1 - \cos 2\theta}{2} \tag{1.24}$$

gives

$$\sigma'_{yy} = \frac{\sigma_{xx}(1 - \cos 2\theta)}{2} \tag{1.25}$$

The remaining component of stress is

$$\sigma'_{yx} = \sigma'_{xy} = \frac{\sigma_{xx} \sin 2\theta}{2} \tag{1.26}$$

A general expression for the stresses on a plane of any orientation in three dimensions can be written in terms of the direction cosines of the normal to the plane and the direction of action of the stress (Sines, 1969). Consider the plane of Figure 1.4 with a plane having a normal **r** with direction cosines a_{ir} to the axes $xyz = x_1x_2x_3$. The stress normal to this plane is

$$\sigma_{rr}' = \sum_{ij} a_{ir}\, a_{jr}\, \sigma_{ij} \tag{1.27}$$

For a shear stress acting on this plane in the direction **s** with direction cosines a_{js} the expression is

$$\sigma_{rs}' = \sum_{ij} a_{ir}\, a_{js}\, \sigma_{ij} \tag{1.28}$$

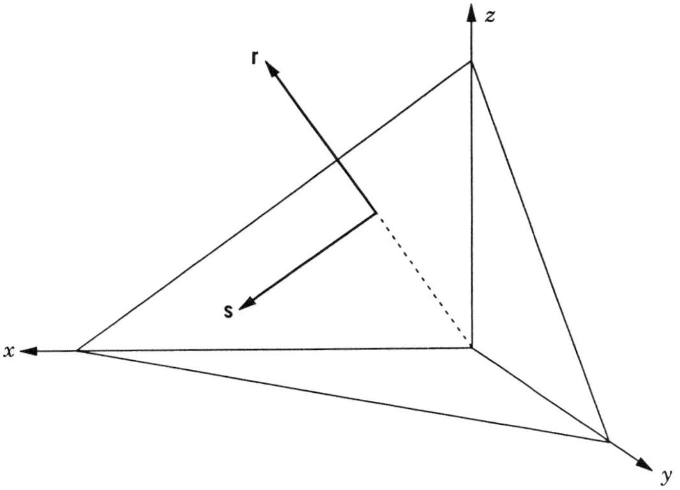

Figure 1.4 Plane with normal and direction in the plane.

1.4 BIAXIAL STRESS

An important stress condition results when force is applied in two noncollinear directions. Consider the bar again and add a perpendicular force F_y. The equations for the stress components in a rotated system can be worked out as above. However, the results can be obtained from above just by interchanging x and y and bearing in mind the sign convention for shear stress. The results for the stress components in the $x'y'$ system produced by σ_{yy} alone are

$$\sigma'_{xx} = \frac{\sigma_{yy}(1 - \cos 2\theta)}{2} \tag{1.29}$$

$$\sigma'_{yy} = \frac{\sigma_{yy}(1 + \cos 2\theta)}{2} \tag{1.30}$$

$$\sigma'_{xy} = -\frac{\sigma_{yy}\sin 2\theta}{2} \tag{1.31}$$

$$\sigma'_{yx} = -\frac{\sigma_{yy}\sin 2\theta}{2} \tag{1.32}$$

The total stress in the rotated coordinate system is obtained by adding the stresses caused by σ_{xx} and σ_{yy} to obtain

$$\sigma'_{xx} = \frac{(\sigma_{xx} + \sigma_{yy}) + (\sigma_{xx} - \sigma_{yy}) \cos 2\theta}{2} \tag{1.33}$$

$$\sigma'_{yy} = \frac{(\sigma_{xx} + \sigma_{yy}) - (\sigma_{xx} - \sigma_{yy}) \cos 2\theta}{2} \tag{1.34}$$

$$\sigma'_{xy} = \frac{(\sigma_{xx} - \sigma_{yy}) \sin 2\theta}{2} \tag{1.35}$$

$$\sigma'_{yx} = \frac{(\sigma_{xx} - \sigma_{yy}) \sin 2\theta}{2} \tag{1.36}$$

One might wonder if this result is completely general because the effect of a shear stress, σ_{xy}, in the original coordinate system has not been considered. However, it can be shown that for any stress system in two dimensions an orientation of the coordinates can be found for which there are only two tensile stresses, for which the shear stresses are zero. In three dimensions the corresponding result is also true; that is, there is a setting of the three axes for which the tensile stresses take their extreme values and the shear stresses are all zero. These axial directions are called principal directions, the corresponding planes are called principal planes, and the three corresponding stresses are called the principal stresses. Returning to two dimensions we conclude that the above result is general provided that the original coordinate system has been chosen to give only these two tensile stresses. Thus the above equations can be used to discuss the variation of all the stress components as the primed system of coordinates is rotated.

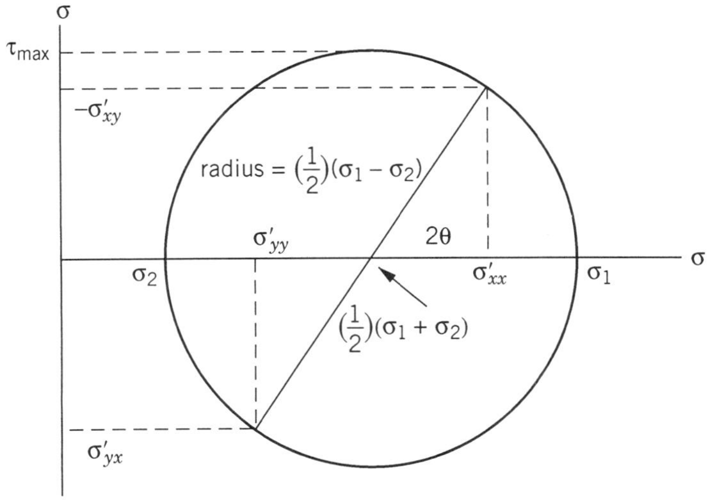

Figure 1.5 Mohr circle for biaxial tension.

This discussion is facilitated by the Mohr circle construction (Sines, 1969; Courtney, 1990). In this construction the abscissa is tensile stress σ, and the ordinate is shear stress, usually written as τ for the Mohr circle. A special convention applies to shear stresses in the Mohr circle construction: shear stresses causing a clockwise rotation are taken as positive and those causing a counterclockwise rotation are taken as negative (Sines, 1969). The above four equations describe a circle in this plot as shown in Figure 1.5. One can consider this circle in two ways: (1) for a situation in which the initial axes are principal axes and (2) for initial axes that are not principal axes.

In the first situation the initial axes are principal axes $\sigma_{xx} = \sigma_1, \sigma_{yy} = \sigma_2$, and $\sigma_{xy} = \sigma_{yx} = 0$, where the subscripts 1 and 2 are used to denote principal stresses. The center of the circle is on the σ axis at $(\sigma_1 + \sigma_2)/2$. The radius of the circle is $(\sigma_1 - \sigma_2)/2$. The circle intercepts the σ axis at σ_1 and σ_2. To obtain the stress in the primed system of coordinates rotated by an angle θ, a diameter of the circle rotated through 2θ is drawn. This diameter intersects the circle at two points. The intersection point from the end of the diameter above the σ axis has an abscissa value of σ'_{xx} and an ordinate value of $\tau'_{xy} = -\sigma'_{xy}$, where τ is conventionally used with the Mohr circle to designate shear stress. The intersection of the diameter located below the axis has an abscissa value of σ'_{yy} and an ordinate value of $\tau'_{yx} = \sigma'_{yx}$.

In the second situation the initial axes are not principal axes so that $\sigma'_{xx} \neq \sigma_1$, $\sigma'_{yy} \neq \sigma_2$, and $\sigma'_{xy} = \sigma'_{yx} \neq 0$. It is useful to have the Mohr circle parameters expressed in terms of stresses given in the nonprincipal axis system. Two points on the Mohr circle can then be constructed with coordinates $(\sigma'_{xx}, -\sigma'_{xy})$ and $(\sigma'_{yy}, \sigma'_{yx})$ when the normal sign convention for shear stresses (positive shear corresponds to positive force on positive plane or negative force on negative plane) is used. Alternatively and equivalently, the Mohr convention for the sign of shear stress (clockwise rotation corresponds to positive shear) may be used and the two points are $(\sigma'_{xx}, \tau_{xy} = -\sigma'_{xy})$ and $(\sigma'_{yy}, \tau_{yx} = \sigma'_{yx})$. The center of the Mohr circle is on the σ axis at $(\sigma'_{xx} + \sigma'_{yy})/2$ and the radius is given by

$$\text{radius} = \left[\left(\frac{\sigma'_{xx} - \sigma'_{yy}}{2}\right)^2 + (\sigma'_{xy})^2\right]^{1/2} \tag{1.37}$$

It is interesting to examine the stresses given by the Mohr circle construction (or equivalently by direct calculation from the above equations) for special situations. Consider the case of a bar under simple tension σ_1. The center of the circle is at $\sigma_1/2$ and the radius is also $\sigma_1/2$. The circle intersects the axis at σ_1 and at 0 corresponding to $\sigma_{xx} = \sigma_1$ and $\sigma_{yy} = 0$ in the original axes. As the primed coordinate system is rotated, the shear stresses increase in absolute value and reach a maximum absolute value at $2\theta = 90°$ with $\sigma'_{xy} = \sigma_1/2$. At this orientation the two tensile stresses are $\sigma'_{xx} = \sigma_1/2$ and $\sigma'_{yy} = \sigma_1/2$. Note that this is NOT the same stress state as is produced by two equal tensile stresses in perpendicular directions, as shown next.

As a further example consider the application of equal values of σ_{xx} and σ_{yy}. If their common value is σ_1, the center of the Mohr circle is at $\sigma = \sigma_1$ and the radius is zero. The Mohr circle is just a point. As the primed coordinate system rotates, the values of σ'_{xx} and σ'_{yy} remain equal to σ_1 and the shear stresses τ'_{xy} and τ'_{yx} remain equal to zero for all values of θ.

1.5 STRESS INVARIANTS

When stresses are transformed from one coordinate system to a rotated system, there are three quantities that remain constant. These three stress invariants are

$$J_1 = \sigma_{xx} + \sigma_{yy} + \sigma_{zz} = \sigma_1 + \sigma_2 + \sigma_3 \tag{1.38}$$

$$J_2 = -(\sigma_{xx}\sigma_{yy} + \sigma_{yy}\sigma_{zz} + \sigma_{zz}\sigma_{xx} - \sigma^2{}_{xx} - \sigma^2{}_{yz} - \sigma^2{}_{zx})$$

$$= -(\sigma_1\sigma_2 + \sigma_2\sigma_3 + \sigma_3\sigma_1) \tag{1.39}$$

$$J_3 = \sigma_{xx}\sigma_{yy}\sigma_{zz} +$$

$$2\sigma_{xy}\sigma_{yz}\sigma_{zx} - \sigma_{xx}\sigma^2{}_{yz} - \sigma_{yy}\sigma^2{}_{xz} - \sigma_{zz}\sigma^2{}_{xy} = \sigma_1\sigma_2\sigma_3 \tag{1.40}$$

The isostatic pressure (taking compression as positive) is given by

$$p = -\frac{J_1}{3} \tag{1.41}$$

1.6 STRESS DEVIATOR

It is useful to separate the stress into the components causing dilation without change of shape (pressure) and components causing distortion without change in volume. For example, to a first approximation isostatic pressure alone causes transformation in transformation-toughened zirconia. Plastic deformation is not caused by isostatic pressure so that it is sometimes useful to subtract the isostatic stress and consider only the remaining stresses. The stress deviator is defined by

$$\sigma^*{}_{ij} = \sigma_{ij} - J_1/3 \text{ for } i = j \tag{1.42}$$

and

$$\sigma^*{}_{ij} = \sigma_{ij} \text{ for } i \neq j \tag{1.43}$$

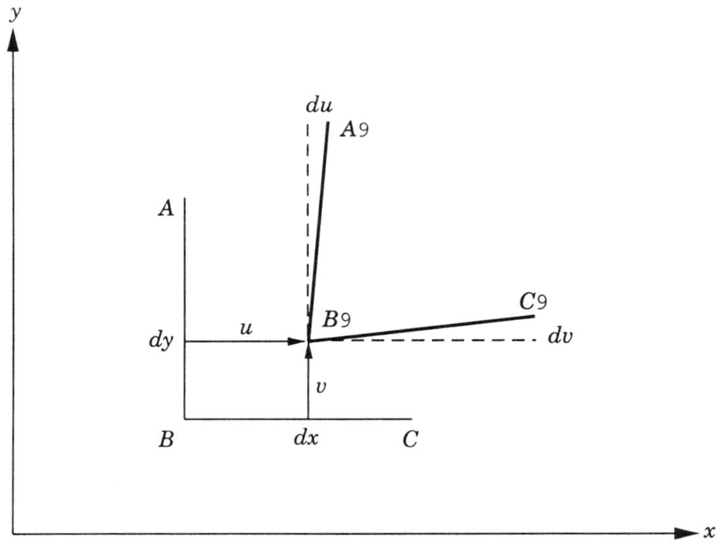

Figure 1.6 Definition of strain.

1.8 STRAIN

Consider again the bar in Figure 1.1 of length L under tensile stress σ. The tensile strain is defined in terms of the change δL in the total length L by

$$\varepsilon = \frac{\delta L}{L} \tag{1.44}$$

To define the complete set of strain components requires consideration of the deformation of a body under arbitrary stress. We begin with the two-dimensional case for simplicity. Using an xy coordinate system as shown in Figure 1.6 allows the strain at a point to be defined in terms of the local deformation rather than the overall deformation. We fix attention on the deformation of the corner of a square ABC. The point at B with coordinates (x, y) is displaced to B'. Let $u(x, y)$ and $v(x, y)$ be the displacements in the x and y directions, respectively, of the point originally at (x, y). The displacement in the x direction of a point C at a distance dx from (x, y) is $u(x + dx, y) = u(x, y) + (\partial u/\partial x)dx$. Then the strain is

$$\varepsilon_{xx} = \frac{u(x+dx,y) - u(x,y)}{dx} = \frac{u(x,y) + \dfrac{\partial u}{\partial x}dx - u(x,y)}{dx} = \frac{\partial u}{\partial x} \tag{1.45}$$

Under σ_{xx} a point at (x, y) not only moves in the u direction but also in the y direction as shown in Figure 1.6. The y component of tensile strain is analogous to the x component and is

$$\varepsilon_{yy} = \frac{\partial v}{\partial y} \tag{1.46}$$

The definition of the shear strain is a little more complicated. The point C is displaced to the point C'. The straight lines CB and AB are rotated as well as displaced to become $C'B'$ and $A'B'$, respectively. Rigid body rotation is not of interest to definition of strain and so we develop an expression for the change of shape of the square. Consider how much the original right angle between CA and CB is changed and define β as this change. Taking this change of angle as equal to its tangent for small deformations gives

$$\beta = \frac{dv}{dx} + \frac{du}{dy} = \frac{\frac{\partial v}{\partial x}dx}{dx} + \frac{\frac{\partial u}{\partial y}dy}{dy} = \frac{\partial v}{\partial x} + \frac{\partial u}{\partial y} \tag{1.47}$$

The xy component of shear strain is defined as half the angle β, so that

$$\varepsilon_{xy} = \frac{\beta}{2} = \frac{1}{2}\left(\frac{\partial v}{\partial x} + \frac{\partial u}{\partial y}\right) \tag{1.48}$$

The fourth component of strain in two dimensions is the shear strain given by

$$\varepsilon_{yx} = \frac{1}{2}\left(\frac{\partial v}{\partial x} + \frac{\partial u}{\partial y}\right) \tag{1.49}$$

Evidently $\varepsilon_{xy} = \varepsilon_{yx}$.

In three dimensions, a point at (x, y, z) will be displaced by stress through the corresponding distances u, v, w. Going from two to three dimensions increases the total number of strain components from four to nine by the additions of the components ε_{zz}, ε_{xz}, ε_{zx}, ε_{yz}, and ε_{zy}. Here again $\varepsilon_{xz} = \varepsilon_{zx}$ and $\varepsilon_{yz} = \varepsilon_{zy}$. Going from two to three dimensions thus increases the number of independent components from three to six. The defining equations for the three new independent components are

$$\varepsilon_{zz} = \frac{\partial w}{\partial z} \tag{1.50}$$

$$\varepsilon_{xz} = \frac{1}{2}\left(\frac{\partial u}{\partial z} + \frac{\partial w}{\partial x}\right) \tag{1.51}$$

$$\varepsilon_{yz} = \frac{1}{2}\left(\frac{\partial w}{\partial y} + \frac{\partial v}{\partial z}\right) \tag{1.52}$$

Using the numerical notation for the axes gives the strain components written as a matrix

$$\varepsilon = \begin{pmatrix} \varepsilon_{11} & \varepsilon_{12} & \varepsilon_{13} \\ \varepsilon_{21} & \varepsilon_{22} & \varepsilon_{23} \\ \varepsilon_{31} & \varepsilon_{32} & \varepsilon_{33} \end{pmatrix} \tag{1.53}$$

where

$$\varepsilon_{ij} = \varepsilon_{ji} \tag{1.54}$$

It was noted earlier that for any set of stress components there are three orthogonal directions, called the principal directions for stress, for which the shear stresses are all zero; the corresponding tensile stresses are called the principal stresses. An analogous result holds for strain. For any set of strain components there are three orthogonal directions, called the principal directions for strain, for which the shear strains are zero; the corresponding tensile strains are called the principal strains. For an elastically isotropic body the principal directions for stress and the principal directions for strain coincide. For an elastically anisotropic body this is not necessarily so.

The Mohr circle construction can be used in two dimensions for strain as for stress. In three dimensions, if one of the principal strain directions is known, the Mohr circle can be used to represent the strain in the perpendicular plane. For the more general case of finding a stress component in a rotated coordinate system in terms of the stress components in an initial coordinate system, the same transformation equations can be used for strain as for stress. That is, equations (1.27) and (1.28) can be used with stress replaced by strain.

Another set of results that can be taken over from stress and used for strain is the stress invariants and the definition of stress deviator. The corresponding strain invariants and strain deviator are given by substituting strain for stress in equations (1.39) through (1.43).

CHAPTER 2

TYPES OF MECHANICAL BEHAVIOR

2.1 INTRODUCTION

With the concepts of stress and strain established, one can describe the general types of mechanical behavior of solids. The intention in this chapter is to present the major categories of mechanical behavior in terms of their general characteristics in order to provide an overview and give perspective to the material to follow. The approach here is to follow the historical tradition and discuss types of mechanical behavior in terms of the overall macroscopic response of a solid to stress, considering the solid as a homogeneous body. Real solids are usually inhomogeneous and the overall behavior is the result of detailed processes that take place on an atomic scale. Full and accurate descriptions of various types of mechanical behavior often require further development of concepts and mathematical tools to treat these detailed processes. These concepts and tools will be treated in later chapters for types of mechanical behavior pertinent to ceramics.

The response of a solid to an applied stress is to deform and, if the stress exceeds the strength, to fail. Mechanical behavior categories are generally defined with regard to the modes of deformation and failure. The individual types of mechanical behavior discussed below often closely approximate the real behavior of solids, at least over some range of time and temperature. However, a real solid may have several types of mechanical behavior taking place at once so that the overall behavior is more complex than any one of the somewhat idealized types discussed below would indicate. Nevertheless, these idealized types of behavior permit quantitative treatments of great usefulness.

2.2 ELASTICITY AND BRITTLE FRACTURE

A relatively simple type of mechanical behavior, and one of great importance for ceramics, is elastic deformation and brittle failure in tension. This type of behavior is illustrated in Figure 2.1, which shows the strain (approximately instantaneous and time independent) as a function of the applied stress. Elastic deformation is defined by full and instantaneous recovery of strain if the stress is removed. Many, but not all, ceramics at room temperature are elastic to a good approximation. Others, such as MgO single crystals, show residual deformation after the load is removed. Even ceramics such as polycrystalline aluminum oxide that are generally considered elastic at room temperature may show small amounts of permanent deformation when strain is measured very accurately. Such effects can be important when accurate dimensional stability, such as in a gyroscope, is required. The precision elastic limit is said to be exceeded when a permanent strain of 10^{-6} is exceeded.

Elastic strain need not be a linear function of stress. Curve A in Figure 2.1 is drawn with distinct nonlinearity to emphasize this fact. When examined closely, all solids show some degree of nonlinearity in their stress–strain curve. However, for many solids including ceramics the nonlinearity is quite small so that a linear relation between stress and strain is assumed for most purposes, as is done here. The behavior of geological materials under very high pressures is an example requiring consideration of nonlinear elastic effects.

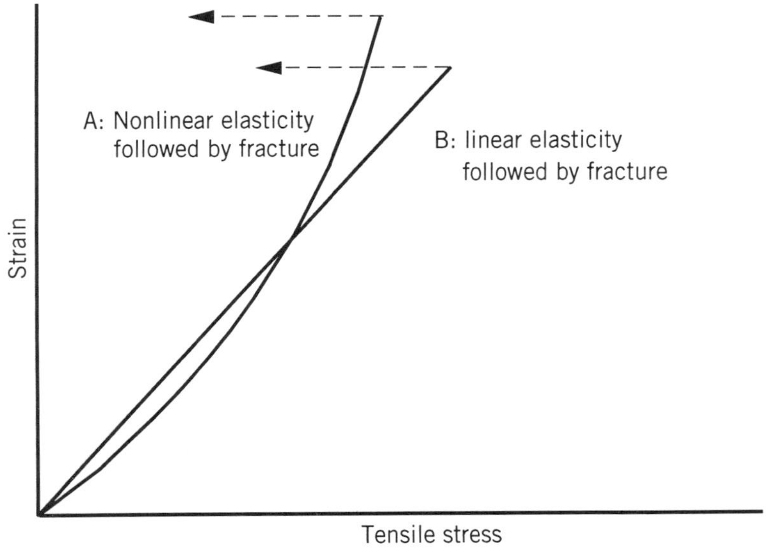

Figure 2.1 Elastic behavior.

The curves of Figure 2.1 show dotted arrows indicating failure at the highest stress reached. For ceramics under uniform stress at or near room temperature this failure is usually sudden and complete.

Much of the design and use of ceramics is based on the conceptually simple model of elastic and brittle behavior described above. In subsequent chapters linear elasticity and linear elastic fracture mechanics will be developed extensively to describe the mechanical behavior of ceramics at relatively low temperatures. Despite the apparent simplicity of this concept, its application to materials with defects and complex microstructures is a considerable undertaking. Indeed, there are still unsolved problems. However, the body of successful theory provides very useful guidance.

Another feature is observed if the linear curve of Figure 2.1 is measured on a series of nominally identical ceramic specimens. As shown in Figure 2.2 the elastic deflection repeats quite accurately, but the maximum tensile stress that the sample will sustain varies considerably. This is one of several facts pointing to flaws as the source of failure. Different specimens have flaws with different degrees of effectiveness in reducing strength. This fact is of great importance to the practical use of ceramics, to the practical processing of ceramics, and to the theory of the strength of ceramics. The field of fracture mechanics deals with the behavior of flaws under stress and is a central subject of this book.

So far in the discussion in this chapter it has tacitly been assumed that any effect of stress is instantaneous. This assumption can be used to provide a good

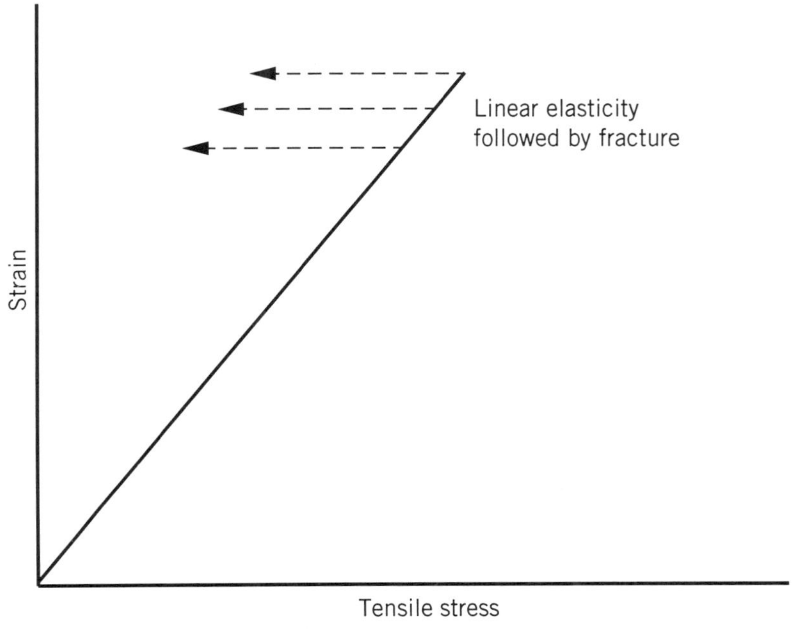

Figure 2.2 Fracture of samples under the same loading rates.

description of behavior for conditions of fairly rapid loading (e.g., tens of seconds to maximum load) and for relatively short time behavior in general (e.g., tens of seconds total time under load) provided the stresses are not too high. At very high strain rates and very high stresses corresponding to ballistic loading, other effects must be considered; we exclude this region of impact behavior from consideration although ceramics are sometimes used as armor (Ferguson and Rice, 1971). Within the range of loading accessible with a normal testing machine, some time-dependent effects must be considered, however. Thus cracks can have a time-dependent response to stress, namely, slow, subcritical crack growth, which shows itself as an effect on strength.

This time-dependence effect in the mechanical behavior of ceramics becomes apparent if nominally identical samples are tested in certain ranges of temperature and/or corrosive environments by being placed under various constant stress values below the stress required for immediate fracture. As shown in Figure 2.3, each specimen typically breaks after a time. This behavior is termed stress rupture in general. The time to failure increases rapidly as the stress value is decreased so that a log–log plot is commonly used. Sometimes the curve shows an indication of leveling out, indicating a stress level below which the specimen life is effectively infinite. Several mechanisms can be at work depending on the material and the temperature. For the case under consideration here of slow crack growth this behavior is termed static fatigue. Static fatigue is often associated with stress-assisted environmental attack at the tip of cracks. This is an important effect that must be taken into account in the use of ceramics in structures intended to carry a load for a long time. A methodology for describing this effect quantitatively and for predicting lifetimes will be described later.

The same static fatigue effect can manifest itself in another way if stress–strain curves are measured out to failure over a wide range of loading rates. As

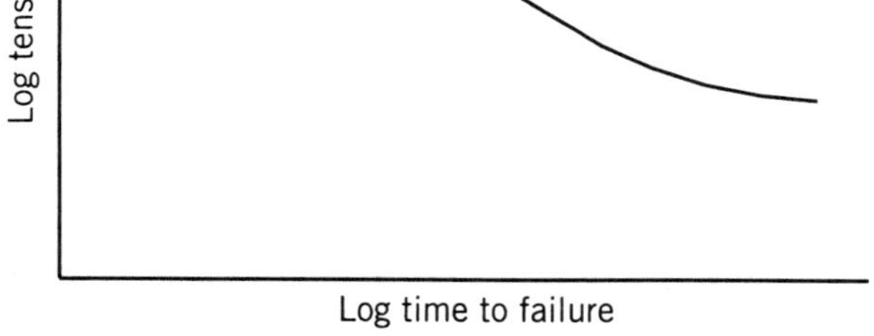

Figure 2.3 Stress-rupture plot.

shown in Figure 2.4, the strength is progressively lower as the smaller loading rates are used.

So far we have considered stress–strain curves for solids that undergo only small amounts of elastic deformation followed by fracture. Another class of behavior experienced by some materials involves permanent deformation that is not recovered when the stress is removed. There is always some elastic deformation in addition to the permanent deformation, but we omit the elastic component from the following descriptions for simplicity. An important case of permanent deformation is viscous deformation of a liquid. Glass and fine-grained polycrystalline ceramics at high temperatures behave like highly viscous liquids so that consideration of liquid-like behavior is appropriate. Figure 2.5 shows that for a liquid the response to an applied shear stress is a shear strain rate that depends on the stress value. The simple case where the shear strain rate is proportional to shear stress (Newtonian viscosity) is shown. More complex relationships are possible and are important in many liquids including ceramic slips but will not be pursued here.

Another type of permanent deformation is plastic deformation. This deformation can be considered to be time independent even though it is based on dynamic processes whose rates depend on stress. This is because the time elapsing in stress–strain tests is generally long enough for the dynamic processes to reach completion. Figure 2.6 shows a typical stress–strain curve for a plastically deforming material. If the test is interrupted at some point and the stress is removed, the elastic portion of the strain is recovered but the plastic

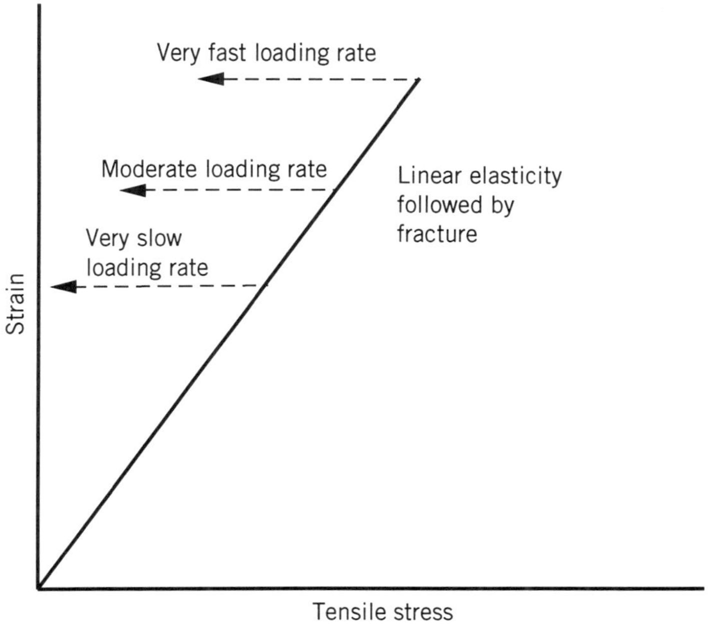

Figure 2.4 Fracture of samples under different loading rates.

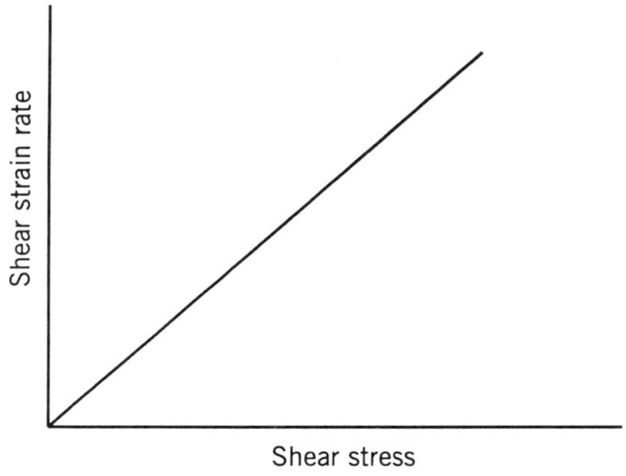

Figure 2.5 Viscous deformation.

component is not. The tensile yield strength is generally defined as the stress that upon removal leaves a permanent strain of 0.002 as indicated by the straight line in Figure 2.6. Plastic deformation commonly occurs in metals at room temperature but usually takes place in ceramics only at high temperatures, although localized plastic deformation does occur even in ceramics in special circumstances such as in hardness testing.

A final type of permanent deformation needing consideration in ceramics is creep. Here permanent deformation begins when stress is applied, slows down in rate, but continues for a long time before ultimate failure. Figure 2.7 shows this type of behavior. These two stages, termed transient and steady-state creep, generally occur in polycrystalline ceramics at low stresses and sufficiently high temperature. In metals there is often a final period of increas-

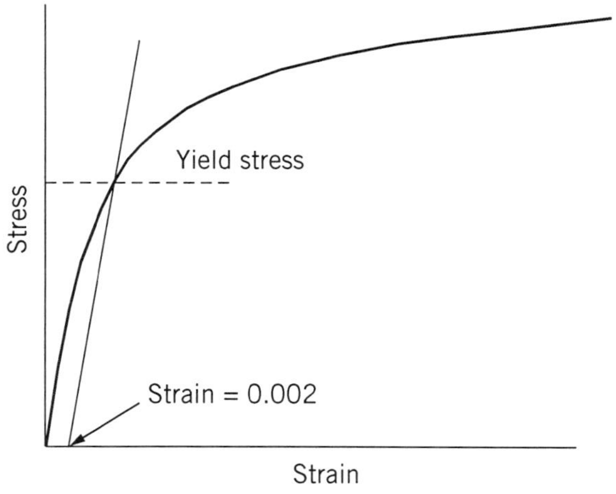

Figure 2.6 Plastic deformation.

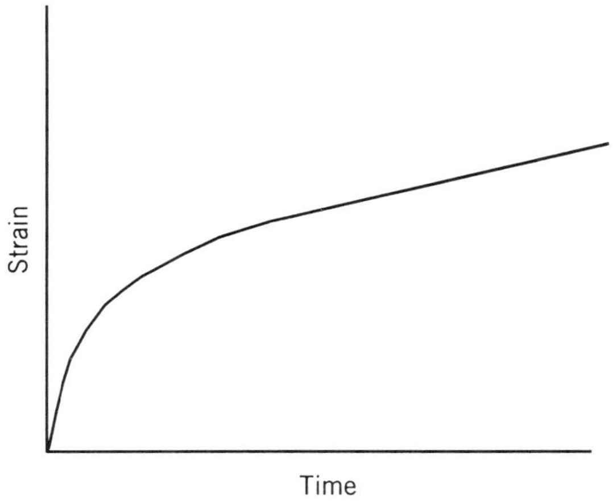

Figure 2.7 Creep.

ing creep rate before failure, termed third-stage or tertiary creep, but in ceramics failure often occurs with little or no third stage. Creep and creep-assisted failure are the primary processes limiting structural use of ceramics at high temperature and are accordingly considered in Chapters 19 and 20.

CHAPTER 3

ELASTICITY

3.1 INTRODUCTION

Linear elastic behavior is commonly assumed for ceramics under load at temperatures up to about one-half their absolute melting point. This is usually a good approximation, although there are exceptions, such as single crystal magnesium oxide for which appreciable bulk plastic deformation occurs at room temperature. Linear elasticity is usually assumed when calculating the stresses caused by loads applied to bodies and so is at the heart of design with brittle ceramics. The basic equations of linear elasticity for elastically isotropic material are accordingly briefly reviewed.

Nonisotropic elasticity (anisotropic elasticity) can occur in polycrystalline ceramics with preferred orientation and, in fact, does sometimes occur in hot pressed ceramics, although the anisotropy produced by hot pressing may be small and is sometimes ignored. Two types of ceramics in which a large degree of elastic anisotropy often occurs are single crystals and composites with oriented reinforcing fibers. Because these are important types of ceramics, the basic equations of anisotropic elasticity are reviewed.

Both the values of the elastic constants and their temperature dependence are related to the interatomic potentials. A few results are reviewed to provide physical insight. Some useful approximate relations for temperature dependence and some correlations with other properties are presented.

3.2 ELASTICITY OF ISOTROPIC BODIES

Consider a body under a single component of stress σ_{xx}. Hooke's law of elasticity states that the strain is proportional to the stress. The strain components for an elastically isotropic body resulting from σ_{xx} are

$$\varepsilon_{xx} = \frac{\sigma_{xx}}{E} \tag{3.1}$$

$$\varepsilon_{yy} = -\nu\, \varepsilon_{xx} = -\nu \frac{\sigma_{xx}}{E} \tag{3.2}$$

$$\varepsilon_{zz} = -\nu\, \varepsilon_{xx} = -\nu \frac{\sigma_{xx}}{E} \tag{3.3}$$

Here E is Young's modulus of elasticity and ν is Poisson's ratio defined by $\nu = -\varepsilon_{yy}/\varepsilon_{xx}$.

An idea of the size of elastic deformations in ceramics can be obtained by considering a rod of polycrystalline aluminum oxide having randomly oriented grains with $E = 4.1 \times 10^5$ MPa and $\nu = 0.21$. If a tensile stress of $\sigma_{xx} = 500$ MPa is applied, the resulting strains for polycrystalline aluminum oxide are

$$\varepsilon_{xx} = 0.0012 = 0.12\%$$

$$\varepsilon_{yy} = \varepsilon_{zz} = -0.00025 = -0.025\%$$

Here the minus sign indicates that the strains are at a right angle to the extension and are compressive. The maximum elastic strain that can be achieved in a polycrystalline ceramic prior to failure is typically 0.1% or so. Ceramic fibers and bulk single crystals specially prepared to be nearly flaw-free may have elastic tensile strains at fracture as great as 1%.

Consider next the application of a shear stress $\sigma_{xy} = \sigma_{yx}$. Hooke's law states that

$$\varepsilon_{xy} = \varepsilon_{yx} = \frac{\sigma_{xy}}{\mu} \tag{3.4}$$

Here μ is used to designate the shear modulus of elasticity (sometimes termed the rigidity modulus) to distinguish from G, which, although used in many books to designate the shear modulus, is used in this book for the mechanical energy release rate defined in Chapter 5.

There are only two independent moduli of elasticity for an isotropic material. The shear modulus is related to Young's modulus and Poisson's ratio by

$$\mu = \frac{E}{2(1+\nu)} \tag{3.5}$$

Another important elastic modulus is the bulk modulus B defined by $B = -V\, \partial P/\partial V$. The bulk modulus is related to Young's modulus and the shear modulus by

$$B = \frac{E\mu}{3(3\mu - E)} \tag{3.6}$$

In the above example with aluminum oxide under 500 MPa tensile stress, the maximum shear stress will occur on planes at 45° and will be half the tensile stress or 250 MPa. For polycrystalline aluminum oxide with $\mu = 1.6 \times 10^5$ MPa, this gives a shear strain of $\varepsilon_{xy} = 0.0016 = 0.16\%$.

Omitting the subscripts, the above equations are the familiar expressions for isotropic bodies; $\varepsilon = \sigma/E$ for tensile strain in the direction of applied tensile stress, $\varepsilon = -\nu\sigma/E$ for the corresponding lateral strain, and $\varepsilon = \sigma/\mu$ for shear strain caused by applied shear stress. The subscripts must be retained to deal with important classes of ceramics that are not isotropic. However, the double subscript tensor notation for stress and strain components can be somewhat simplified by going to a single subscript matrix notation. Because this is the notation commonly used for anisotropic elasticity, it is taken up in the next section.

Other elastic moduli have been defined to describe the elastic behavior of isotropic materials. Elastic moduli defined for isotropic materials include Young's modulus E, the shear modulus μ, Poisson's ratio ν, the bulk modulus B, and λ; the pair μ, λ are called the Lame constants. Interrelationships between these elastic moduli are summarized in Table 3.1.

Values for Young's modulus and the shear modulus for pore-free polycrystalline ceramics are given in Table 3.2. These values are in some cases averages of measurements from several authors as given in more detail elsewhere (Wachtman, 1969).

TABLE 3.1 Expressions for Various Isotropic Elastic Moduli in Terms of Various Pairs of Moduli

	Independent Pair of Moduli			
Moduli	E, μ	B, μ	B, ν	λ, μ
E	E	$9B\mu/(3B + \mu)$	$3B(1 - 2\nu)$	$\mu(3\lambda + 2\mu)/(\lambda + \mu)$
μ	μ	μ	$3B(1 - 2\nu)/2(1 + \nu)$	μ
B	$E\mu/3(3\mu - E)$	B	B	$\lambda + (2\mu/3)$
ν	$(E/2\mu) - 1$	$\frac{3B - 2\mu}{6B + 2\mu}$	ν	$\lambda/(2\lambda + \mu)$
λ	$\frac{(E - 2\mu)\mu}{3\mu - E}$	$B - (2\mu/3)$	$3\nu B(1 + \nu)$	λ

TABLE 3.2 Elastic Moduli of Polycrystalline Ceramics

Substance	Young's Modulus (10^{10} Pa)	Shear Modulus (10^{10} Pa)
Al_2O_3	40.28	16.30
Dy_2O_3	17.05	—
Er_2O_3	18.63	—
MgO	31.09	13.34
ThO_2	26.10	10.06
TiO_2	28.42	11.15
Y_2O_3	13.83	—
ZnO	12.35	4.56

3.3 MATRIX NOTATION FOR STRESSES, STRAINS, AND ELASTIC CONSTANTS

The general form of Hooke's law expresses each elastic strain as a linear component of each of the stresses (Nye, 1957):

$$\varepsilon_{ij} = \sum_{k,l=1}^{k,l=3} s_{ijkl}\,\sigma_{kl} \tag{3.7}$$

where the s_{ijkl} are the elastic compliances in tensor notation. The stresses can equivalently be expressed as linear combinations of the strains according to

$$\sigma_{ij} = \sum_{k,l=1}^{k,l=3} c_{ijkl}\,\varepsilon_{kl} \tag{3.8}$$

where the c_{ijkl} are the elastic stiffness constants in tensor notation. Note that compliance is represented by the parameter s, and stiffness by c, and not the other way as you might expect.

In the matrix notation the stresses are defined by

$$\sigma_1 = \sigma_{11},\ \ \sigma_2 = \sigma_{22},\ \ \sigma_3 = \sigma_{33}$$

$$\sigma_4 = \sigma_{23} = \sigma_{32},\ \ \sigma_5 = \sigma_{13} = \sigma_{31},\ \ \sigma_6 = \sigma_{12} = \sigma_{21} \tag{3.9}$$

and the strains in single-subscript (reduced or matrix) notation are defined by

$$\varepsilon_1 = \varepsilon_{11},\ \ \varepsilon_2 = \varepsilon_{22},\ \ \varepsilon_3 = \varepsilon_{33}$$

$$\varepsilon_4 = 2\varepsilon_{23} = 2\varepsilon_{32},\ \ \varepsilon_5 = 2\varepsilon_{13} = 2\varepsilon_{31},\ \ \varepsilon_6 = 2\varepsilon_{12} = 2\varepsilon_{21} \tag{3.10}$$

Note that these shear strains in reduced notation are twice the size of the tensor components of shear strain; the shear strains in reduced notation are often termed engineering shear strains and designated by γ_i. Note further that this use of the term engineering strain is different from the context of using "engineering strain" to distinguish from "true strain"; see Section 17.18 for a discussion of the latter.

In matrix form Hooke's law then takes the form

$$\begin{pmatrix} \varepsilon_1 \\ \varepsilon_2 \\ \varepsilon_3 \\ \varepsilon_4 \\ \varepsilon_5 \\ \varepsilon_6 \end{pmatrix} = \begin{pmatrix} s_{11} & s_{12} & s_{13} & s_{14} & s_{15} & s_{16} \\ s_{21} & s_{22} & s_{23} & s_{24} & s_{25} & s_{26} \\ s_{31} & s_{32} & s_{33} & s_{34} & s_{35} & s_{36} \\ s_{41} & s_{42} & s_{43} & s_{44} & s_{45} & s_{46} \\ s_{51} & s_{52} & s_{53} & s_{54} & s_{55} & s_{56} \\ s_{61} & s_{62} & s_{63} & s_{64} & s_{65} & s_{66} \end{pmatrix} = \begin{pmatrix} \sigma_1 \\ \sigma_2 \\ \sigma_3 \\ \sigma_4 \\ \sigma_5 \\ \sigma_6 \end{pmatrix} \tag{3.11}$$

or in summation form

$$\varepsilon_i = \sum_{j=1}^{j=6} s_{ij}\ \sigma_j \tag{3.12}$$

Here the conversion of elastic compliances from four subscripts $ijhl$ to two subscripts mn is done by replacing ij equals 11, 22, 33, 23 or 32, 31 or 13, and 12 or 21 by m equals 1, 2, 3, 4, 5, and 6 respectively. The same holds for replacing hl by n. This leads to $s_{jhl} = s_{mn}$ when both m and n are 1, 2, or 3; $2s_{ijkl} = s_{mn}$ when either m or n is 4, 5, or 6; $4s_{ijkl} = s_{mn}$ when both m and n are 4, 5, or 6 (Nye, 1957).

The stresses can be expresses in terms of the strains by

$$\begin{pmatrix} \sigma_1 \\ \sigma_2 \\ \sigma_3 \\ \sigma_4 \\ \sigma_5 \\ \sigma_6 \end{pmatrix} = \begin{pmatrix} c_{11} & c_{12} & c_{13} & c_{14} & c_{15} & c_{16} \\ c_{21} & c_{22} & c_{23} & c_{24} & c_{25} & c_{26} \\ c_{31} & c_{32} & c_{33} & c_{34} & c_{35} & c_{36} \\ c_{41} & c_{42} & c_{43} & c_{44} & c_{45} & c_{46} \\ c_{51} & c_{52} & c_{53} & c_{54} & c_{55} & c_{56} \\ c_{61} & c_{62} & c_{63} & c_{64} & c_{65} & c_{66} \end{pmatrix} = \begin{pmatrix} \varepsilon_1 \\ \varepsilon_2 \\ \varepsilon_3 \\ \varepsilon_4 \\ \varepsilon_5 \\ \varepsilon_6 \end{pmatrix} \tag{3.13}$$

or in summation form

$$\sigma_i = \sum_{j=1}^{j=6} c_{ij}\ \varepsilon_j \tag{3.14}$$

The same rule for conversion from four to two subscripts holds here. However, $c_{ijkl} = c_{mn}$ for all values of m and n.

There are six independent components of stress and six independent components of strain leading to 36 coefficients in the elasticity equations. However, it can be shown (Nye, 1957; Hearmon, 1961) that the elastic compliance and elastic stiffness matrices are symmetric; that is, $s_{ij} = s_{ji}$ and $c_{ij} = c_{ji}$. In the most general case there are accordingly a total of 21 independent coefficients whether expressed as compliances or stiffnesses.

One can choose to describe the elastic behavior of single crystals by using the two elastic moduli defined for isotropic materials—Young's modulus and the shear modulus. However, these quantities will vary with orientation for single crystals. The relation between these two elastic moduli and the single crystal stiffnesses can be seen by considering the meaning of the moduli. The reciprocal of Young's modulus gives the tensile strain in the direction of an applied tensile stress. Taking the x_1 direction as the direction of application of tensile stress then gives

$$E^{-1} = s_{11} \tag{3.15}$$

Similarly, a shear stress in the yz plane produces a shear strain in the yz plane so that the shear modulus for this situation is given by

$$\mu^{-1} = s_{44} \tag{3.16}$$

To obtain either of these elastic moduli for any other crystal direction requires considering a rotated coordinate system and transforming the compliance values. The procedure is to use the four-index form of the compliances (e.g., $s_{11} = s_{1111}$) and to work out the transformations of these. The task is straightforward but tedious and the resulting equations are quite lengthy for lower symmetry crystals such as sapphire (Wachtman, Tefft, Lam, and Apstein, 1960).

3.4 EFFECT OF SYMMETRY ON ELASTIC CONSTANTS

The existence of symmetry reduces the number of independent elastic constants. The 32 possible crystal point group symmetries fall into 11 categories in terms of the effect of crystal symmetry on elastic properties. The results are given in detail by Nye (1957). Only two cases of special interest will be summarized here: cubic symmetry and elastic isotropy.

Crystals with cubic symmetry have three independent elastic constants and are not elastically isotropic. One might think by analogy with electrical, thermal, and optical properties (for which cubic symmetry ensures isotropic behavior) that cubic crystals would be elastically isotropic. The reason that this is not so is that the tensor compliances and stiffnesses are fourth rank tensors while the analogous resistivity, dielectric constant, and thermal expansion are second rank tensors. Application of symmetry operations in any one of the cubic point groups leads to the relations

$$s_{11} = s_{22} = s_{33}$$

$$s_{12} = s_{13} = s_{23}$$

$$s_{44} = s_{55} = s_{66} \tag{3.17}$$

All the other compliances are required to be zero by cubic symmetry. The same relationships hold with the s_{ij} replaced by c_{ij}.

The compliance and stiffness matrices for cubic symmetry are thus

$$\begin{pmatrix} s_{11} & s_{12} & s_{12} & 0 & 0 & 0 \\ s_{12} & s_{11} & s_{12} & 0 & 0 & 0 \\ s_{12} & s_{12} & s_{11} & 0 & 0 & 0 \\ 0 & 0 & 0 & s_{44} & 0 & 0 \\ 0 & 0 & 0 & 0 & s_{44} & 0 \\ 0 & 0 & 0 & 0 & 0 & s_{44} \end{pmatrix} \text{and} \begin{pmatrix} c_{11} & c_{12} & c_{12} & 0 & 0 & 0 \\ c_{12} & c_{11} & c_{12} & 0 & 0 & 0 \\ c_{12} & c_{12} & c_{11} & 0 & 0 & 0 \\ 0 & 0 & 0 & c_{44} & 0 & 0 \\ 0 & 0 & 0 & 0 & c_{44} & 0 \\ 0 & 0 & 0 & 0 & 0 & c_{44} \end{pmatrix}$$

The stiffness components are related to the compliances for cubic symmetry by $c_{11} = (s_{11} + s_{12})/[(s_{11} - s_{12})(s_{11} + 2s_{12})]$, $c_{12} = -s_{12}/[(s_{11} - s_{12})(s_{11} + 2s_{12})]$, and $c_{44} = 1/s_{44}$ (Nye, 1957).

An elastically isotropic material can be regarded as a cubic material with an additional relation between the elastic constants. Specifically, for elastic isotropy in a cubic crystal

$$s_{12} = \frac{2s_{11} - s_{44}}{2} \tag{3.18}$$

The compliance matrix for an elastically isotropic material written in terms of Young's modulus, Poisson's ratio, and the shear modulus is

$$\begin{pmatrix} 1/E & -\nu/E & -\nu/E & 0 & 0 & 0 \\ -\nu/E & 1/E & -\nu/E & 0 & 0 & 0 \\ -\nu/E & -\nu/E & 1/E & 0 & 0 & 0 \\ 0 & 0 & 0 & 1/\mu & 0 & 0 \\ 0 & 0 & 0 & 0 & 1/\mu & 0 \\ 0 & 0 & 0 & 0 & 0 & 1/\mu \end{pmatrix}$$

Elastic constants for some ceramic single crystals with cubic symmetry are given in Table 3.3 (Nye, 1957; Wachtman, 1969). More complete tabulations were given by Hearmon in the Landolt–Bornstein tables (Hearmon, 1966).

3.5 ORIENTATION DEPENDENCE OF ELASTIC MODULI IN SINGLE CRYSTALS AND COMPOSITES

The orientation dependence of Young's modulus in a cubic crystal is given by

$$E^{-1} = s'_{11} = s_{11} - [2(s_{11} - s_{12}) - s_{44}][\sin^2\theta\cos^2\theta + \tfrac{1}{4}\sin^4\theta\sin^2 2\varphi] \qquad (3.19)$$

Here θ and φ are the usual polar angles; that is, θ is the angle between the x_1 axis and the chosen direction and φ is the angle between x_2 and the projection of the chosen direction on the x_2x_3 plane.

Crystals with orthorhombic symmetry have nine independent compliances as follows: $s_{11}, s_{22}, s_{33}, s_{12}, s_{13}, s_{23}, s_{44}, s_{55}$, and s_{66}. Such crystals are not often dealt with, but fiber-reinforced composites with continuous, aligned fibers arranged in a square or rectangular array in the plane perpendicular to the fibers have the same elastic symmetry as an orthorhombic crystal and are spoken of as having orthotropic symmetry (Hearmon, 1961; Chawla, 1993). If the arrangement of the aligned fibers in the transverse plane is either random or hexagonal, the elastic symmetry of the composite is the same as that of a hexagonal crystal; that is, five independent elastic constants are required.

Crystals with trigonal symmetry fall into two categories from the point of view of elastic behavior: those with point group symmetry 3 or $\bar{3}$ have seven independent elastic constants while those with symmetry 32, $3m$, or $\bar{3}m$ have six independent elastic constants. The stable form of aluminum oxide, corundum = alpha alumina, falls in the last group. Expressions for the orientation dependence of elastic moduli are available (Wachtman, Tefft, Lam, and Apstein, 1962).

3.6 VALUES OF POLYCRYSTALLINE MODULI IN TERMS OF SINGLE CRYSTAL CONSTANTS

It is of interest to express the elastic moduli of an isotropic polycrystalline body in terms of the single crystal elastic constants. This is more difficult than might appear at first. Application of a strain to a polycrystalline body causes

TABLE 3.3 Single Crystal Elastic Constants (entry in units of 10^{10} N/m^2) and Compliances (in units of 10^{-12}m^2/N) for Some Ceramic Crystals with Cubic Symmetry

Substance	c_{11}	s_{11}	c_{44}	s_{44}	c_{12}	s_{12}
C (diamond)	102.0	1.12	49.2	2.07	25.0	−0.22
MgO	29.6	3.99	15.6	6.41	9.51	−0.97
ThO_2	36.7	3.13	7.97	12.5	10.6	−0.70
UO_2	39.5	2.96	6.41	15.6	12.1	−0.70

strains in the individual grains that lead to forces on neighboring grains. Each grain is constrained by its neighbors in a complex way. There appears to be no exact solution for this problem. Instead, various authors have derived upper and lower bounds for the values of the bulk modulus and the shear modulus of polycrystalline bodies. For crystals of any symmetry it has been shown that

$$E_v \geq E \geq E_r \tag{3.20}$$

$$\mu_v \geq \mu \geq \mu_r \tag{3.21}$$

where

$$E_v = \frac{(A - B + 3C)(A + 2B)}{2A + 3B + C} \tag{3.22}$$

$$\mu_v = \frac{A - B + 3C}{5} \tag{3.23}$$

$$3A = c_{11} + c_{22} + c_{33}$$

$$3B = c_{23} + c_{13} + c_{12}$$

$$3C = c_{44} + c_{55} + c_{66} \tag{3.24}$$

$$\frac{1}{E_r} = \frac{3A' + 2B' + C'}{5} \tag{3.25}$$

$$\frac{1}{\mu_r} = \frac{4A' - 4B' + 3C'}{5} \tag{3.26}$$

$$3A' = s_{11} + s_{22} + s_{33}$$

$$3B' = s_{23} + s_{13} + s_{12}$$

$$3C' = s_{44} + s_{55} + s_{66} \tag{3.27}$$

The quantities E_v and μ_v are termed the Voigt moduli. They result from assuming that the strain is uniform in a polycrystalline aggregate and averaging the stress. The quantities E_r and μ_r are termed the Reuss moduli. They result from assuming that the stress is uniform and averaging the strain (Hearmon, 1961).

3.7 VARIATION OF ELASTIC CONSTANTS WITH LATTICE PARAMETER

Crystals with the same structure show a systematic variation of elastic constants with lattice parameter. This variation can be understood quite well in terms of the relatively simple Born model of ionic crystals and is instructive. In the Born model the energy U of an ionic crystal is assumed to be the sum of two terms according to

$$U = -\frac{Q_1}{r} + \frac{Q_2}{r^n} \tag{3.28}$$

Here the first term arises from the electrostatic attraction and the second from the repulsive force. A simple approach can be used to treat c_{44}, the elastic constant for shear (Gilman, 1963; Davidge, 1979). The force corresponding to a small displacement dr is minus the derivative of the energy with respect to r and the area per atom is approximately r_0^2, where the subscript 0 indicated the equilibrium value of r for the unstressed crystal. The stress is then approximately

$$\sigma \approx \frac{dU}{dr}\frac{1}{r_0^2} \tag{3.29}$$

The elastic constant is

$$c = \frac{d\sigma}{d\varepsilon} = \frac{d\sigma}{(dr)/r_0} \tag{3.30}$$

Then the elastic constant is

$$c = r_0\frac{d\sigma}{dr} = \frac{1}{r_0}\frac{d^2U}{dr^2} \tag{3.31}$$

For shear deformation of ionic crystals the second term in the energy can be neglected. This gives

$$c_{44} \approx \text{constant}\ /r_0^4 \tag{3.32}$$

The alkali halides obey this relation reasonably well.

A somewhat less approximate yet still simple treatment can be given for the bulk modulus (Born and Huang, 1954; Wachtman, 1969). The energy is assumed to be given by

$$U = -\frac{Q_1}{r^m} + \frac{Q_2}{r^n} \tag{3.33}$$

For ionic crystals $m = 1$. From thermodynamics the pressure is given by

$$p = -\frac{dU}{dV} \tag{3.34}$$

and the bulk modulus by

$$B = -V\frac{dp}{dV} \tag{3.35}$$

For ionic crystals these equations give (using V = constant $\times\ r^3$ so that $dr/dV = r/3V$)

$$p = (-\frac{Q_1}{r} + \frac{nQ_2}{r^n})/3V \tag{3.36}$$

and

$$B = p + (-\frac{Q_1}{r} + \frac{n^2Q_2}{r^n})/9V \tag{3.37}$$

Using the condition $p = 0$ at $r = r_0$ gives

$$\frac{nQ_2}{r_0^n} = \frac{Q_2}{r_0} \tag{3.38}$$

so that

$$B = p(1 + n/3) + (n - 1)Q_1/9rV \tag{3.39}$$

For small values of p the first term can be neglected so that the bulk modulus is predicted to vary as $1/r_0 V_0 \approx 1/r_0^4$. This is the same r_0 variation that was obtained above by a simpler argument for the shear modulus.

Anderson (1966b, 1995) has shown that this predicted $1/r_0^4$ variation is obeyed by the bulk modulus of alkali halides, certain carbides, and the Si family of elements. However, he found that the bulk modulus of oxides instead follows a variation of approximately $1/r_0^9$. He suggested that the bulk modulus of oxides is determined approximately simply by the volume per oxygen atom regardless of whether this volume is the result of a particular chemistry, pressure, or temperature.

3.8 VARIATION OF ELASTIC CONSTANTS WITH TEMPERATURE

Elastic constants of single crystals generally decrease slowly with increasing temperature (Wachtman, 1969). As the temperature is decreased toward abso-

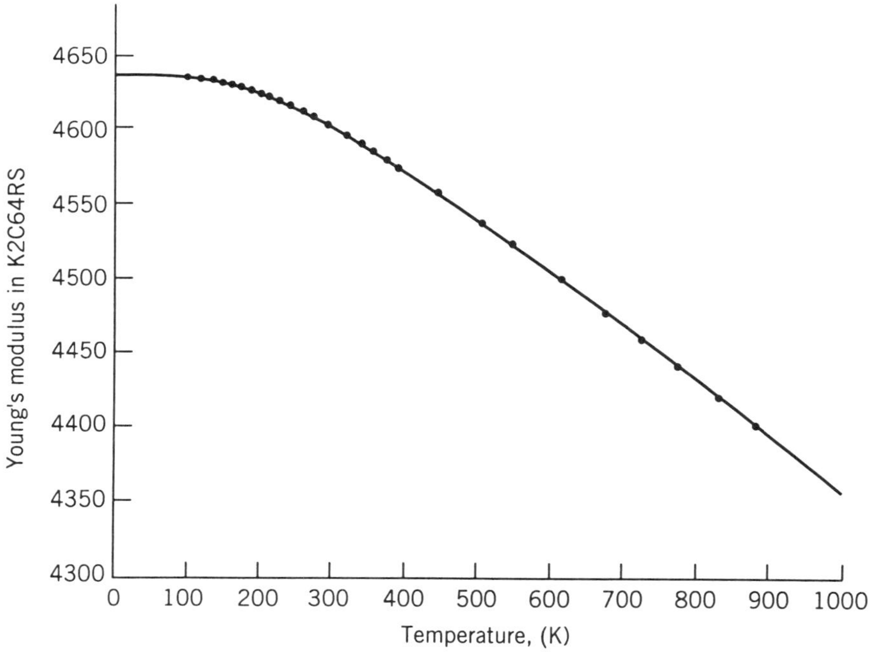

Figure 3.1 Temperature dependence of Young's modulus for an aluminum oxide single crystal. (From Wachtman, Tefft, Lam, and Apstein, 1961.)

lute zero, the slope of the curve of elastic constant as a function of temperature approaches zero as required by the third law of thermodynamics. As the temperature increases toward the Debye Θ the slope approaches a constant for many ceramics. An empirical equation was proposed (Wachtman, Tefft, Lam, and Apstein, 1961) that fit the data well for several oxides. This equation written for Young's modulus is

$$E = E_0 - bT \exp\left(\frac{-T_0}{T}\right) \tag{3.40}$$

where E_0 is the value of Young's modulus at absolute zero and b and T_0 are empirical constants. T_0 was found to be about half the Debye temperature. A plot of this equation fitted to data for an aluminum oxide single crystal is shown in Figure 3.1. Anderson (1966b) gave a derivation justifying the form of

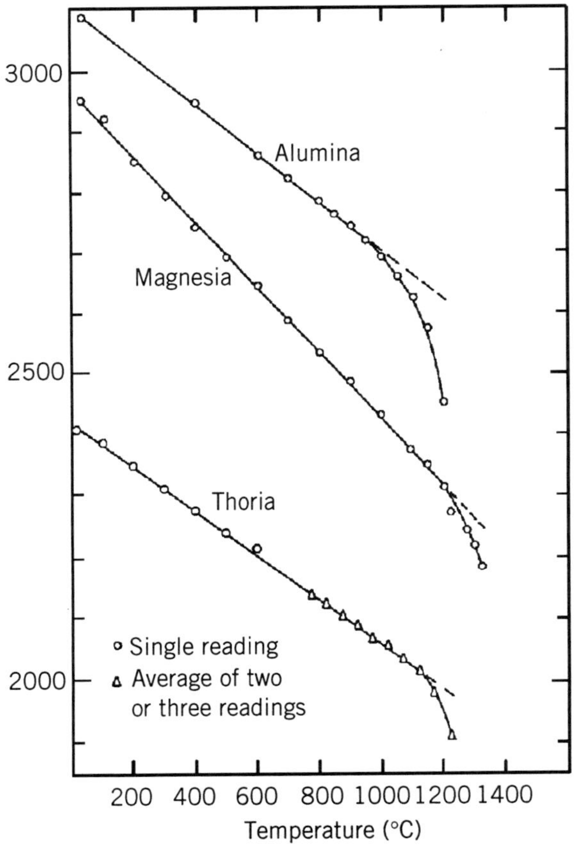

Figure 3.2 Temperature dependence of Young's modulus of polycrystalline ceramics. (From J. B. Wachtman, Jr., and D. G. Lam, Jr., "Young's Modulus of Various Refractory Materials as a Function of Temperature," *J. Am. Ceram. Soc.*, Vol. 42, 254–260, 1959. Reprinted by permission of the American Ceramic Society.)

this equation and providing an interpretation of the parameters. Hillig (1993) has recently discussed estimating the values of mechanical properties including elastic moduli at high temperature.

For polycrystalline ceramics (Figure 3.2) this variation with temperature holds, but an additional effect is superimposed (Wachtman and Lam, 1959). At some temperature characteristic of the grain boundary phase the measured elastic moduli decrease rapidly from the linear decrease characteristic of material free of grain boundaries. The additional rapid decrease is attributed to grain boundary softening and sliding and is not really an elastic effect. A useful rule of thumb is that for common ceramics between room temperature and the onset of grain boundary softening the decrease in either Young's modulus or the shear modulus is about 1% per hundred kelvins of temperature increase. More detailed data are given by Wachtman (1969). The extensive and ground-breaking work of Anderson on pressure–volume–temperature relationships for geophysical and ceramic materials is summarized in his book (Anderson, 1995).

CHAPTER 4

STRENGTH OF DEFECT-FREE SOLIDS

4.1 INTRODUCTION

Both theory and experiment show that defect-free solids are much stronger than the solids usually encountered. This fact points to the necessity for understanding the behavior of various types of defects in solids under stress in order to understand mechanical behavior other than elastic deformation. Before turning to the study of defects in relation to mechanical behavior, it is useful to understand the behavior to be expected of ideal solids.

The relationship of elastic constants to the energy of a solid was discussed in Chapter 3. The force resulting from a strain was shown to be related to the first derivative of the energy and the elastic constants to the second derivative. This approach can be extended to estimate the strength of a solid in tension or in shear.

4.2 THEORETICAL STRENGTH IN TENSION

To estimate the strength of a defect-free solid in tension, a treatment due to Orowan (1949) can be used (Davidge, 1979; Kelly and MacMillan, 1986). His method approximates the stress–strain curve by a sine function. This representation is not exact but has the basic qualitative features needed. To justify the use of a sine function approximation, consider the interatomic potential for ions separated by distance x usually represented (Kittel, 1968) by

$$U = -\frac{A}{x} + B\exp(-x/C) \tag{4.1}$$

Here the first term is the Coulomb attraction and the second is the repulsion. This function is plotted in Figure 4.1, in which the parameters A, B, and C have been chosen to give values characteristic of sodium chloride: that is, a binding energy per ion pair of about 7.9 eV and an interatomic spacing of about 2.8 angstroms. The force, given by the derivative of this function, is plotted in Figure 4.2 with the sign of the force changed so that a positive ordinate value indicates a restoring force. This figure shows zero force at the equilibrium interatomic spacing and a steeply rising repulsive force for a decrease in atomic separation. As the atomic separation is increased, the force becomes negative (indicating an attraction). As the separation increases, the magnitude of the attraction increases to a maximum and then decreases with further separation.

Orowan assumed a sine function for the stress normal to the planes as a function of the separation between two atomic planes and determined the parameters in his sine function by fitting this function to Young's modulus and to the interfacial energy. He took the equilibrium separation to be $x = a_0$ and the separation for maximum restoring force to occur at $x = a_0 + a/2$ and wrote

$$\sigma = \sigma_0 \sin \frac{\pi}{a}(x - a_0) \tag{4.2}$$

where a is a measure of the range of interatomic forces and σ_0 the maximum restoring force. Then Young's modulus, E, is given by

$$E = \frac{d\sigma}{dx/a_0} = \sigma_0 \, \pi \frac{a_0}{a} \cos \frac{\pi}{a} (x - a_0) \tag{4.3}$$

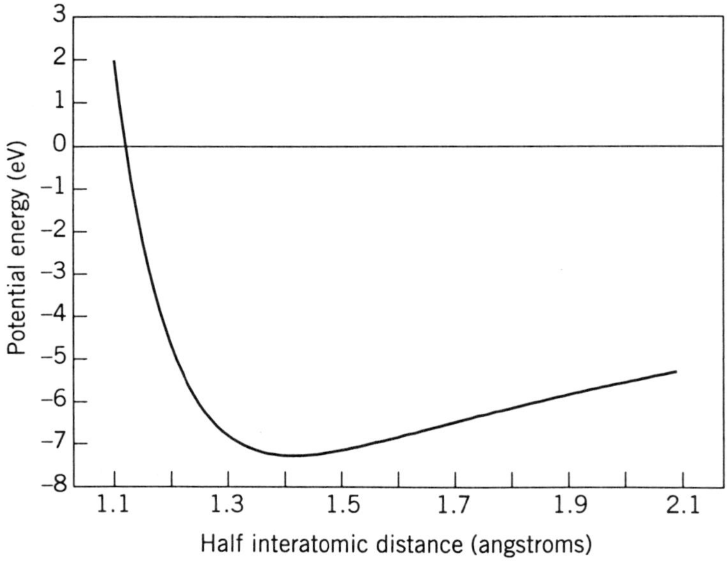

Figure 4.1 Potential energy.

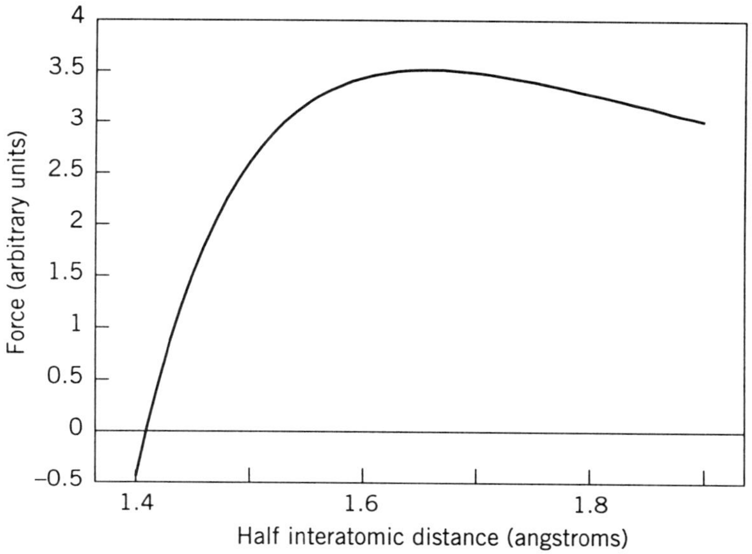

Figure 4.2 Interatomic force.

Taking $x = a_0$ gives the Young's modulus value at zero strain and leads to

$$\sigma_0 = \frac{E}{\pi}\left(\frac{a}{a_0}\right) \tag{4.4}$$

To determine a, Orowan took the energy, γ, to create a unit area of new surface to be the integral of the force times distance over the range of force given by the sine representation of the interatomic forces and wrote

$$2\gamma = \int_{a_0}^{a_0+a} \sigma\, dx = \int_{a_0}^{a_0+a} \sigma_0 \sin \frac{\pi}{a}(x-a_0)\, dx = \frac{\sigma_0\, a}{\pi}\left[-\cos \frac{\pi}{a}(x-a_0)\right]_{a_0}^{a_0+a} = 2\sigma_0 \frac{a}{\pi} \tag{4.5}$$

Here the factor 2 is introduced to account for the fact that two unit areas of new surface are created when a crystal is separated by one unit distance. Rearranging gives

$$a = \frac{\pi\,\gamma}{\sigma_0} \tag{4.6}$$

The theoretical breaking stress is the value of the stress at the maximum of the sine function and so is

$$\sigma_{\text{theoretical}} = \sigma_0 = \frac{E}{\pi}\left(\frac{a}{a_0}\right) \tag{4.7}$$

Substituting $a = \pi\gamma/\sigma_0$ gives

$$(\sigma_0)^2 = \frac{\gamma E}{a_0} \tag{4.8}$$

and

$$\sigma_{\text{theoretical}} = \sigma_0 = \left(\frac{E\gamma}{a_0}\right)^{1/2} \tag{4.9}$$

If values of E and γ are available this equation can be used to estimate the theoretical breaking strength of solids. Table 4.1 summarizes several values compiled by Kelly and MacMillan (1986).

The Orowan treatment can also be used to give an estimate in terms of Young's modulus alone by estimating the surface energy. This is done by using

$$\gamma = \frac{\sigma_0\, a}{\pi} = \frac{E}{\pi}\left(\frac{a}{a_0}\right)\frac{a}{\pi} = \frac{E}{a_0}\left(\frac{a}{\pi}\right)^2 \tag{4.10}$$

Taking a approximately equal to a_0 gives

$$\gamma = \frac{Ea_0}{\pi^2} \tag{4.11}$$

Using this estimate for γ in the equation for $\sigma_{\text{theoretical}}$ gives

$$\sigma_{\text{theoretical}} = \left(\frac{E\gamma}{a_0}\right)^{1/2} = \left(\frac{E}{a_0} \times \frac{Ea_0}{\pi^2}\right)^{1/2} = \frac{E}{\pi} \tag{4.12}$$

Kelly and MacMillan (1966) discuss various attempts to calculate the theoretical strength more accurately and conclude that for ionic crystals Orowan's estimate is too high by a factor of about 2. Other estimates lead to an often-used rule of thumb that the theoretical strength in tension is about $0.1E$ (Davidge, 1979).

Values estimated for several ceramics and other materials from Orowan's equation are given in Table 4.1.

Whiskers of sapphire (single crystal alumina) have exhibited strength values around 1.5×10^4 MPa and carefully polished bulk crystals gave values up to 0.68×10^4 MPa (Kelly and MacMillan, 1986). Strength values for a good polycrystalline alumina in normal use vary but are typically in the range 275–425 MPa. Evidently the theoretical treatments are correct in predicting breaking strengths above 3×10^4 MPa, but in practical bulk ceramics flaws lower the strength by a factor of about 100 or more from the theoretical value. Strengths as high as $E/5$ have been reported in silica fibers with pristine surfaces.

TABLE 4.1 Values of the Theoretical Breaking Strength from Equation (4.9)

Material	Direction	E (GPa)	γ (J/m^2)	$\sigma_{theoretical}$ (GPa)
α-Iron	<111>	132	2	30
Si	<111>	188	1.2	32
NaCl	<100>	44	.25	6.3
MgO	<100>	245	1.2	37
Al_2O_3	<0001>	460	1	46

Source: Kelly and MacMillan (1986).

4.3 THEORETICAL STRENGTH IN SHEAR

Another important consideration is the theoretical shear strength of solids. Frenkel (1926; summarized by Kelly and MacMillan, 1986) considered that a plane of atoms in a crystal moves rigidly over an adjacent plane and takes a sine force law analogous to that used in the Orowan strength derivation. That is, for a displacement x he assumed that the restoring shear stress τ is given by

$$\tau = k \sin 2\pi x/b \tag{4.13}$$

where b is the distance in the shear direction (i.e., not necessarily along a crystallographic axis) to bring the crystal into registry again. The shear modulus μ is given by

$$\mu = \frac{d\tau}{dx/h} = \frac{2\pi kh}{b} \cos 2\pi x/b \tag{4.14}$$

where h is the distance between planes being sheared. For small displacements $\mu = 2\pi kh/b$. The maximum shear stress occurs when the sine is 1 so that the theoretical shear stress is

$$\tau_{max} = k = b\frac{\mu}{2\pi h} \tag{4.15}$$

For a face-centered cubic material such as many metals $b = a/\sqrt{6}$ and $h = a/\sqrt{3}$, where a is the lattice parameter. This gives

$$\tau_{max} = \frac{\mu}{(2\pi)\sqrt{2}} \approx \frac{\mu}{9} \tag{4.16}$$

For single crystals the appropriate shear constant must be used as described by Kelly and MacMillan (1986). A more refined calculation by Mackenzie as summarized by Kelly and MacMillan leads to different values of τ/μ for different materials. A selection of their results is summarized in Table 4.2.

TABLE 4.2 Values of the Theoretical Shear Stress

Material	Appropriate μ	τ_{max}/μ	τ_{max}	τ_{max}
	(10^4 MPa)		(10^4 MPa)	(10^6 psi)
Diamond	50.5	0.24	12.1	17.6
Fe	6.0	0.11	0.66	1.0
Al_2O_3	14.7	0.115	1.69	2.5

Source: Kelly and MacMillan (1966).

These results on theoretical tensile strength and theoretical shear strength are of great importance to the study of mechanical properties. First, the comparison with values for perfect crystals validates the theoretical tensile strength values, say, within a factor of 2. The much lower values found in most practical material directs attention to the study of flaws (in particular microcracks) and offers hope of achieving much higher strengths if flaws can be eliminated or their behavior modified to increase strength. Much of the design and processing of ceramic materials is directed toward these goals.

The high theoretical shear strength values contrasts with much lower values that are found for ductile metals. This in turn directs attention to another defect, the dislocation, and its behavior in deformation including slip and twinning as subsequently discussed in Chapter 17. Much of the design and processing of metallic alloys is directed toward controlling the behavior of dislocations. For ceramics, dislocations exist yet most polycrystalline ceramics typically show brittle behavior at temperatures below about one-half their melting point. The student needs first to understand how dislocations contribute to plasticity in metals through slip and sometimes twinning and then to understand what limits their motion in ceramics at moderate temperatures and what special effects limit their effectiveness in providing plasticity in ceramics at moderate temperatures in crystal structures in which considerable dislocation mobility is possible.

Consideration of flaws and dislocations and how their interaction with other defects and microstructure affects mechanical behavior is an underlying theme of this book.

CHAPTER 5

LINEAR ELASTIC FRACTURE MECHANICS

5.1 INTRODUCTION

The strength of polycrystalline ceramics is usually much less than that of a corresponding perfect single crystal, as noted earlier. Understanding the strength of real polycrystalline ceramics requires understanding the behavior of microcracks under stress. A completely successful theory of the behavior of microcracks on an atomic level still does not exist, although progress is being made. Linear elastic fracture mechanics is a very useful theory that treats a crack in a continuous body while avoiding the detail of what happens on an atomic scale. This theory is introduced in stages in this chapter using a minimum amount of mathematical complexity. Basic references include Sneddon and Lowengrub (1969), Kelly and MacMillan (1986), Broek (1987), Lawn (1993), and the Fracture Mechanics of Ceramics series (Bradt, Hasselman, and Lange, 1973a, 1973b, 1978a, 1978b; Bradt, Evans, Hasselman, and Lange, 1983a, 1983b; Bradt, Hasselman, Munz, Sakai, and Shevchenko, 1992).

5.2 STRESS CONCENTRATIONS

Stress concentrations are usually discussed in terms of the effect of the thought experiment of introducing a flaw into a body that was originally under uniform stress. Consider a sample initially under a uniform tensile stress, σ. If an elliptical flaw whose maximum length is perpendicular to the tensile stress is introduced, the stress will change, especially near the end of the flaw (Inglis, 1913). Figure 5.1 shows a two-dimensional view of a sample with a flaw of length $2c$ in the x direction and a dimension $2b$ in the y direction (i.e., in the direction of the initial tensile stress). The size of the flaw in Figure 5.1 is exaggerated; $2c$ is assumed to be small compared to the sample width.

The equation of such an ellipse is

$$\frac{x^2}{c^2} + \frac{y^2}{b^2} = 1 \tag{5.1}$$

The radius of curvature at the end of the major axis of the ellipse is given by

$$\rho = b^2/c \tag{5.2}$$

The stress distribution around the flaw is complex (Timoshenko and Goodier, 1951; Kelly and MacMillan, 1986) but a key result can be stated simply. The stress σ_{yy} in the initial stress direction at the end of the crack is given by

$$\sigma_{yy} = \sigma\left(1 + \frac{2c}{b}\right) = \sigma\left[1 + 2(c/\rho)^{1/2}\right] \tag{5.3}$$

It is interesting to note that the latter form of this equation contains only two characteristics of the elliptical crack: the total length and the radius of curvature at the end of the crack perpendicular to the initial stress. These two factors give approximately the correct result for cracks of other shapes.

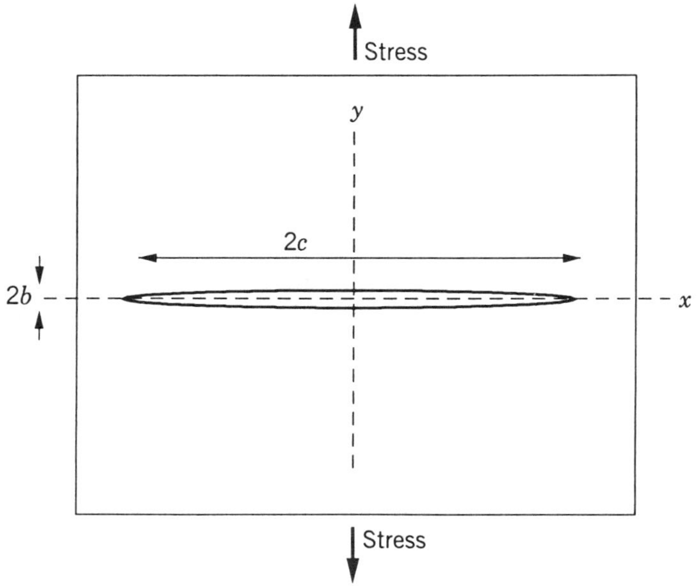

Figure 5.1 An elliptical crack.

To a good first approximation only the length perpendicular to the applied stress and the radius of curvature at the end of that length determine the maximum stress.

It is evident that for long thin cracks the stress concentration can be great and so account for the fact that practical strengths are much less than the theoretical strength.

5.3 THE GRIFFITH THEORY OF FRACTURE OF A BRITTLE SOLID

The treatment above gives the stress at the tip of a smoothly rounded crack of known geometry. One might attempt to proceed by developing a theory to give the stress at the tip of a very sharp crack and set this equal to the theoretical stress as a criterion for fracture. This presents difficulties because very sharp cracks bring atomic dimensions into play for which it is unreasonable to apply continuum elasticity. Griffith (1920, 1924) avoided this difficulty by developing a theory of brittle fracture based on energy considerations. This type of treatment remains the basis of fracture mechanics. The essential ideas of his treatment are given in the following paragraphs. The discussion here is in modern terms rather than in the original terms. An interesting discussion of the development of knowledge of the strength of glass following Griffith's papers is given by Holloway (1983).

Griffith based his treatment of fracture on energy conservation. The first law of thermodynamics states that

$$dQ = dU + dW$$

where U is the internal energy of the specimen (dU is positive for an increase in internal energy), dW is the mechanical work done by the system on its environment (dW is negative when work is done by the external system on the specimen), and dQ is the heat flowing into or out of the specimen (dQ is positive when heat flows into the specimen). We assume rapid crack propagation so that little time is available for heat flow to or from the specimen and so inquire into the terms dU and dW when a crack propagates from a half-length c to $c + dc$ in a specimen under constant applied stress σ corresponding to constant applied force P.

Griffith assumed that the internal energy was the sum $U_e + U_s$, where U_e is the elastically stored energy in the specimen, and U_s is the surface energy of the crack. He took the critical stress for fracture to be the value for which the energy changes associated with crack growth just balance, giving

$$\frac{d(U_e + U_s + W)}{dc} = 0 \tag{5.4}$$

which is the Griffith energy balance condition.

This is a reasonable assumption for a mechanical system but strictly speaking it is only a necessary condition. At present, this condition will be taken as necessary and sufficient and the question of sufficiency will be considered later. For simplicity a plate of thickness t is assumed to be under uniform tensile stress. A crack through the body and of length $2c$ is then introduced. The problem is to write expressions for the elastically stored energy and the surface energy.

The surface energy expression is straightforward and is

$$4\gamma ct \tag{5.5}$$

The total area of new surface created by a crack of length $2c$ is $4ct$ and γ is the surface energy per unit area. It is convenient to take the surface energy U_s per unit length of crack in the thickness dimension and so drop the thickness factor t, giving

$$U_s = 4\gamma c \tag{5.6}$$

Fortunately, there is a general result from the theory of elasticity that for any linearly elastic body undergoing crack extension under constant load the work done by the external loading system as a result of the additional deflection caused by crack growth is equal to minus twice the change in elastically stored energy associated with the crack growth deflection; that is,

$$dW = -2dU_e \tag{5.7}$$

Before going on to insert this relation into the Griffith energy balance condition and proceeding with the fracture stress derivation, we give a proof for simple constant tensile loading. We consider the work done on a specimen and the elastic energy stored in the specimen under constant force with a small central crack of length $2c$ that extends a distance dc. As a preliminary it is interesting to consider the specimen without a crack. As shown in Figure 5.2 the application of a force that increases from zero to P_0 causes a proportional elastic deflection reaching the value u_0. For this case

$$\delta U_e(c{=}0, P{\rightarrow}P_0) = \frac{1}{2}P_0u_0 = -\delta W(c{=}0, P{\rightarrow}P_0) \tag{5.8a}$$

Now consider a specimen with crack length c loaded to $P = P_0$ followed by an infinitesimal crack extension dc under constant load and an accompanying infinitesimal specimen deflection du. The external load does work on the specimen during the crack growth so that the energy of the external loading system changes during crack growth by

$$\delta W(c{\rightarrow}c{+}dc, P{=}P_0) = -P_0du \tag{5.8b}$$

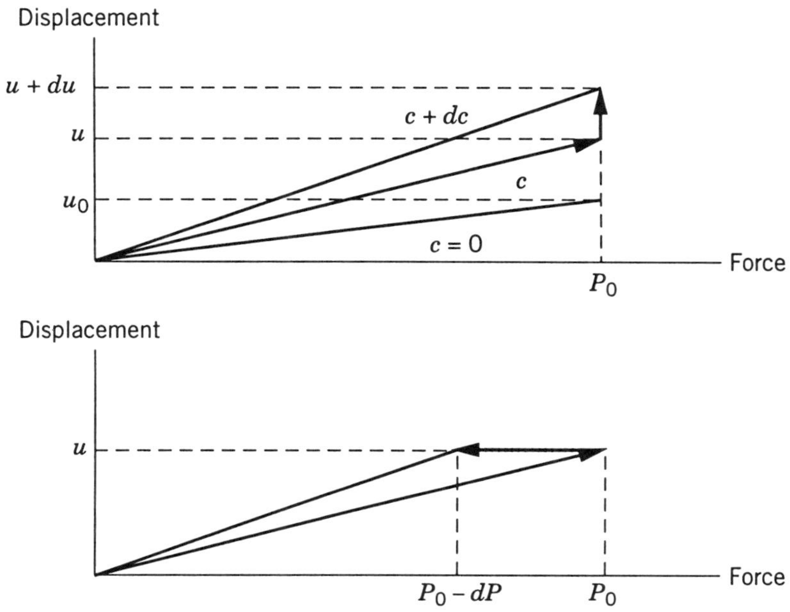

Figure 5.2 Force and displacement during crack growth. Upper portion: constant force. Lower portion: constant displacement.

The change in elastic stored energy in the specimen is

$$dU_e(c \to c+dc, P=P_0) = U_e(c, P=P_0) - U_e(c+dc, P=P_0) \tag{5.9}$$

or

$$dU_e(c \to c+dc, P=P_0) = \frac{1}{2}P_0(u+du) - \frac{1}{2}P_0 u = \frac{1}{2}P_0 du \tag{5.10}$$

so that

$$\delta W(c \to c+dc, P=P_0) = -2dU_e(c \to c+dc, P=P_0) \tag{5.11}$$

which is the relation stated for the general case above. Note that the elastically stored energy increases during crack growth under constant load.

Using the above relation, the Griffith energy balance condition for crack growth under constant load becomes

$$\frac{dU_e}{dc} = \frac{dU_s}{dc} \tag{5.12}$$

To apply this condition to the crack in a body under stress an equation for the elastically stored energy is needed. Griffith (1920) based on Inglis (1913) calculated this quantity for a thin plate (corresponding to the condition of plane stress) and obtained

$$U_e = \frac{\pi c^2 \sigma^2}{E} \tag{5.13}$$

For a thick plate (corresponding to plane strain) the result is

$$U_e = \frac{\pi(1 - \nu^2) c^2 \sigma^2}{E} \tag{5.14}$$

Substituting these expressions into the Griffith crack propagation condition gives the Griffith equations for the fracture stress of a brittle solid under plane stress,

$$\sigma_f = \left(\frac{2E\gamma}{\pi c} \right)^{\frac{1}{2}} \tag{5.15}$$

and under plane strain,

$$\sigma_f = \left(\frac{2E\gamma}{\pi(1 - \nu^2)c} \right)^{\frac{1}{2}} \tag{5.16}$$

The case of infinitesimal crack propagation in a specimen constrained to constant displacement leads to the same Griffith equations. As shown in Figure 5.2 the specimen is initially loaded to P_0, causing a deflection u_0 as before. The specimen is then held at constant displacement while the crack propagates an amount dc. The load drops from P_0 to $P_0 - dP$. No work is done on the specimen; that is,

$$\delta W(c \rightarrow c + dc, u = \text{constant}) = 0 \tag{5.17}$$

Hence Griffith's energy balance condition for crack growth during constant displacement takes the form

$$-\frac{dU_e}{dc} = \frac{dU_s}{dc} \tag{5.18}$$

In this case of constant displacement the elastically stored energy decreases as the crack extends so that the left-hand side of this equation is positive, allowing the equation to be satisfied. The same Griffith equations for σ_f follow as before.

5.4 STRESS AT THE CRACK TIP—AN ESTIMATE

The Griffith equations are based on energy considerations and therefore provide a necessary condition for fracture to occur as noted above. The question remains whether the stress is actually high enough to break a bond at the crack tip when the Griffith condition is met so that this condition is also a sufficient condition for fracture. The following approximate argument leads to the plausible conclusion that it is. The Inglis condition for stress concentration can be combined with the Orowan estimate of theoretical strength to give an expression for fracture stress and the latter can be compared with the Griffith condition.

The Orowan estimate for the theoretical stress was developed in Chapter 4 and is

$$\sigma_{\text{theoretical}} = \left(\frac{E\gamma}{a_0}\right)^{\frac{1}{2}} \tag{5.19}$$

The maximum tensile stress at the tip of a sharp crack is given by the Inglis expression as

$$\sigma_{\max} = 2\sigma\left(\frac{c}{\rho}\right)^{\frac{1}{2}} \tag{5.20}$$

σ is the uniform tensile stress applied to the specimen. For fracture

$$\sigma_{\max} = \sigma_{\text{theoretical}} \tag{5.21}$$

and the applied stress is the fracture stress σ_f, giving

$$\sigma_f = \left(\frac{E\gamma}{4c}\frac{\rho}{a_0}\right)^{\frac{1}{2}} \tag{5.22}$$

The crack tip radius for an atomically sharp crack is about half the interatomic spacing a_0, giving

$$\sigma_f = \left(\frac{E\gamma}{8c}\right)^{\frac{1}{2}} \tag{5.23}$$

This stress is actually lower than the Griffith stress so that the Griffith condition can be taken to be a sufficient as well as a necessary condition. This is the conclusion adopted in the literature; Lawn (1993) gives an argument in support of this conclusion.

5.5 CRACK SHAPE IN BRITTLE SOLIDS

Accurate calculation of the crack shape in solids remains a challenging problem (Kelly and MacMillan, 1986; Lawn, 1993). It is important, however, to keep in mind the qualitative features that are expected of real cracks in stressed solids as the approximate models of cracks used in fracture mechanics are considered. Figure 5.3 shows a sketch of the shape expected for a real crack under stress. On the left-hand side the uncracked material is bonded and has the normal interatomic spacing slightly expanded by elastic deformation. Far to the right the opposite surfaces of the crack are effectively beyond the range of interatomic forces and are free of force. In the intermediate region atoms on one crack surface experience an attractive force from atoms on the opposite surface. The crack opens smoothly and it is somewhat arbitrary where the crack edge is taken.

This picture of a crack differs from the fracture mechanics model of a crack. The latter assumes a well-defined edge and generally assumes no forces between the opposite faces of a crack behind the edge. The fracture mechanics model has the advantage of allowing a continuum elasticity treatment of stresses in an intermediate region that is near, but not too near, the crack tip. One way of interpreting the success of fracture mechanics is to argue that whatever the true state of affairs is in the region of atomic bonding for the true crack, it will be the same in two specimens of the same material provided the stresses in the intermediate region are the same. It can then be assumed that fracture will occur whenever a critical stress state is reached in the intermediate region. Fracture mechanics allows a rigorous connection to be established between the stresses in this intermediate region and the external forces acting on a specimen. Thus fracture mechanics allows any loading geometry to be treated in terms of the assumed critical state of affairs in the intermediate region. Fracture mechanics thus has great practical value as well a providing insight.

5.6 THE IRWIN FORMULATION OF FRACTURE MECHANICS—THE STRESS INTENSITY FACTOR

Linear fracture mechanics as developed by Irwin and many others can be introduced in a manner analogous to the way that fracture stress and the

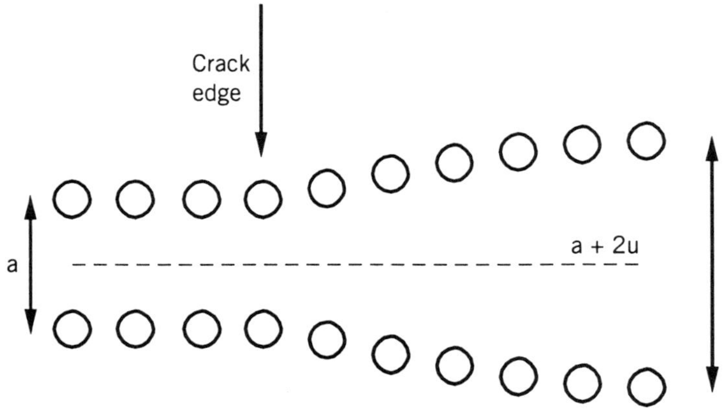

Figure 5.3 Atomic planes on either side of crack.

Griffith condition were discussed. An important new feature is that by concentrating attention on stresses in an intermediate region, the stress-based and energy-based treatments can be shown to be equivalent. Irwin (1957) utilized a solution by Westergaard (1939) for the stresses near the tip of a crack in the xz plane with its edge defined by x = constant. For any system of distant loads the stresses near the crack tip can be expressed in terms of polar coordinates r, θ, z and expanded in a power series in r. There are three types of crack opening to be considered corresponding to a separation of the crack faces in the y, x, and z directions and termed, respectively, mode I, mode II, and mode III crack opening. The first term of the series expansion of stress as a function of r dominates for small values of r. The stresses σ_{ij} and the displacements u_i have the form

$$\sigma_{ij,m} = \frac{K_m}{(2\pi r)^{1/2}} f_{ij,m}(\theta) \tag{5.24}$$

$$u_{i,m} = \frac{K_m}{(2\pi r)^{1/2}} f_{i,m}(\theta) \tag{5.25}$$

Here the subscripts i and j take the values 1, 2, 3 and indicate the coordinate directions. The subscript m takes the values I, II, III and indicates the mode of crack opening. The complete functions are given by Lawn (1993). The tensile stresses parallel to the direction of crack advance, perpendicular to the crack plane, and parallel to the crack front are, respectively,

$$\sigma_{xx,I} = \frac{K_I}{(2\pi r)^{1/2}} \cos\frac{\theta}{2} \left[1 - \sin\frac{\theta}{2} \sin\frac{3\theta}{2}\right] \tag{5.26}$$

$$\sigma_{yy,I} = \frac{K_I}{(2\pi r)^{1/2}} \cos\frac{\theta}{2} \left[1 + \sin\frac{\theta}{2} \sin\frac{3\theta}{2}\right] \tag{5.27}$$

$$\sigma_{zz} = \nu' (\sigma_{xx} + \sigma_{yy}) \tag{5.28}$$

For plane stress $\nu' = 0$ and for plane strain $\nu' = \nu$.

The equations for displacements in the x and y directions are

$$u_{x,I} = \frac{K_I}{2E}\left[\frac{r}{2\pi}\right]^{1/2} (1+\nu)\left[(2\kappa - 1)\cos\frac{\theta}{2} - \cos\frac{3\theta}{2}\right] \tag{5.29}$$

$$u_{y,I} = \frac{K_I}{2E}\left[\frac{r}{2\pi}\right]^{1/2} (1+\nu)\left[(2\kappa + 1)\sin\frac{\theta}{2} - \sin\frac{3\theta}{2}\right] \tag{5.30}$$

where

$$\kappa = \frac{3-\nu}{1+\nu} \quad \text{(for plane stress)} \tag{5.31}$$

and

$$\kappa = 3 - 4\nu \quad \text{(for plane strain)} \tag{5.32}$$

and where ν is Poisson's ratio. The crack profile in mode I loading is interesting in itself and is used in various theories for toughening. The crack profile is

$$u_{y,I}(\theta{=}\pi) = \frac{4K_I}{E}\left[\frac{r}{2\pi}\right]^{1/2} \quad \text{for plane stress} \tag{5.33}$$

$$u_{y,I}(\theta{=}\pi) = \frac{4K_I}{E}\left[\frac{r}{2\pi}\right]^{1/2} \quad \text{for plane stress} \tag{5.33}$$

and

$$u_{y,I}(\theta{=}\pi) = \frac{4K_I}{E}\left[\frac{r}{2\pi}\right]^{1/2}(1 - \nu^2) \quad \text{for plane strain} \tag{5.34}$$

The factor K depends on the applied loading and the specimen geometry and can be worked out for any given type of specimen and loading. The radial and angular dependence are separable factors. The reason for the remarks above about considering stresses in an intermediate region now become apparent. The distance r cannot be taken too small. Continuum elasticity cannot be expected to apply at atomic dimensions and the divergence of the stress equation if r is allowed to approach zero indicates the inadequacy of the treatment for extremely small r values. If r is allowed to increase too much, higher order terms become appreciable in the power series. However in the intermediate region three parameters, K_{I}, K_{II}, and K_{III} (with dimensions stress $\times$ length$^{1/2}$) completely specify the stress for any type of external loading as far as the effect of crack propagation is concerned. For ceramics, it is usually assumed that mode I behavior is overridingly important so that the other K values are ignored. This is perhaps justified for polycrystalline specimens by the argument that flaws of all orientations will occur and that failure will take place from the largest flaw normal to the greatest tensile stress for which K_{I} is the stress intensity factor driving stress. Fracture control by K_{I} is assumed here but the behavior of a flaw under compound loading leading to simultaneous action of K_{I} and K_{II} is discussed later. Under this assumption any combination of specimen shape, external loading, and crack configuration leads to a value of K_{I} acting at the tip of the crack and so a single number characterizes the combined effect of all these factors on the driving force for crack propagation. Calculating this number for a particular combination of specimen shape, loading configuration and amount, and crack configuration may be a difficult task in practice but it is just a problem in continuum elasticity that can be solved for each case by numerical methods with computer assistance. For special specimen shapes, loads, and crack configurations there are simple solutions. Some of the most commonly used of these are summarized in Chapter 6.

The central assumption of linear elastic fracture mechanics is that for a given material there is some critical value of K_I, usually termed K_{IC}, at which the crack is in equilibrium and above which rapid crack propagation will take place. The fracture condition in this formulation is taken as the equilibrium point so that the fracture condition becomes

$$K_I = K_{IC} \tag{5.35}$$

The quantity on the left is the applied stress intensity factor and that on the right is a property of the particular material under consideration. This equation is the analog of setting the applied stress at the tip of an elliptical crack equal to the theoretical stress. In the present formulation the parameter K_{IC} is usually not calculated in terms of other properties (as was done with the theoretical strength) but is measured. The idea is that K_{IC} can be measured for a particular material using a convenient specimen and loading system and the resulting value can then be used in design with the same material for more complex shapes and loads. This scheme works well up to a point and this view will be taken for now. A note of caution is in order concerning the fact that some ceramics show a dependence of K_{IC} on crack length. This is termed R-curve behavior and will be discussed later. Another important departure from this simple scheme is that some ceramics show slow crack propagation at K_I values below K_{IC}. Again, this will be taken up later. Setting aside these complications for the time being permits building up the basic treatment, which is very useful and which can subsequently be modified to take into account these complications.

5.7 THE IRWIN FORMULATION OF FRACTURE MECHANICS—THE ENERGY RELEASE RATE

The Griffith treatment of fracture was formulated in terms of energy. The Irwin formulation can also be expressed in terms of energy and related to the formulation in terms of stress intensity factor. As in the case of the Griffith treatment the portion of the energy change that is available to drive any processes associated with creation of new crack surfaces is considered. In Irwin's formulation the amount of energy available per unit of new crack surface, A, from the mechanical work and change of internal energy is termed G, the energy release rate (with dimensions of energy/area, not energy/time), and is

$$G = -\left(\frac{dW}{dA} + \frac{dU_e}{dA}\right) \tag{5.36}$$

This is equivalent to the differentiations in the Griffith energy balance condition as previously stated; there the differentiation was with respect to $c(dA = t\,dc)$, where we had taken $t = 1$. The Griffith energy balance condition can be restated in terms of a crack resistance "force" (again with dimensions of energy/area) R as

$$G = R \tag{5.37}$$

Note that this definition does not make any assumption as to how the energy required for crack propagation is divided into portions going into new thermodynamic surface energy and into other processes such as local plastic deformation (Evans and Langdon, 1976). For the case of all the energy required being surface energy, the crack resistance force is

$$R = \frac{dU_s}{dA} = 2\gamma \tag{5.38}$$

The energy release rate can be expressed in terms of the internal energy by using the relations derived in Section 5.2 as follows. For crack propagation under constant load $dW = -2\,dU_e$ so that

$$G = \frac{dU_e}{dA} \quad \text{(constant load)} \tag{5.39}$$

while for crack propagation under constant displacement $dW = 0$ so that

$$G = -\frac{dU_e}{dA} \quad \text{(constant displacement)} \tag{5.40}$$

Since the magnitude of G is the derivative of the elastic energy with respect to area in both cases, G is often termed the strain energy release rate despite the fact that the strain energy actually increases during crack propagation under constant load.

The relation between G and K_I can be derived by taking the case of a crack in a specimen with fixed displacement so that $dW = 0$ (Irwin, 1957; Jayatilaka, 1979). The result is independent of the condition assumed but this assumption simplifies the calculation. Then

$$G \equiv -\frac{dU_e}{dA} = \lim_{\Delta x \to 0} \frac{2}{\Delta x} \int_0^{\Delta x} \frac{\sigma_{yy}(\theta = 0)}{2} u_y(\theta = \pi)\,dx \tag{5.41}$$

Here the factor 2 in front of the integral arises from the fact that both surfaces of the crack contribute to the energy as they move against the stress. The factor of 2 in the denominator inside the integral arises from the proportionality of stress and displacement as follows. At each value of x in the integration, the expression for the work done per unit area is

$$\int_0^{\sigma_{max}} \sigma \, du \tag{5.42}$$

Since the stress is proportional to the displacement, one can take $u = u_{max}$ σ/σ_{max} so that the integral can be written

$$\int_0^{\sigma_{max}} \sigma \, du = \frac{u_{max}}{\sigma_{max}} \int_0^{\sigma max} \sigma \, d\sigma = \frac{1}{2} \sigma_{max} u_{max} \tag{5.43}$$

The σ value in the G integral is thus the value before the crack opens and so is

$$\sigma_{yy}(r = x, \theta = 0) = \frac{K_I}{(2\pi r)^{1/2}} \tag{5.44}$$

but the u_y value in the G integral is the value after the crack opens and so is

$$u_y(r = \Delta x - x, \theta = \pi) = \frac{K_I}{E}\left(\frac{\Delta x - x}{2\pi}\right)^{\frac{1}{2}} (1 + \nu)(\kappa + 1) \tag{5.45}$$

The energy release rate is then

$$G = \frac{1}{\Delta x} \frac{K_I^2}{2E\pi} (1 + \nu)(\kappa + 1) \int_0^{\Delta x} \left(\frac{\Delta x - x}{x}\right)^{\frac{1}{2}} dx \tag{5.46}$$

Using the substitution $w = x/\Delta x$ leads to a definite integral given in tables; alternately, the subsequent substitution $w = \cos^2 \theta$ gives a form that is directly integrable. Either gives

$$G = \frac{K_I^2}{4E} (1 + \nu)(\kappa + 1) \tag{5.47}$$

or

$$G = \frac{K_I^2}{E} \quad \text{(for plane stress)} \tag{5.48}$$

$$G = \frac{K_I^2}{E}(1 - \nu^2) \quad \text{(for plane strain)} \tag{5.49}$$

The contributions to G by K_{II} and K_{III} are often treated similarly (e.g., Lawn, 1993), leading to the equations

$$G = \frac{K_{\mathrm{I}}^2}{E} + \frac{K_{\mathrm{II}}^2}{E} + \frac{(1+\nu)K_{\mathrm{III}}^2}{E} \quad \text{(for plane stress)} \tag{5.50}$$

$$G = \frac{(1-\nu^2)K_{\mathrm{I}}^2}{E} + \frac{(1-\nu^2)K_{\mathrm{II}}^2}{E} + \frac{(1+\nu)K_{\mathrm{III}}^2}{E} \quad \text{(for plane strain)} \tag{5.51}$$

However, the direction of crack propagation in mode II is not necessarily in the plane of the crack, so that this procedure is questionable and therefore the above equations for G in terms of the K's may not be accurate as far as K_{II} is concerned (Jayatilaka, 1979). This delicate point is generally not pertinent to ceramics because the latter are generally taken to be governed by K_{I} alone. The idea of additivity of different contributions to G or K_{I} in ceramics is, however, of great importance, as will be seen in later discussions of toughening.

The energy release rate as developed above can now be used to define an empirical "fracture surface energy," γ_f, that can be much larger than the thermodynamic surface energy. At the beginning of the discussion of energy release rates, the thermal contribution in the first law of thermodynamics was taken to be zero for crack propagation. There is evidence for radiated energy in the phenomenon of acoustic emission and there may be other means of energy dissipation, as in some degree of plasticity. All these energy contributions are collected into the fracture surface energy so that the crack resistance force R is defined by

$$R = 2\gamma_f \tag{5.52}$$

This gives a modified Griffith equation for fracture:

$$\sigma_f = \left(\frac{2E\gamma_f}{\pi c}\right)^{\frac{1}{2}} \quad \text{(for plane stress)} \tag{5.53}$$

The measurement of fracture surface energies is discussed in Chapter 6 and the use of fracture mechanics concepts forms a principal theme in the subsequent treatment of the design of microstructures and the design for safe life service. We anticipate a result from that chapter to show the equivalence of the Griffith and Irwin formulations of fracture mechanics. For uniform tensile stress on a specimen, elastic theory gives for the applied stress intensity factor for a thin specimen $K = Y\sigma c^{1/2}$, where Y is a dimensionless factor that depends on the geometry of the crack and is of order one. For fracture

$$K_{IC} = Y\sigma_f c^{1/2} \tag{5.54}$$

Reverting to γ rather than γ_f for ease in comparing with earlier equations, the Irwin fracture condition is

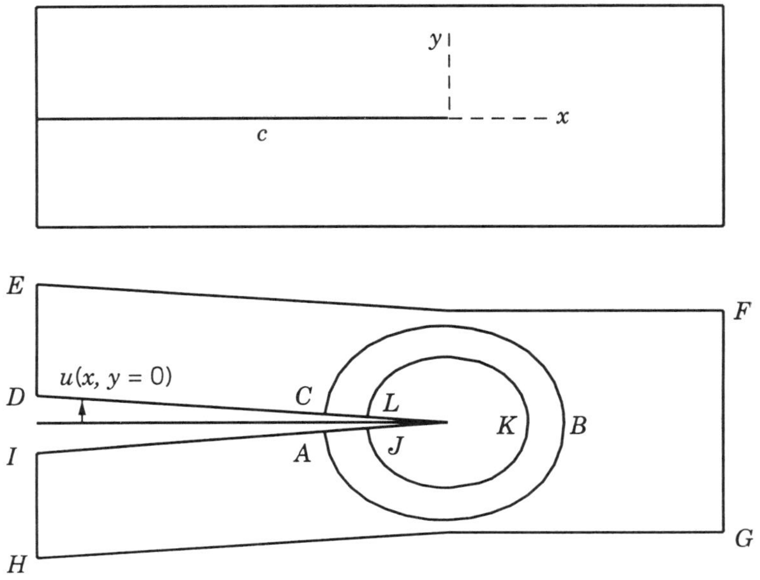

Figure 5.4 J-Integral contours.

$$R = 2\gamma = G = \frac{K_{IC}^2}{E} = \frac{(Y\sigma_f c^{1/2})^2}{E} \tag{5.55}$$

This gives

$$\sigma_f = \left(\frac{2\gamma E}{Y^2 c}\right)^{1/2} \tag{5.56}$$

For a through crack under plane stress, $Y = \pi^{1/2}$, giving

$$\sigma_f = \left(\frac{2\gamma E}{\pi c}\right)^{1/2} \tag{5.57}$$

which is the Griffith condition for plane stress.

5.8 THE J-INTEGRAL

The Irwin strain energy release rate, G, is defined for linearly elastic behavior. It can be generalized to nonlinear elasticity by defining a quantity J that is equal to G for linear elastic behavior (J. Rice, 1968a, 1968b; Broek, 1987; Lawn, 1993). This quantity also permits the derivation of a relation that is useful in treating toughening by partial bridging of cracks and so it is briefly discussed here. Furthermore, the J-integral can be used to formulate a condition for the initiation of cracking in materials that undergo plastic deformation, but care is needed in this case and the reader is referred to texts dealing with fracture involving plastic deformation for information on this subject (e.g., Broek, 1987).

The fundamental idea underlying the J-integral approach to fracture is still that of energy conservation from the first law of thermodynamics applied to the case where only mechanical energy changes are involved, as in the Griffith and Irwin theories. As in the case of the Griffith theory, we consider an edge crack of length c extending through a specimen and take the thickness to be of unit length. The problem can then be treated in two-dimensions as before. We consider any portion of a two-dimensional specimen bounded by a closed curve and consider the changes in mechanical energy as a crack extends infinitesimally. An example of such a contour is *ABCDEFGHIA* shown in Figure 5.4. Changes in the mechanical energy of the portion of the specimen within the contour will depend only on changes in the strain energy per unit volume, U, and the combination of the force vector per unit area acting on the surface, $\mathbf{F}$ (termed the traction), with the displacement vector of the surface, $\mathbf{u}$. In terms of these quantities the mechanical energy per unit length perpendicular to the plane of the contour and within the contour is

$$U_M = \int_A U dA - \int_S \mathbf{F} \cdot \mathbf{u}\, ds \tag{5.58}$$

where

$$F_i = \sum_i \sigma_{ij} n_j \tag{5.59}$$

and the n_j are the components of the unit vector normal to the contour at the point under consideration. Note that the first integral is over the area and the second is a line integral taken along the curve bounding the area. Rice showed that the energy release rate could be expressed as a line integral that he defined as J. That is,

$$-\frac{dU_M}{dc} = \int_S [\, U dy - \mathbf{F} \cdot \frac{\partial \mathbf{u}}{\partial x} ds] \equiv J \tag{5.60}$$

This integral can be shown to be zero for any contour, such as *ABCDEFGHIA*, that is taken to enclose a portion of the material free of body and surface forces and that does not enclose a singularity. Here we have been careful to take points A and C sufficiently far from the crack tip to be outside the crack cohesion zone so that the crack surfaces CD and IA are free of traction. We can take a second contour *ABCLKJA* still remaining outside the cohesion zone and write J as the sum of four contributions from successive parts of the contour:

$$J_{ABCLKJA} = J_{ABC} + J_{CL} + J_{LKJ} + J_{JA} = 0 \qquad (5.61)$$

The crack before the opening displacements is shown in the upper portion of Figure 5.4 and is described by $y = 0$ along $x = 0$ to $x = -c$, where c is the crack length so that $dy = 0$ in J_{CL} and J_{JA}. Also, for the portions of the crack surface CL and AJ the traction is zero so that

$$J_{CL} = J_{JA} = 0 \text{ and } J_{ABC} = -J_{LKJ} \qquad (5.62)$$

If we adopt the convention that the J-integral around a crack tip should be taken in a counterclockwise sense we wish to have J_{JKL}. Reversing the direction of integration along a portion of the contour changes the sign of the integral so that

$$J_{JKL} = -J_{LKJ} \qquad (5.63)$$

and we finally have

$$J_{ABC} = J_{JKL} \qquad (5.64)$$

Thus the contour can be drawn along any path from one face of the crack to the other as long as the end points on the contour are outside the cohesion zone; that is, with this constraint of the end points the J-integral is path-independent. The J-integral will be used to obtain an important result for partially bridged cracks in the next section.

5.9 CRACKS WITH INTERNAL LOADING

So far we have considered the crack faces to be free of traction apart from the short cohesion zone near the crack tip. The applied stress intensity factor arises from loads applied to the external surface of the specimen in this case. When traction on the face of a crack exists (as is the case with a crack filled with liquid under pressure, for example), we expect an additional contribution to the applied stress intensity factor. An important type of internally loaded crack results when the crack is partially bridged by fibers or grains that are elastically loaded and so exert a closure stress on the crack faces and lead to toughening of the ceramic. We shall discuss this in more detail in later chapters and here simply summarize some of the fundamental relations that underlie treatment of this form of toughening.

If a tensile stress $\sigma_{yy} = \sigma(x)$ acts on each face of the crack surface, where x is the distance from the crack tip, it can be shown (Sneddon and Lowengrub, 1969; Lawn, 1993) that the contribution to the stress intensity factor for a straight crack is

$$K = 2\left(\frac{c}{\pi}\right)^{\frac{1}{2}} \int_0^c \left[\frac{\sigma(x)}{(c^2 - x^2)^{1/2}}\right] dx \tag{5.65}$$

and that for a penny-shaped crack of radius c it is

$$K = \frac{2}{(\pi c)^{1/2}} \int_0^c \frac{r\sigma(r)}{(c^2 - r^2)^{1/2}} dr \tag{5.66}$$

For loads that are applied directly to the crack surfaces (as in the case of internal hydraulic loading), these equations can be applied directly. One would like to apply these formulas to partially bridged cracks and treat the resulting K as a contribution to the critical stress intensity factor. However, for the case of cracks partially bridged by elastic fibers or grains, the closure stress at a point depends on the crack opening $2u$ at that point. This in turn depends on the entire stress distribution over the crack surface. The result is that, before the appropriate one of the two equations above can be used to calculate the K contribution from the closure stress, two equations for this stress must be solved simultaneously by numerical means or simplifying approximations must be made by assuming a crack opening profile. Both procedures have been done by Marshall and Evans (1986) for the case of a fully bridged crack and their results will be discussed in Chapter 10. To illustrate the problem we anticipate some of their steps here. For a volume fraction V_f of fibers with Young's modulus E_f and radius r that require an interfacial sliding stress τ in a matrix with Young's modulus E_m with a crack opening displacement at x of $2u(x)$, the closure stress on the faces of the crack is a function of u given by

$$\sigma(x) = 2\left[\frac{u\tau V_f^2\, E_f(1 + E_f V_v) / [E_m(1-V_f)]}{r}\right]^{1/2} \tag{5.67}$$

However, the crack opening displacement for a crack under both a remote uniform stress σ_∞ and a closure stress $\sigma(x)$ is given in general by

$$u(x) = \frac{4(1-\nu^2)c}{\pi E_c} \int_x^1 \frac{1}{(s^2 - x^2)^{1/2}} \int_0^s \frac{[\sigma_\infty - \sigma(x)]t\,dt}{(s^2 - t^2)^{1/2}} ds \tag{5.68}$$

where s and t are normalized position coordinates. These are the equations that must be solved simultaneously for $\sigma(x)$ before the K equation can be used.

Fortunately, the J-integral can be used to derive a relationship for the contribution to the resistance to crack propagation for partially bridged cracks that requires knowledge of the stress as a function of crack opening displacement and the value of the crack opening displacement itself at only one point. We consider a crack with a bridged region near the tip giving rise to a stress $\sigma(x)$. Turning again to Figure 5.4, we consider the contour LKJ and shrink it to the surface of the crack itself; that is, the contour then runs from L along the upper surface of the crack to the tip and back to J along the lower surface. Also, the points L and J are taken at the end of the bridging zone. Along the upper and lower crack surfaces $dy = 0$ so that

$$J = 2\int_0^{u_L} p(x)\frac{du}{dx}\,dx \tag{5.69}$$

where p is stress. Recognizing that

$$x = x(u) \text{ so } p(x) = p(u) \tag{5.70}$$

leads to

$$J = 2\int_0^{u_L} p(u)\,du \tag{5.71}$$

This result allows the toughening due to bridging (the J value from bridging) to be calculated from knowledge of the displacement and the bridging force as a function of displacement at the end of the bridging zone rather than requiring complete knowledge of these quantities over the whole bridged portion of the crack length.

5.10 FAILURE UNDER MULTIAXIAL STRESS

Throughout most of the present book the treatment of brittle fracture is focused on cracks under simple tensile stress perpendicular to the crack plane. This is justified for many situations of interest to ceramic engineers, as in uniaxial tensile or flexural testing of polycrystalline ceramics. Here, even though cracks are presumably found in all orientations with respect to the stress axis, for a random distribution of crack orientations failure should occur from a crack at the maximum end of the crack size distribution that is oriented perpendicular to the uniaxial stress axis. However, situations involving multiaxial loading can arise so that the behavior of cracks under compound loading should be studied. The behavior of a crack under mixed mode loading is treated by Jayatilaka (1979) and Broek (1987). The treatment of mixed mode loading is complex and remains somewhat unsatisfactory. We limit ourselves to treatment of two-dimensional mixed mode loading and content ourselves with showing how the most general two-dimensional case can be reduced to consideration of an inclined crack in a tensile stress and to presenting the results of the simplest mixed mode fracture criterion.

In the most general case of two-dimensional loading of a central crack in a body, there are three stress components—σ_{xx}, σ_{yy}, and $\sigma_{xy} = \sigma_{yx}$. From the Mohr circle construction one can find an inclined coordinate system in which there are no shear components. In this principal axis system the crack will be inclined to the principal stress axes. Taking the angle of inclination to the axis for the principal stress σ_1 to be β, the stress intensity factors expressed in the principal stress axis system from this principal stress for a penny-shaped crack of radius c are

$$K_I = \sigma_1 (\pi c)^{1/2} \sin^2 \beta \tag{5.72}$$

$$K_{II} = \sigma_1 (\pi c)^{1/2} \sin \beta \cos \beta \tag{5.73}$$

Each of the two principal axes produces applied stress intensity factors K_{I} and K_{II}; that is, mixed mode I and mode II loading of the crack. The contributions of each of the principal stresses can be added for each of these two modes so that the problem reduces to the treatment of a simple tension acting on an inclined crack.

As noted earlier in this chapter, the general energy release rate is

$$G_t = G_I + G_{II} + G_{III} = \frac{1-\nu^2}{E}\left(K_I^2 + K_{II}^2 + \frac{K_{III}^2}{1-\nu}\right) \tag{5.74}$$

Fracture is assumed to occur when the energy release rate exceeds the energy consumption rate. For the combination of mode I and mode II but not mode III, this condition becomes

$$\frac{E}{1-\nu^2} G_t = K_I^2 + K_{II}^2 = \text{constant} = K_{IC}^2 \tag{5.75}$$

where it is assumed that the crack propagates only in its initial plane so that the constant can be set equal to K_{IC}^2. The failure criterion then becomes

$$\left(\frac{K_I}{K_{IC}}\right)^2 + \left(\frac{K_{II}}{K_{IC}}\right)^2 = 1 \tag{5.76}$$

which gives the locus of fracture in the $K_{\mathrm{I}}, K_{\mathrm{II}}$ plane as a circle. It is more likely that $K_{\mathrm{I}}C \neq K_{\mathrm{II}}C$, giving

$$\left(\frac{K_I}{K_{IC}}\right)^2 + \left(\frac{K_{II}}{K_{IIC}}\right)^2 = 1 \tag{5.77}$$

which gives the locus of fracture in the K_I, K_{II} plane as an ellipse.

However, it is commonly observed that cracks placed under mixed mode loading extend at an angle to their original planes. Two criteria for such propagation under mixed mode loading have been proposed: the maximum principal stress criterion and the strain energy density criterion (Broek, 1987). The two criteria give very similar results although neither criterion is free from some objection. Experimental examples given by Broek fit the maximum principal stress criterion reasonably well but also are in fair agreement with the above simple ellipse result despite the experimental fact that cracks under mixed mode loading are observed to extend at an angle to their original plane and so do not fit the assumption made in deriving the above simple result. The reader is referred to Jayatilaka (1979) and Broek (1987) for further details.

One might suppose that the issue of an appropriate failure criterion for ceramics under multiaxial loading could be settled by experiment and attempts have been made to do so. Thiemeier and Bruckner-Foit (1991) considered several failure criteria: mode I failure, coplanar energy release rate, maximum hoop stress factor, minimum strain-energy density, maximum noncoplanar energy release rate, and an empirical criterion of Richard. They developed predictions made with these criteria for the relation of strength in four-point bending to strength measured in a ring-on-ring biaxial flexure test and compared the results to experiments on aluminum nitride. All criteria with the exception of mode I failure, coplanar energy release rate, and minimum strain-energy density for plane strain gave reasonable agreement with experiment. However, the scatter in the data did not permit a unique failure criterion to be determined. The issue of the statistical treatment of fracture under multiaxial stress is discussed in Section 7.11.

Mixed mode fracture toughness has been studied in polycrystalline alumina by Suresh, Shih, Morrone, and O'Dowd (1990) using a four-point bending technique with an asymmetrical notch. They found that the mode II fracture toughness was comparable to the mode I toughness. Combining these results with an earlier study by Suresh and Tschegg (1987), they concluded that the mode III toughness was 2.3 times higher than the mode I value.

CHAPTER 6

MEASUREMENTS OF ELASTICITY, STRENGTH, AND STRESS INTENSITY FACTORS

6.1 INTRODUCTION

This chapter presents a tabulation of the commonly used methods for measurement of elastic moduli, strength and fracture toughness of ceramics. Many tabulations are available (Jayatilaka, 1979; Young, 1989; Lawn, 1993). The reader embarking on a testing program would be well advised to review recent work on developing standards for mechanical and related tests; these are briefly discussed in Chapter 22 and listed in Table 22.2 (G. D. Quinn, personal communication, 1995).

The focus is first on elasticity and strength tests for which nominally uncracked specimens are used. That is, no artificial crack is introduced. The sensitivity of ceramics to surface cracks and the possibility of introducing surface cracks in sawing and grinding the specimen to accurate dimensions are special concerns. The surface condition is usually made as close as possible to that to be used in service when the intention is to produce strength data for design. For research on strength improvement theory, microstructure design, and processing, the surfaces are prepared to be as free as possible of cracks not characteristic of the general microstructure. Although this class of test with nominally uncracked specimens can be used for both elastic moduli and strength determinations, the accuracy of elastic moduli determination is typically a few percent at best. An alternative approach using resonance or ultrasonic methods gives much better accuracy and is summarized in Section 6.7.

The second focus is on fracture toughness test methods in which a crack is deliberately introduced and a critical stress intensity factor or equivalently a fracture surface energy is measured. Measurement in ductile materials such as many metals presents special problems and requires special test procedures. The tests considered here are those appropriate to ceramics characterized by brittle fracture. Three conditions are required for successful fracture toughness testing (Jayatilaka, 1979): (1) the specimen geometry and loading system must allow calculation of the applied stress intensity factor; (2) the crack length and load at the beginning of rapid crack propagation must be accurately determined; and (3) the initial crack must be sharp. ASTM Committee C-28 is developing a standard on fracture toughness testing; the proposed standard recommends three methods: the single edge precracked beam method, the chevron notch method, and the surface crack in flexure method (G. D. Quinn, personal communication, 1995). All are discussed in the present chapter.

For metals the development of a zone of plastic deformation near the crack tip must be taken into account. For most ceramics at room temperature this complication does not exist. However, ceramics at high temperatures do show plastic effects so that plasticity considerations are briefly mentioned here. A characteristic parameter for the size of the plastic zone is $(K_{IC}/\sigma_y)^2$, where K_{IC} is the critical stress intensity factor and σ_y is the yield stress. When the yield stress is difficult to measure directly, it is estimated by taking $\sigma_y = H/3$, where H is the hardness. $(K_{IC}/\sigma_y)^2$ is typically about 4 mm for a strong metal but is only about 0.1 mm for a brittle material. Both the crack length and the specimen thickness should exceed $2.5(K_{IC}/\sigma_y)^2$ (Jayatilaka, 1979).

The second category of test methods (i.e., using specimens with deliberately introduced cracks) is also useful for tests in which a chosen value of stress intensity factor below the critical stress intensity factor is to be applied. Such test methods are useful, for example, in studying slow crack propagation.

The test methods covered in this chapter are appropriate for bars. Test techniques for optical fibers have been summarized by Matthewson (1994).

6.2 TENSILE TESTS

A straight elastic bar of uniform cross section, A, is assumed. The loads (tensile or compressive) are applied at the ends uniformly distributed over the cross section. In practice these conditions often apply only to the central portion of a longer specimen with larger ends to which the loads are applied. The y axis is taken along the axis of the bar as shown in Figure 6.1. Tensile stress is denoted by σ and shear stress by τ as a function of the angle θ between the axis of the bar and the normal to the plane on which the stress acts. P is the applied load, E is Young's modulus and ν is Poisson's ratio.

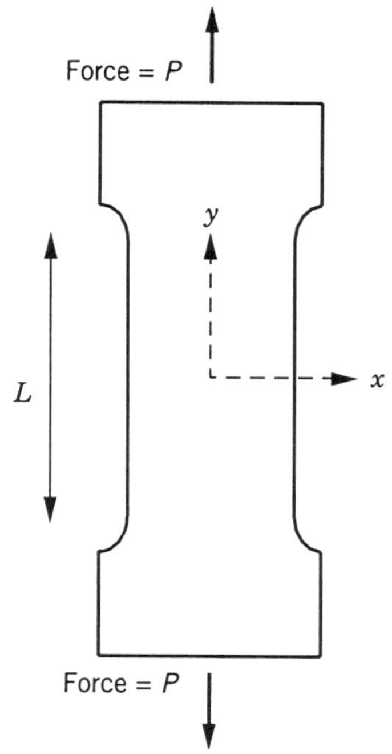

Figure 6.1 Schematic of a tensile test.

The resulting stresses do not depend on the elastic properties and are (Young, 1989)

$$\sigma_{yy} = \sigma\,(\theta = 0) = \frac{P}{A} \tag{6.1}$$

$$\sigma\,(\theta) = \frac{P}{A}\cos^2\theta \tag{6.2}$$

The maximum value of $\sigma(\theta)$ occurs when $\theta = 0$ and is P/A.

$$\tau\,(\theta) = \frac{P}{2A}\sin 2\theta \tag{6.3}$$

The maximum value of $\tau(\theta)$ occurs when $\theta = 45°$ or $135°$ and is $P/2A$.

The strains in general depend on the elastic and plastic behavior of the material. For purely elastic behavior the strains are

$$\varepsilon_{yy} = \frac{\sigma}{E} \tag{6.4}$$

$$\varepsilon_{xx} = \varepsilon_{zz} = -\nu\varepsilon_{yy} \tag{6.5}$$

For linear elastic strain, the strain energy per unit volume is

$$U_e = \frac{1}{2}\frac{\sigma^2}{E} \tag{6.6}$$

The tensile strength as measured in a tensile test is simply the stress P/A at the instant of fracture. The most common experiment is to program a testing machine to increase the displacement (typically for a screw-driven testing machine) or the load (typically for a hydraulic testing machine) at a constant rate with time. Strength measured in this way at a rapid strain rate is often termed the instantaneous strength.

The tensile test has several advantages in principle. It produces a value of stress not dependent on the response of the material provided necking does not occur. It subjects the test section of the specimen to a uniform stress, which facilitates studies of the probability of failure. However, the tensile test requires very accurate loading to eliminate a component of bending. The specimens are complex and expensive. Several special forms of tensile specimens and associated testing equipment have been developed in an attempt to minimize these problems.

The phenomenon of slow crack propagation (treated in Chapter 8) causes the measured strength to decrease with decreasing loading rate. Measurements of strength as a function of loading rate can be analyzed to give information on slow crack propagation.

Another loading pattern used in tensile testing is the stress rupture test in which a constant load corresponding to a tensile stress below the instantaneous strength is applied. The time to failure is determined as a function of the stress. The time to failure increases as the stress is decreased and eventually becomes too large to measure. This type of test can also be used to study slow crack propagation. The resulting behavior is sometimes termed static fatigue. At high temperatures other effects associated with creep can also cause stress-rupture behavior.

Still another type of loading pattern is used to study cyclic fatigue (termed simply as fatigue in the metallurgical literature). The load is typically cycled between a maximum value below the instantaneous strength and some chosen minimum stress value. This type of test is important for metallic alloys, which often display a decrease in strength as the number of fatigue cycles is increased; the effect is related to local plasticity in metals. Cyclic fatigue was once thought unlikely in most ceramics at room temperature because of the lack of plastic deformation, but strong cyclic fatigue effects are now known to occur in many ceramics at room temperature and are treated in Chapter 14.

Tensile tests on specimens containing deliberate flaws can also be used to measure fracture mechanics properties of a ceramic: the critical stress intensity factor or equivalently the fracture surface energy. Some common specimen configurations are summarized in Table 6.1 and then the tests are discussed.

The simplest situation is a crack extending through the thickness of a specimen and of length $2c$ in the center of the width. The applied stress intensity factor resulting from a tensile stress $\sigma = \sigma_{yy}$, which is uniform across the specimen at a large distance from the crack, is given (Jayatilaka, 1979) by

TABLE 6.1 Values of Y in $K_I = Y\sigma c^{1/2}$

Type of Flaw	Value of Y
Small central, double-ended through crack with total length $2c$	$\pi^{1/2} = 1.772$
Central double-ended through crack with total length $2c$ in specimen of width w	$[(w/c) \tan (p\, c/w)]^{1/2}$
Small edge through crack with total length c	$1.12\pi^{1/2} = 1.9851$
Small central, circular crack with radius c	$2/\pi^{1/2} = 1.128$

$$K_I = \pi^{1/2}\, \sigma\, c^{1/2} = 1.772\, \sigma\, c^{1/2} \text{ (through crack of length } 2c) \quad (6.7)$$

when the length of the crack $2c$ is small compared to the width of the specimen. If the length of the crack is an appreciable fraction of the specimen width, w, the result is

$$K_I = \sigma \left[w \tan \left(\frac{\pi c}{w} \right) \right]^{1/2} \text{ (through crack of length } 2c) \quad (6.8)$$

For small values of c/w this becomes

$$K_I = \pi^{1/2}\, \sigma\, c^{1/2} \left[1 + \frac{\pi^2 c^2}{6w^2} \right] \text{ (through crack of length } 2c) \quad (6.9)$$

If a through crack is located with one end on the surface, the stress intensity factor will be larger than if the same crack is located within the specimen. For a crack of total length c ending on the surface of a specimen whose width is great compared to the crack length,

$$K_I = 1.12\, \pi^{1/2}\, \sigma\, c^{1/2} = 1.98\, \sigma\, c^{1/2} \text{ (edge crack of length } c) \quad (6.10)$$

However, the centrally located crack discussed above was taken to have a total length $2c$. If this through crack instead ends on the surface, the stress intensity factor would be

$$K_I = 1.12\, \pi^{1/2}\, \sigma\, (2c)^{1/2} = 2.81\, \sigma\, c^{1/2} \text{ (edge crack of length } 2c) \quad (6.11)$$

Thus a through crack perpendicular to the stress and ending at the surface will have a stress intensity factor $2.80/1.77 = 1.59$ times as great as the same crack distant from the surface. This suggests that a specimen containing many randomly distributed cracks of the same size would usually fail at the surface. Many ceramics and glasses do fail from crack propagation beginning from cracks at the surface. However, surfaces may not be characteristic of the volume of a sample so that failure may begin from flaws within a sample despite the greater stress-concentrating effect of surface flaws. Also, cracks in ceramics are rarely through cracks so that the above simple picture does not apply strictly.

Another crack geometry of interest is that of a "penny-shaped" crack within a specimen as shown in Figure 6.2. This crack shape is not used in testing but is encountered in failure analysis using fractography; the latter subject is discussed in Chapter 16. The crack is taken to be oriented perpendicular to the applied stress and to be a circle of radius c located far from any surface. For this geometry (Jayatilaka, 1979; Lawn, 1993)

$$K_I = 2\,\pi^{-1/2}\,\sigma\, c^{1/2} = 1.128\,\sigma\, c^{1/2} \tag{6.12}$$

The tensile test can be used to measure the critical stress intensity factor of a ceramic. If a flaw of known size can be introduced and the fracture stress is measured, the stress intensity factor at fracture can be calculated from the above equations. The equations for the critical stress intensity factor are commonly expressed in terms of a crack of total length c rather than $2c$. There are two types of tensile specimens for critical stress intensity factor determination: the edge-cracked tensile specimen and the compact tension specimen (Evans and Langdon, 1976; Tada, Paris, and Irwin, 1985; Murakami, 1986).

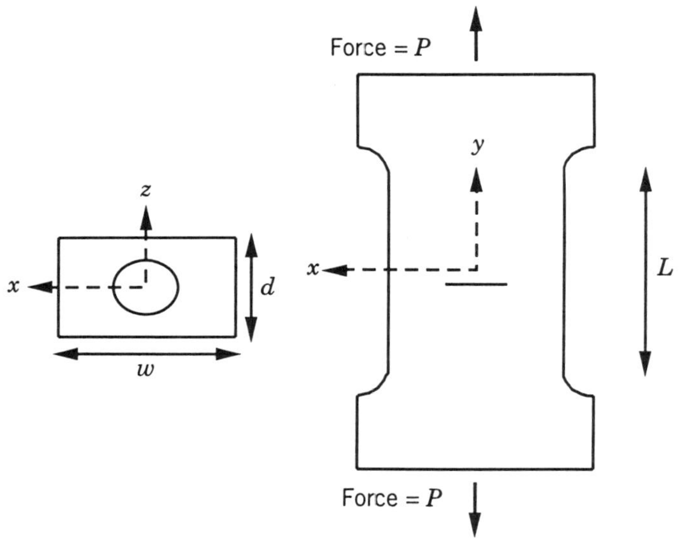

Figure 6.2 Tensile test with penny-shaped crack.

The edge-cracked test uses a tensile specimen, shown in Figure 6.3, under total load P with width w and thickness d containing an edge crack of total length c extending through the thickness. This crack experiences an applied stress intensity factor

$$K_I = \frac{YP}{wd} c^{1/2} \tag{6.13}$$

where Y is a numerical factor given by

$$Y = 1.99 + 0.41\left(\frac{c}{w}\right) + 18.70\left(\frac{c}{w}\right)^2 - 38.48\left(\frac{c}{w}\right)^3 + 53.85\left(\frac{c}{w}\right)^4 \tag{6.14}$$

The compact tension specimen shown in Figure 6.4 is again a bar of width w and thickness d with an edge crack of total length c extending through the thickness. However, the load P is not applied over the end of the specimen as in the usual tension test but is instead applied at the edge of the specimen containing the crack. Also, the total length of the specimen $2h$ is relatively small, $2h/w < 1$. The applied stress intensity factor is given by

$$K_I = \frac{YP}{wd} c^{1/2} = Y \sigma c^{1/2} \tag{6.15}$$

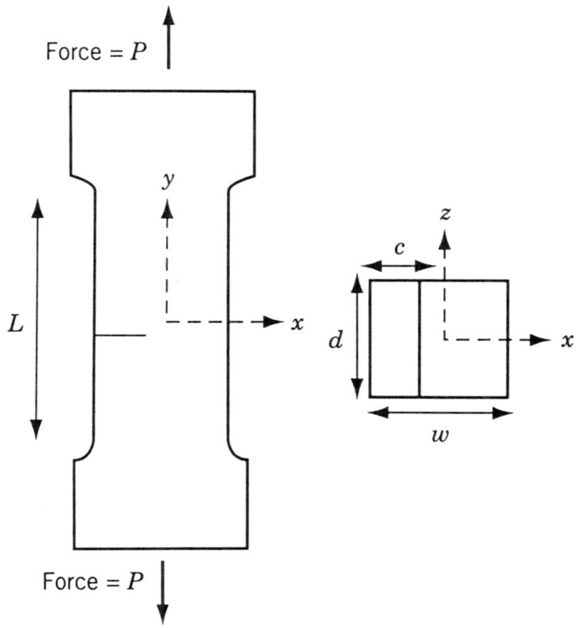

Figure 6.3 Tensile test with edge-cracked specimen.

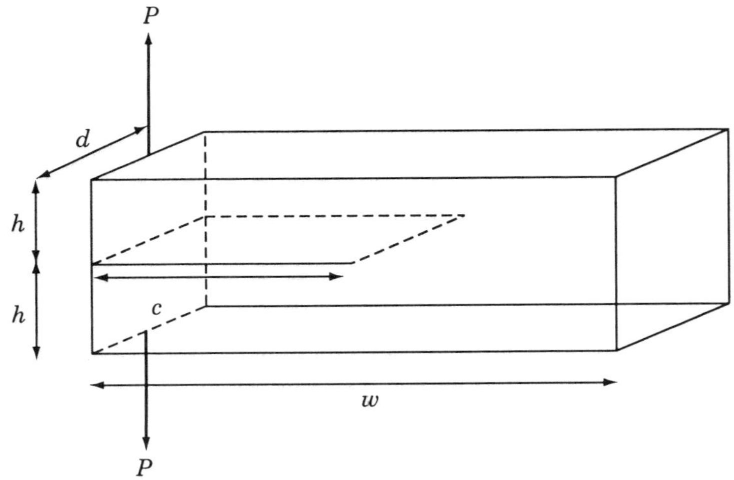

Figure 6.4 Tensile test with compact tensile specimen.

For this specimen Y depends for a given h/w on the value of c/w. For $h/w = 0.4$, $Y_f = 1.42$ at $c/w = 0.3$ and increases smoothly to $Y_f = 2.2$ at $c/w = 0.65$.

Tension tests are often used for measurements of critical stress intensity factors for metals. However, tensile testing of ceramics is difficult as mentioned. Accordingly, bending tests and other specialized tests (some of which are described later) are commonly used to determine critical stress intensity factors of ceramics.

Tensile (and compressive) testing is especially important for studying creep of ceramics at high temperatures. Creep of ceramics in bending leads to complex behavior because the creep rate of ceramics in tension is typically greater than in compression at the same stress level, leading to a shift of the neutral axis in bending tests (Jakus and Wiederhorn, 1988). Although a method of analyzing tests of creep on bending to give creep rate in tension has been developed (Hollenberg, Terwilliger and Gordon, 1971; Chuang, 1986; Chuang and Wiederhorn, 1988), the procedure is difficult and creep testing in tension, when possible, is generally considered much to be preferred. A method for creep testing in tension at temperatures up to 1500°C has been developed (Carroll, Wiederhorn, and Roberts, 1989) that uses relatively simple specimens and equipment.

6.3 FLEXURE TESTS

The relative simplicity of flexural (bend) testing and the low cost of the required specimen have made this test very popular for ceramics despite several drawbacks in interpretation of the resulting data. These drawbacks include nonuniform stress distribution and a stress distribution that changes with increasing strain if some plastic deformation takes place. Special problems accordingly arise in connection with failure probability analyses and with experiments involving creep. These will be discussed in the appropriate chapters. In the present chapter attention is focused on the use of bend tests for linearly elastic, brittle materials.

Bars of rectangular or circular cross section are generally used for bend tests on ceramics. However, it is useful to recognize that some of the basic treatment of bars in flexure apply to bars of arbitrary cross section and to state these results before specializing to a rectangular cross section. Figure 6.5 shows a bar of trapezoidal cross section under the action of a bending moment. The cross section of the bar is shown in an end view on the right and the bending of the bar is shown in a side view on the left of the figure. The upper surface of the bar is in compression and the lower in tension, as shown in the side view. Somewhere between the upper surface and the lower surface is a line EF such that EE is unchanged in length by the bending but is bent into the arc of a circle of radius R. EF is termed the neutral axis. Consider an element of area dA at a distance y from the neutral axis. The radius of curvature at y is $R + y$. The strain ϵ in GG is

$$\varepsilon = \frac{GG - EE}{EE} = \frac{(R + y)\theta - R\theta}{R\theta} = \frac{y}{R} \tag{6.16}$$

That is, the strain is proportional to the distance from the neutral axis. From Hooke's law

$$\varepsilon = \frac{\sigma}{E} \tag{6.17}$$

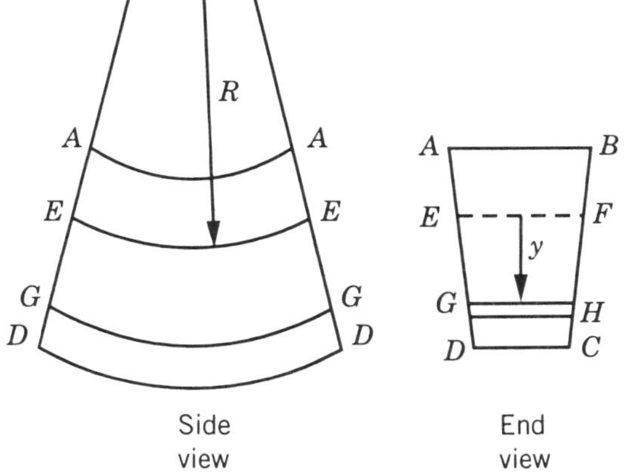

Figure 6.5 Flexure of a beam.

so that

$$\sigma = \frac{E}{R} y \tag{6.18}$$

Thus the stress is proportional to the distance from the neutral axis. The total force on the cross section *ABCD* must be zero so that

$$\int \sigma \, dA = 0 \tag{6.19}$$

giving

$$\int y \, dA = 0 \tag{6.20}$$

Thus the neutral axis passes through the centroid of the section. The bending moment *M* is defined by

$$M = \int \sigma y \, dA \tag{6.21}$$

so that

$$M = \frac{E}{R} \int y^2 \, dA = \frac{EI}{R} \tag{6.22}$$

where the second moment of area, *I*, is defined by

$$I = \int y^2 \, dA \tag{6.23}$$

The stress in the bar is then given by

$$\sigma = \frac{M}{I} y = \frac{E}{R} y \tag{6.24}$$

These general expressions can be used to calculate the stress in bars under various loading conditions. Often the load on a bar is applied in such a manner that the bending moment varies along the length of the bar so that the radius of curvature is not constant, as shown in Figure 6.5, but instead varies along the length of the bar. The elastic solutions taken from the literature for three- and four-point bending are given below.

The case of a beam of rectangular cross section supported at two points and loaded at the center of the span is illustrated in Figure 6.6. This configuration is usually termed three-point bending. Note that the "points" are

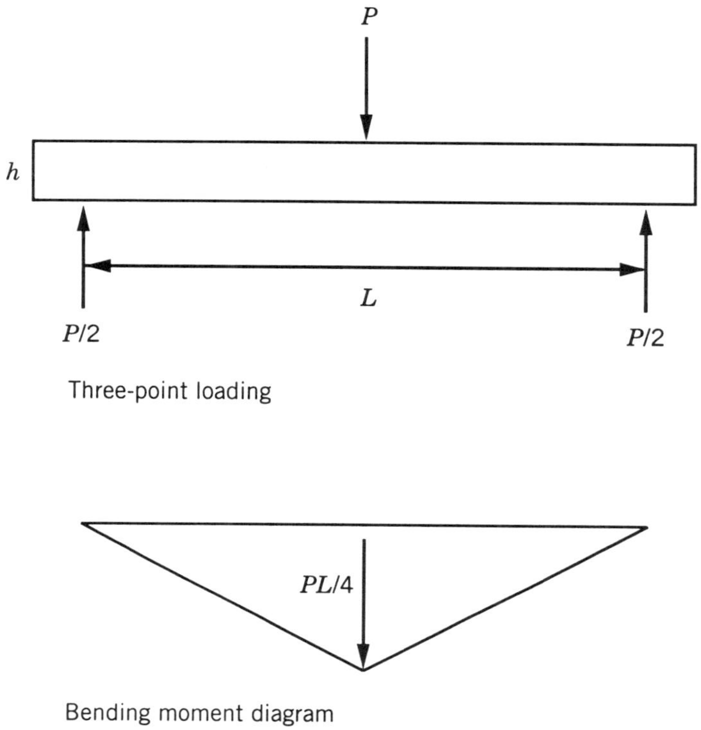

Figure 6.6 Three-point bending and moment diagram.

actually line contacts extending across the top and bottom of the specimen. The neutral axis is halfway between the upper and lower surfaces so that the maximum tensile and compressive stresses are equal in absolute value and will be given by

$$\sigma_{max} = \frac{M}{I}\frac{h}{2} \tag{6.25}$$

Integration over the cross section for a bar of rectangular cross section of width d and height h and for a rod or radius a gives

$$I = \frac{h^3 d}{12} \text{ (for rectangular bar) and } I = \frac{\pi a^2}{4} \text{ (for rod)} \tag{6.26}$$

and the bending moment at the center is due to a force $P/2$ acting at a distance $L/2$ giving

$$M = \frac{PL}{4} \tag{6.27}$$

so that

$$\sigma_{max} = \frac{3PL}{2h^2d} \text{ (for rectangular bar)} \tag{6.28}$$

More detailed treatment shows that the concentrated load causes a wedging action and when this effect is taken into account the result is

$$\sigma_{max} = \frac{3PL}{2h^2d}\left(1 - \frac{4h}{3\pi L}\right) \text{ (for rectangular bar)} \tag{6.29}$$

The deflection of the bar with respect to the support points can be measured and used to determine Young's modulus. Letting δ represent this deflection, the deflection at the center is given by

$$\delta_c = \frac{PL^3}{48EI} \tag{6.30}$$

As shown in Figure 6.6, the bending moment varies from zero at the support points to $PL/4$ at the center. Thus only a very small portion of the specimen is under the maximum stress or near it. For this reason four-point bending rather than three-point bending is often preferred and is commonly used in testing ceramics to obtain design data. R. W. Rice (personal communication, 1994) has emphasized that the three-point bending test has a useful place in studies on the development of ceramics because its ease of use facilitates scoping effects of microstructural variation on strength.

A four-point bend specimen is shown in Figure 6.7. The central portion of the bar (between the two upper loads) is in uniform bending. The bending moment in the central portion is constant and is equal to $PD/2$. Note that the total load is taken here as P rather than $2P$ as in some references. The maximum stress (i.e., the stress in the outer fiber) is constant over the central portion and is

$$\sigma_{max} = \frac{3PD}{h^2d} \tag{6.31}$$

The deflection of the center measured with respect to the support points is

$$y_c = \frac{PD}{48EI}(3L^2 - 4D^2) \tag{6.32}$$

Flexure tests with four-point loading are the most widely used technique for strength measurements of ceramics over the temperature range in which creep is not appreciable. Typically quarter-point ($D = L/4$) or third-point ($D = L/3$)

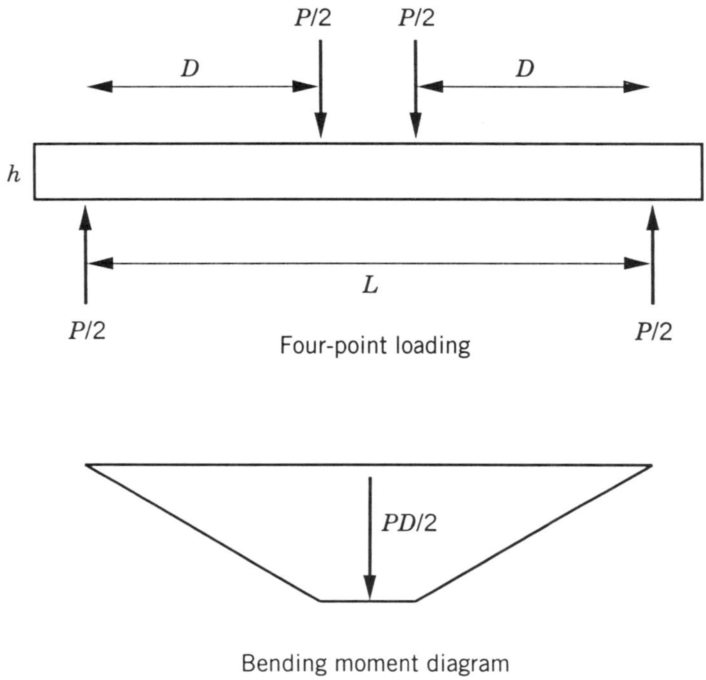

Figure 6.7 Four-point bending and moment diagram.

loading is used. The use of this test including standards for test procedures and possibilities for error was reviewed by Quinn and Morrell (1991) and a discussion was also given by Baratta (1982).

Flexural tests with precracked specimens are commonly used to determine the fracture surface energy of ceramics. One of the simplest is the precracked bend bar in three-point or four-point bending. The latter case is shown in Figure 6.8 with an edge crack extending a distance c from the tension surface. The critical stress intensity factor for three-point bending (i.e., $D = L/2$) is given by

$$K_{IC} = Y \frac{3PL}{2h^2 d} c^{1/2} \tag{6.33}$$

where P in the breaking load and the numerical factor for the case $L/h = 4$ is given by (Evans and Langdon, 1976)

$$Y = 1.93 - 3.07\left(\frac{c}{h}\right) + 13.66\left(\frac{c}{h}\right)^2 - 23.98\left(\frac{c}{h}\right)^3 + 25.22\left(\frac{c}{h}\right)^4 \tag{6.34}$$

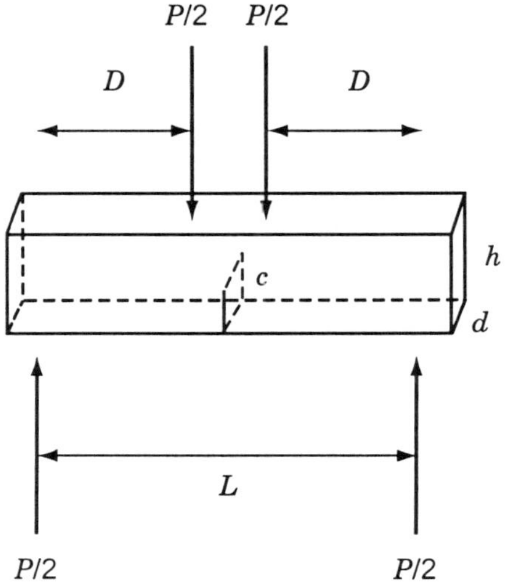

Figure 6.8 Bending test with edge-cracked specimen.

For four-point bending of an edge-cracked specimen, Srinivasan and Seshadri (1981) give

$$K_{IC} = Y \frac{3Pc}{h^2 d} c^{1/2} \tag{6.35}$$

with

$$Y = 1.99 - 2.47\left(\frac{c}{h}\right) + 12.97\left(\frac{c}{h}\right)^2 - 23.17\left(\frac{c}{h}\right)^3 + 24.8\left(\frac{c}{h}\right)^4 \tag{6.36}$$

Equations for this type of fracture toughness testing are given in several places (Srawley and Brown, 1965; Brown and Srawley, 1966).

Care is necessary in the use of edge-notched specimens for K_{IC} determination. The value calculated is commonly found to decrease with decreasing slot width and is believed to finally level off below some critical value. Munz, Bubsey, and Shannon (1980) found good agreement between critical stress intensity values measured by the chevron notch test (see below) and the single-edge-notched beam in bending, provided the slot was not wider than 70 μm.

The above test methods generally assume elastic behavior up to the point of failure. When irreversible behavior occurs such as plastic deformation, phase changes, or some combination of matrix cracking, fiber breaking, and fiber pullout in composites, testing becomes more difficult and complex. A J-integral approach is sometimes used in such situations. The technique for plastic metals is summarized, for example, by Broek (1987). A J-integral testing approach for ceramic–matrix composites has been described by Hashida, Li, and Takahashi (1994).

A relatively simple technique for determining fracture toughness in mixed mode I and mode II loading has been presented by Suresh, Shih, Morrone, and O'Dowd (1990).

Another important form of bend test used to determine the critical stress intensity factor utilizes a specimen with a chevron notch (Munz, Bubsey, and Shannon, 1980; Munz, Himsolt, and Eschweiler, 1981; Sakai and Yamasaki, 1983). A beam of original cross section depth h and width w is notched as shown in Figure 6.9. The load–displacement curves produced by bending chevron-notched brittle specimens differ from the curves produced by bending straight-through-notched brittle specimens. For the latter the load increases linearly with displacement up to the point of failure. For the chevron-notched specimens the load–displacement curve is initially linear but passes through a maximum. Failure occurs at higher displacements and smaller loads than the maximum load. This behavior results from the fact that a crack can propagate some distance from the tip of the solid web of material and come to a stable position. The critical stress intensity factor is given by

$$K_{IC} = \frac{2DP_{\max}}{wh^{3/2}}\left(3.08 + 5.00\frac{c}{h} + 8.33\left(\frac{c}{h}\right)^2\right)\left[1 + 0.007\left(\frac{L(L-2D)}{h^2}\right)\right]\left[\frac{b-c}{h-c}\right] \tag{6.37}$$

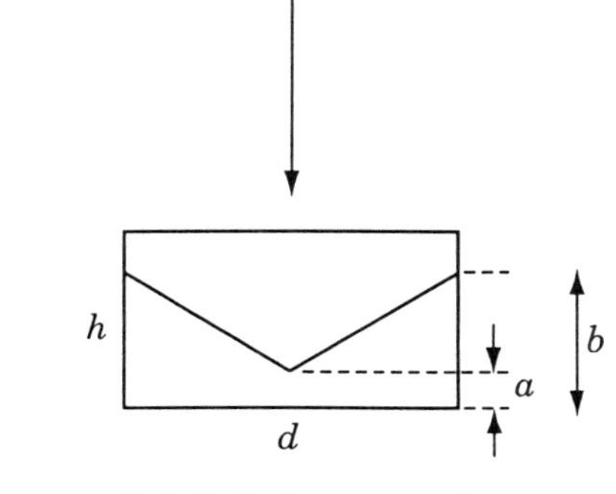

Figure 6.9 Bending test with chevron-notched specimen.

Three other types of test are of considerable importance for ceramics and are reviewed in the following sections: the cantilever beam test, the double torsion test, and the indentation test.

6.4 THE CANTILEVER BEAM TEST

The cantilever beam test, shown in Figure 6.10, uses a specimen similar to the compact tension test but differs in that the "width" is now the length and is large compared to the height, $2h$. The stress intensity factor is

$$K_{\mathrm{I}} = \left[\frac{Pc}{\left(d^2h^3\right)^{1/2}}\right]\left(3.47 + 2.43\frac{h}{c}\right) \tag{6.38}$$

If a constant load P is used, the applied stress intensity factor increases as the crack length increases so that the crack is unstable once propagation has begun.

A variation on the double cantilever test, the constant moment double cantilever beam test, is shown in Figure 6.11. This test employs an applied moment, M, rather than an applied load. The applied stress intensity factor is given by

$$K_{\mathrm{I}} = \frac{M}{(Id)^{1/2}} \tag{6.39}$$

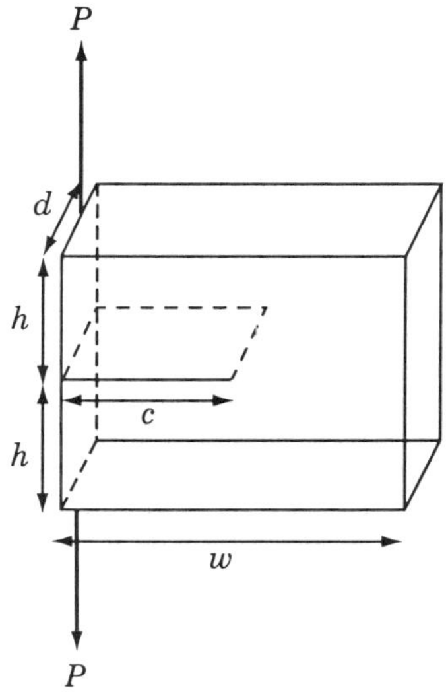

Figure 6.10 Double cantilever beam test.

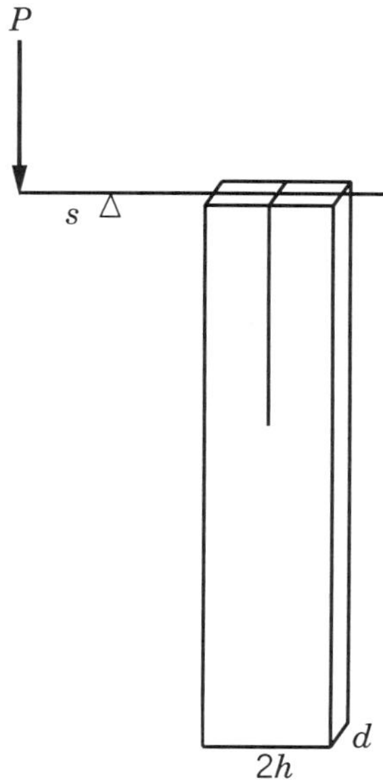

Figure 6.11 Constant moment double cantilever beam.

where M is the bending moment and I is the moment of inertia of a single arm. For the lever arrangement shown, the bending moment is

$$M = Ps \tag{6.40}$$

Taking the total depth of the specimen as $2h$, so that the depth of a single arm is h, gives a moment of inertia

$$I = \frac{h^3 d}{12} \tag{6.41}$$

The applied stress intensity is then given by

$$K_{\mathrm{I}} = \frac{2\sqrt{3}\, Ps}{d\, h^{3/2}} \tag{6.42}$$

An important property of this test is that the applied stress intensity factor is independent of the crack length. This makes the constant moment double cantilever beam test particularly appropriate for slow crack growth studies.

6.5 THE DOUBLE TORSION TEST

Another test with the property of having a constant applied stress intensity factor independent of the crack length is the double torsion test shown in

Figure 6.12. The double torsion test can be regarded as a four-point bend test specimen, as in Figure 6.7, but having the loads and supports at the edge of the specimen and having a thickness d that is large compared with the depth h. The specimen is notched on the tension surface, leaving a reduced thickness h_n. The applied stress intensity factor is

$$K_{\mathrm{I}} = PD\left[\frac{3(1+\nu)}{L_t h^3 h_n}\right]^{1/2} \tag{6.43}$$

Here ν is Poisson's ratio and L_t is the total length of the specimen.

G. D. Quinn (personal communication, 1995) finds that the notch shown in Figure 6.12 causes the crack to move off center and recommends that a notch not be used. Instead, careful alignment of the specimen should be achieved. In equation (6.43) $h^3 h_n$ is then replaced by h^4.

6.6 THE INDENTATION TEST

Hardness testing is important in itself without reference to its use in toughness testing. The Vickers test is commonly used, but the Knoop indenter with greatly unequal diagonals is also used. G. D. Quinn (personal communication, 1995) has drawn attention to a small but significant difference in the definition of hardness

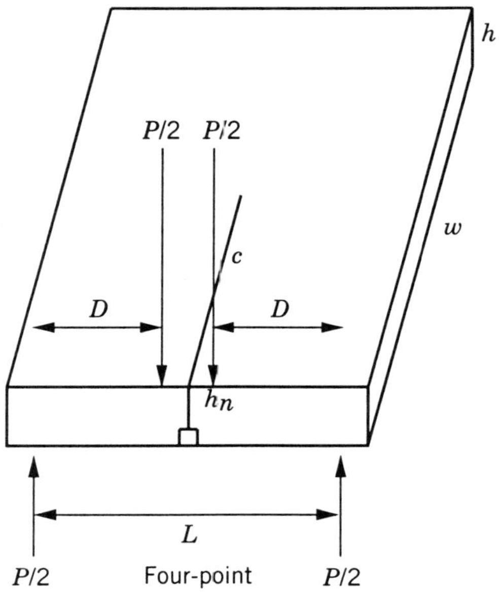

Figure 6.12 Schematic of double torsion test specimen.

in general use in the materials community and that used in the Vickers indentation tests for fracture mechanics measurements. The widely accepted definition in hardness testing is to use the actual area of contact of the four faces of the indenter pyramid leading to a definition of Vickers hardness of $H = 1.8544P/(2a)^2$, where $2a$ is the length of the diagonal of the hardness indent. In contrast, many in the ceramics community use the projected area of the indenter contact onto the original surface plane, leading to a definition of Vickers hardness of $H = 2.000P/(2a)^2$. The definitions differ by 7.9%. Here we follow the use in the ceramics community but warn the reader to check the definitions being used before comparing hardness data from the literature.

The indentation test for toughness can be used in several forms to determine the critical stress intensity factor (Lawn and Wilshaw, 1975; Evans and Charles, 1976; Marshall, Lawn, and Chantikul, 1979). Its use in a material with a toughness that depends on crack length will be discussed in Chapter 10. Only the basic technique for a critical stress intensity factor independent of crack length is summarized here. There are three approaches: direct, indirect, and modified.

The direct approach is to use a load P on a pointed indenter to produce a hardness indent with a diagonal length $2a$ and a system of radial cracks with total length $2c$ extending from the indent as shown in Figure 6.13. A detailed discussion of the types of cracks associated with hardness tests and of the

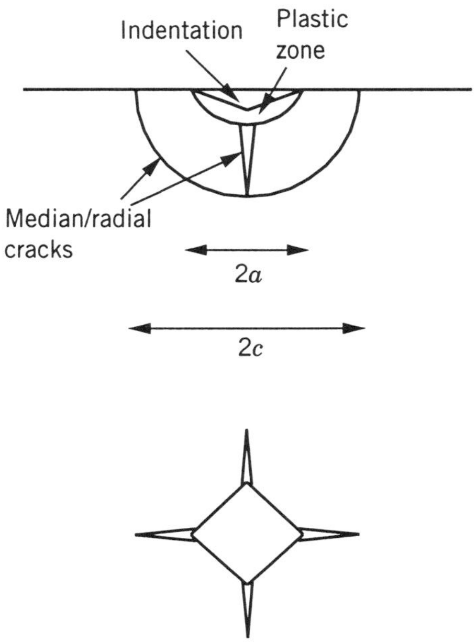

Figure 6.13 Cracking around hardness indentation.

sequence of occurrence is given by Cook and Pharr (1990). The hardness is given by

$$H = \frac{P}{\alpha a^2} \tag{6.44}$$

where α is a numerical factor that depends on the shape of the indenter. For a Vickers indenter $\alpha = 2$.

The critical stress intensity factor is obtained by assuming that the applied stress intensity caused by the load is equal to the critical stress intensity for crack propagation. The result is

$$K_{IC} = \frac{\zeta (E/H)^{1/2} P}{c^{3/2}} \tag{6.45}$$

Here H is the hardness determined at the load P and ζ is a dimensionless constant (Anstis, Chantikul, Lawn, and Marshall, 1981). Studies on many ceramics led to an average value of $\zeta = 0.016 \pm 0.004$.

The indirect approach is to measure the strength after indentation (Chantikul, Anstis, Lawn, and Marshall, 1981). Care should be taken not to allow the crack to elongate by slow crack propagation during the time between indentation and strength measurement. An attempt is usually made by putting a drop of oil on the indent to keep reactive species, such as water, that promote slow crack propagation away from the crack. That is, the crack is kept in inert conditions and the strength σ_m under these conditions is measured, typically by bending. This has led to the use of the term "inert strength" to describe this procedure. The result is

$$K_{IC} = \eta (E/H)^{1/8} \left(\sigma_m P^{1/3}\right)^{3/4} \tag{6.46}$$

Experiments on a number of materials led to an average value of

$$\eta = 0.59 \pm 0.12 \tag{6.47}$$

A modified form of the indentation method developed by Cook and Lawn (1983) combines some aspects of the direct and indirect methods. An indented specimen is broken as in the indirect method. However, in the modified method multiple indents (typically three) are made in the specimen before stressing rather than a single indent as in the original direct method. Failure will occur from the crack associated with one of the indents. The maximum crack length just before failure can be measured from one of the other indents and K_{IC} can be determined from these measurements.

The analysis for the modified method is based on that for the indirect method as follows. A crack of half-length c produced by an indentation load P in a specimen that is subsequently placed under a tensile stress σ will experience an applied stress intensity factor

$$K_a = \chi\, P/c^{3/2} + \psi\, \sigma\, c^{1/2} \tag{6.48}$$

where χ characterizes the intensity of the driving force produced by the indentation and ψ is the crack-geometry parameter Y for the case of crack shapes produced by indentation. Under conditions of no subcritical crack growth (i.e., in an inert environment), the initial crack will extend stably as σ reaches a value such that $K_a = K_C$ because the increase in K_a from the second term will be balanced by a decrease in the contribution from the first term until a critical value of $\sigma = \sigma_m$ corresponding to $c = c_m$ is reached, above which the increase from the second term dominates and the crack propagates unstably. The condition for maximum stable crack growth thus requires both $K_a = K_C$ and $(d\sigma/dc)_{c=c_m} = 0$. Differentiating the first condition with respect to c gives

$$\left(\frac{dK}{dc}\right)_{c=c_m} = -\frac{3}{2}\frac{\chi P}{c_m^{5/2}} + \frac{1}{2}\frac{\sigma_m\, \psi}{c_m^{1/2}} + \psi\left(\frac{d\sigma}{dc}\right)_{c=c_m} c^{1/2} = \frac{dK_C}{dc} = 0 \tag{6.49}$$

Applying $(d\sigma/dc)_{c=c_m} = 0$ gives

$$\sigma_m = \frac{3\,\chi P}{\psi\, c_m^2} \tag{6.50}$$

Substituting equation (6.50) in equation (6.48) gives

$$\sigma_m = \frac{3K_C}{4\psi\, c_m^{1/2}} \tag{6.51}$$

Combining equations (6.50) and (6.51) gives

$$c_m = \left(\frac{4\,\chi\, P}{K_C}\right)^{2/3} \tag{6.52}$$

In the indirect method this equation for c_m is used to eliminate c_m from the σ_m equation, yielding

$$K_C = \left(\frac{256\,\chi\,\psi^3}{27}\right)^{1/4} \left(\sigma_m\, P^{1/3}\right)^{3/4} \tag{6.53}$$

which is equivalent to the result previously quoted in equation (6.46). In the modified method, the equation for σ_m is solved for K_C to give

$$K_C = \left(\frac{4\,\psi}{3}\right)\sigma_m\, c_m^{1/2} \tag{6.54}$$

The advantage of the modified indentation method is that the parameter χ is not required to calculate K_C. Cook and Lawn (1983) measured critical stress intensity values by the double cantilever beam technique and by the modified indentation technique and found a good fit to the equation

$$K_C = A\,\sigma_m\, c_m^{1/2} + B \tag{6.55}$$

with $A = 2.02$ and $B = -0.68$ MPa·m$^{1/2}$. They suggest the use of this equation in conjunction with the modified indentation method to determine critical stress intensity values.

The above methods are based on the assumption of half-penny cracks with radial/median geometry. This assumption has been questioned by Cook and Pharr (1990). Smith and Scattergood (1992) made in situ measurements of crack size as a function of stress and showed that the assumption of a half-penny shape causes the toughness values to be overestimated and produces an apparent R-curve effect.

Use of the indentation test in the indirect form is common and has recently been the subject of an international round robin (Quinn and Gettings, 1994) using an empirical stress intensity factor equation due to Newman and Raju (1981). Twenty laboratories in Europe and the United States participated. The results on silicon nitride were very good: for Norton NC-132 the toughness was 4.50 ± 0.37 MPa·m$^{1/2}$ while for ESK hot isostatic pressed material the value was 4.95 ± 0.55 MPa·m$^{1/2}$. Fourteen laboratories attempted to measure toughness on zirconia with 3 mol % yttria stabilization, but only eight sent results for analysis. The average toughness was 4.36 ± 0.44 MPa·m$^{1/2}$.

6.7 BIAXIAL FLEXURE TESTING

A relatively simple method to apply biaxial tension to the surface of a sample is to use biaxial bending. Several techniques have been used with ceramics (Wachtman, Capps, and Mandel, 1972; Marshall, 1980; Shetty, Rosenfield, Duckworth, and Held, 1983) and the subject has recently been reviewed by de With and Wagemans (1989). The common techniques can be grouped into three categories: ball-on-ring, ring-on-ring, and piston-on-ring (Ritter, Jakus, Batakis, and Bandyopadhyay, 1980). The technique of supporting a thin disk on three balls and loading in the center produces biaxial tension in the center of the lower face and has the advantage that as-fired ceramic specimens can be tested even though they may be slightly warped. This eliminates the expense of machining and, more importantly, allows testing of the surface without any machining damage (Wachtman, Capps, and Mandel, 1972). Also, edge effects are eliminated provided that the disk is not too thick. This test has been made an ASTM standard (ASTM, 1978). The advantage of eliminating edge effects is retained in various other biaxial flexure tests.

6.8 ELASTIC CONSTANT DETERMINATION USING VIBRATIONAL AND ULTRASONIC METHODS

It is difficult to achieve good accuracy in determination of elastic constants by static methods; several percent is typical of the best that can be achieved with care. Dynamic methods can achieve an order of magnitude greater accuracy with relative ease. The dynamic methods give adiabatic values and the static methods give isothermal values when sufficient time is allowed for thermal equilibration; the difference is small but should be considered in very accurate work. One interlaboratory study (Dickson and Wachtman, 1971) resulted in an uncertainty of 0.4% for the dynamic Young's modulus and 0.2% for the dynamic shear modulus. These dynamic methods fall into two classes: resonance and ultrasonic wave propagation.

Resonance methods can be used in principle with any shape of solid but are most often used with flexure and torsion of slender bars and cylinders. Resonance methods are described in ASTM Standard C 1198 - 91(ASTM 1991). A rectangular bar of length L, thickness t, width w, mass m, and flexural resonance frequency f_f in the thickness direction has a Young's modulus value given by

$$E = 0.9465 \frac{m f_f^2}{w} \frac{L^3}{t^3} T_1 \tag{6.56}$$

Here T_1 is a correction factor that approaches 1 as the value of t/L approaches zero. For $t/L < 0.05$ one can use

$$T_1 = 1.000 + 6.585\left(\frac{t}{L}\right)^2 \tag{6.57}$$

while for larger values an accurate approximating equation for T_1 given in the ASTM standard must be used. For the same bar vibrating with torsional resonance frequency f_t the shear modulus is given by

$$\mu = \frac{4 L m f_t^2}{wt} \frac{B}{1 + A} \tag{6.58}$$

where

$$B = \frac{w/t + t/w}{4(t/w) - 2.52(t/w)^2 + 0.21(t/w)^6} \tag{6.59}$$

and

$$A = \frac{0.5062 - 0.8776(w/t) + 0.3504(w/t)^2 - 0.0078(w/t)^3}{12.03(w/t) + 9.892(w/t)} \tag{6.60}$$

For a circular cylinder of diameter D, Young's modulus is given by

$$E = 1.6067 \frac{L^3}{D^4} m f_f^2 T_1' \tag{6.61}$$

where for $D/L < 0.05$

$$T_1' = 1.000 + 4.939(D/L)^2 \tag{6.62}$$

and the shear modulus is given by

$$\mu = 16m f_t^2 \frac{L}{\pi D^2} \tag{6.63}$$

Another dynamic method requires the measurement of the velocity of ultrasonic waves typically by using a piezoelectric transducer and measuring the transit time through the specimen. If v_L is the velocity of longitudinal waves and v_t is the velocity of transverse waves, the shear modulus is given by

$$\mu = \rho v_t^2 \tag{6.64}$$

and Poisson's ratio is given by

$$\nu = \frac{1 - 2(v_t/v_L)^2}{2[1 - (v_t/v_L)^2]} \tag{6.65}$$

Young's modulus is obtained from

$$E = 2\mu(1 + \nu) \tag{6.66}$$

CHAPTER 7

STATISTICAL TREATMENT OF STRENGTH

7.1 INTRODUCTION

Measurement of the strength of a series of nominally identical ceramic specimens typically produces considerable scatter in the results. This behavior is easy to explain qualitatively as resulting from a scatter in the size of the cracks responsible for failure. The existence of this scatter has important engineering consequences. First, the strength that can safely be used in design is evidently less than the average measured strength. Second, the probability of failure of a large specimen is greater than a small specimen under the same stress because of the larger probability of having a serious flaw in a larger volume. It is desirable to have some means of describing these effects quantitatively and incorporating this description into a method of design for safe use. Weibull statistics provides such a means and is the subject of this chapter. Weibull statistics is a form of extreme value statistics dealing with a weakest link situation in which the failure of a single element of a specimen causes failure of the whole specimen.

7.2 STATISTICAL DISTRIBUTIONS IN GENERAL

If any quantity, say, the strength σ, is measured repeatedly, say, N times, a series of values will be obtained: $\sigma_1, \sigma_2, \ldots . \sigma_N$. The mean value is

$$\bar{\sigma} = \frac{1}{N}\sum_{i=1}^{N}\sigma_i \qquad (7.1)$$

If the strength can assume only certain fixed values, the distribution function $p(\sigma)$ can be defined by

$$p(\sigma) = \frac{\text{number of occurrences of the value } \sigma}{N} \qquad (7.2)$$

The distribution function for strength can be defined equivalently in terms of the number of strength values falling in successive equal intervals when the values are not limited to certain fixed values. In either case the distribution so defined is normalized; that is,

$$\sum_{i=1}^{N} p(\sigma_i) = 1 \qquad (7.3)$$

or

$$\int_0^{\infty} p(\sigma)d\sigma = 1 \qquad (7.4)$$

for a continuous distribution. The mean can be written in terms of the distribution function as

$$\bar{\sigma} = \sum_{i=1}^{N} \sigma_i\, p(\sigma_i) \qquad (7.5)$$

or

$$\bar{\sigma} = \int_0^{\infty} \sigma\, p(\sigma)d\sigma \qquad (7.6)$$

Other important quantities associated with the distribution are the deviations, the variance, and the standard deviation. The deviation, ϵ, is the amount that a value differs from the mean

$$\varepsilon_i = \sigma_i - \bar{\sigma} \qquad (7.7)$$

The variance, s^2, is defined as

$$s^2 = \frac{1}{N-1}\sum_1^N \varepsilon_i^2 \tag{7.8}$$

$N - 1$ rather than N is used in this definition because one degree of freedom is used in calculating the mean. The standard deviation is defined as the square root of the variance.

The distribution $p(\sigma)$ defined above gives the probability of failure in an interval σ to $\sigma + d\sigma$. The same behavior can be described by $P(\sigma)$, the cumulative probability of failure under all stresses up to the value σ. The two distribution functions are related by

$$P(\sigma) = \int_0^\sigma p(x)\,dx \tag{7.9}$$

or equivalently by

$$p(\sigma) = \frac{dP(\sigma)}{d\sigma} \tag{7.10}$$

The most commonly used distribution for general treatment of experimental data, the Gaussian distribution, is usually described in terms of $p(\sigma)$, but the Weibull distribution, which is generally used for treatment of the statistics of brittle failure, is usually described in terms of $P(\sigma)$.

7.3 THE GAUSSIAN OR NORMAL DISTRIBUTION

A specific distribution often used to describe experimental data is the Gaussian or normal distribution. Written in terms of the mean $\bar{\sigma}$ and the variance s^2 and normalized to give unit probability of failure when integrated over all stresses, the Gaussian distribution is

$$p(\sigma) = \frac{1}{s(2\pi)^{1/2}} \exp\left(-\frac{(\sigma - \bar{\sigma})^2}{2s^2}\right) \tag{7.11}$$

The Gaussian distribution is probably the most widely used statistical distribution in the description of experimental data. It can be used to represent the distribution of strength data for values not too far from the mean, but for small strength values it suffers from the fact that it gives a finite probability of occurrence of negative strength values. That is, it cannot be correct for strength values far below the mean. The Weibull distribution has a more appropriate form for small strength values and also brings in the volume dependence so that the Weibull distribution is used for calculations of failure probability for small stresses and variable volumes.

7.4 THE WEIBULL DISTRIBUTION

The Weibull treatment of failure is in two parts: (1) a weakest link argument leading to a volume dependence for the average strength and (2) the assumption of a distribution function. The weakest link argument is based on the idea that failure at any flaw leads to total failure and that the material is homogeneous in the sense of the distribution of flaws throughout the volume (Weibull, 1939). Consider a material divided into n equal volume elements δV_i under stress σ. Take $P_f(\sigma, \delta V_i)$ as the probability of failure from the stress for the ith volume element. The probability of survival of this element is $1 - P_f(\sigma, \delta V_i)$. Since the stress is taken to be the same for all volume elements and since the specimen is assumed to be homogeneous, all $P_f(\sigma, \delta V_i)$ can be taken the same, say, P_{f1} (σ). For the specimen to survive under the stress, all the volume elements must survive so that the total probability of survival, $1 - P_f(\sigma, V)$, of a specimen of volume $V = n\,\delta V$ under uniform stress is given by

$$1 - P_f(\sigma,V) = [1 - P_{f1}(\sigma)]^n = [1 - \frac{V\,P_{f1}(\sigma)}{V}]^n$$

$$= [1 - \frac{V}{n}\frac{P_{f1}(\sigma)}{\delta V}]^n = [1 - \frac{V\varphi(\sigma)}{n}]^n \qquad (7.12)$$

where we have assumed that as δV decreases $\frac{Pf1(s)}{dV}$ approaches a limit $\varphi(\sigma)$. For n large the limit

$$\lim_{n\to\infty} (1-\frac{x}{n})^n = \exp(-x) \qquad (7.13)$$

so that

$$1 - P_f(\sigma,V) = \exp(-V\varphi(\sigma)) \qquad (7.14)$$

This line of argument shows that the probability of failure of the specimen depends exponentially on the specimen volume times a function $\varphi(\sigma)$ that characterizes the stress dependence of cumulative failure probability per unit volume at the limit of small volume. This weakest link argument does not give a specific form for $\varphi(\sigma)$. Weibull assumed the form

$$\varphi(\sigma) = \left(\frac{\sigma - \sigma_u}{\sigma_0}\right)^m \quad \text{for } \sigma > \sigma_u \tag{7.15}$$

and

$$\varphi(\sigma) = 0 \text{ for } \sigma \leq \sigma_u. \tag{7.16}$$

This gives the Weibull distribution for the probability of survival of the whole specimen, P_s, and the probability of failure of the whole specimen, P_f, as

$$P_s = 1 - P_f = \exp\left[-V\left(\frac{\sigma - \sigma_u}{\sigma_0}\right)^m\right] \text{ for } \sigma > \sigma_u \tag{7.17}$$

and

$$P_s = 1 - P_f = 1 \text{ for } \sigma \leq \sigma_u \tag{7.18}$$

Note that V must be nondimensional if σ_0 has dimensions of stress so that V should be expressed as V/V_0, where V_0 is some chosen unit volume. If V is instead absorbed into σ_0 then σ_0 must have dimensions of stress times $(\text{volume})^{1/m}$. This point is sometimes ignored in calculating the Weibull parameters from a set of data taken on specimens of constant volume. This procedure is equivalent to taking the specimen volume as the unit volume. When Weibull analysis is applied to specimens of various volumes, the definition of V and σ_0 must be recalled and properly taken into account, as discussed later.

Weibull's assumption of a form for the stress dependence of the survival probability allows the treatment to be extended to the situation where the stress varies with position in the specimen. The result is

$$P_s = \exp\left[-\int_V \left(\frac{\sigma - \sigma_u}{\sigma_0}\right)^m dV\right] \tag{7.19}$$

The Weibull distribution function, $P_s(\sigma)$, contains three parameters: m, σ_u, and σ_0. These are generally treated as empirical parameters and determined experimentally. The parameter σ_u is a stress level below which there is zero probability of failure. For ceramics the remote possibility exists of having a very large flaw in a specimen so that the safest assumption is to take $\sigma_u = 0$, as is usually done. This gives the most conservative estimate of survival probability (i.e., it errs on the side of safety if at all). The resulting Weibull distribution is usually termed a two-parameter Weibull distribution.

We shall use the Weibull distribution in the remainder of this chapter to build up an extensive method of predicting the safe working stress for a chosen level of survival probability. The complexity and power of the method should not obscure the fact that it rests on the assumption of the Weibull form of distribution of strength as a function of stress and that this distribution is only an assumption that gives a reasonable fit to the data in many cases. Typically, the Weibull distribution is used to fit a few tens or hundreds of points at most (as in the example of Table 7.3) and is then extrapolated to probability values far outside the range of stresses and probabilities in the experiment (as in Section 7.8). Anticipating later sections, Figure 7.1 fits the Weibull distribution reasonably well for the most part, but several points at low stress values suggest that some other function might fit the data better. Using some other function in place of the Weibull function can lead to very different estimates of the safe working stress for a chosen value of survival probability.

7.5 DETERMINING THE WEIBULL PARAMETERS

The process of determining the Weibull parameters m and σ_0 is easiest to describe using strength values of n specimens determined in a tensile test. The form of the Weibull function for constant stress can then be used. Two procedures are possible: (1) least-squares fitting of a linearized form of the distribution or (2) the method of maximum likelihood. The latter procedure is preferred by some authors but is more complicated so that discussion of this method is deferred until later. The least-squares method is in widespread use and is now discussed. Taking natural logarithms twice of the two-parameter Weibull distribution for tensile testing gives

$$\ln \ln 1/P_s = \ln V + m \ln \sigma - m \ln \sigma_0 \tag{7.20}$$

The Weibull parameters can then be determined by fitting a straight line to $\ln \ln(1/P_s)$ as a function of $\ln \sigma$. The parameter m is simply the slope and σ_0 is related to the intercept on the $\ln \ln(1/P_s)$ axis at $\ln\sigma = 0$ by

$$\ln V - m \ln \sigma_0 = \text{intercept} \tag{7.21}$$

Here the point made earlier about the choice of how to treat V is evident again. Sometimes the volume term is simply ignored for a set of measurements taken on specimens having all the same volume and σ_0 is calculated from

$$\sigma_0 = [\exp(-\mathit{intercept})]^{1/m} \tag{7.22}$$

This procedure is equivalent to taking the volume of the specimens in the test series as the unit volume. If the Weibull parameters so determined are to be applied to specimens of different volume, account must be taken of this implicit choice.

Plots can be made on special Weibull graph paper with ln ln $(1/P_s)$ as ordinate and ln σ as abscissa. A straight line can then be fitted to the data. An easier procedure is to use a computer with a spreadsheet program, calculate ln ln $(1/P_s)$ and ln σ, and then use the regression program to obtain the slope and intercept for the least-squares best fit. The calculation of ln σ is straightforward, but there is some degree of choice in calculating P_s (Davidge, 1979; Bruckner-Foit and Munz, 1989), although for large values of n the different choices give very closely the same values. The general procedure is to put the measured strength values in ascending order and to assign a number beginning with 1 and ending with n. The survival probability usually assigned to the ith strength value is then

$$P_{si} = 1 - \frac{i - 0.5}{n} \tag{7.23}$$

An example of fitting data measured by bend testing to obtain the Weibull parameters for the two-parameter distribution is given later after the adaptation of the Weibull distribution for bend testing has been discussed.

7.6 ADAPTATION TO BEND TESTING

To adapt the two-parameter Weibull distribution to tests in which the stress varies with position, the form of the Weibull distribution with the integral in the exponent must be used:

$$P_s = \exp\left[-\int_V \left(\frac{\sigma}{\sigma_0}\right)^m dV\right] \tag{7.24}$$

One can express the integration in terms of the ratio of the stress at each point to the maximum stress, σ_{max}, in a specimen and loading system of a particular geometry:

$$P_s = \exp\left[-\left(\frac{\sigma_{max}}{\sigma_0}\right)^m \int_V \left(\frac{\sigma}{\sigma_{max}}\right)^m dV\right] = \exp\left[-kV\left(\frac{\sigma_{max}}{\sigma_0}\right)^m\right] \tag{7.25}$$

where k is a dimensionless number dependent on the stress distribution but independent of volume for a given type of stress distribution and is defined by

$$k = \frac{1}{V} \int_V \left(\frac{\sigma}{\sigma_{\max}} \right)^m dV \tag{7.26}$$

To proceed further the stress distribution for a particular specimen and loading system is required. For the tension test the stress is constant over the test length. For a rectangular specimen of length L, width w, and depth d, the integral in P_s is

$$k_t = \frac{1}{V} \int_V \left(\frac{\sigma_t}{\sigma_{t,\max}} \right)^m dV = 1 \tag{7.27}$$

because $\sigma_t = \sigma_{t,\max}$, where σ_t is the uniform tensile stress.

The stress values for the outer fiber for bend tests have been given in Chapter 6. Three-point bending is considered first, ignoring the effect of wedging at the center. Noting that the stress varies in proportion to the distance y from the neutral axis and in proportion to the distance x from the support point gives

$$\sigma_{3b} = \sigma_{3b,\max} \frac{2x}{L} \frac{2y}{h} \tag{7.28}$$

where $\sigma_{3b,\max}$ is the maximum stress in the outer fiber and so corresponds to the measured breaking strength. The integral becomes

$$k_{3b} = \frac{1}{V} \int_V \left(\frac{\sigma_{3b}}{\sigma_{3b,\max}} \right)^m dV = \frac{2}{V} \int_0^{L/2} \left(\frac{2x}{L} \right)^m dx \int_0^{h/2} \left(\frac{2y}{h} \right)^m dy \int_0^d dz \tag{7.29}$$

$$= \frac{1}{2(m+1)^2} \tag{7.30}$$

Now consider the stress values for a particular probability of survival, say, P_{s1}. Let these be $\sigma_{t,1}$ in tension and $\sigma_{3b,\max}$ for three-point bending of the same bar. Setting these survival probabilities equal $[P_{s,t}(\sigma_{t,1}) = P_{s,3b}(\sigma_{3b,\max,1})]$ in the above equations gives

$$\frac{\sigma_{3b,\max,1}}{\sigma_{t,1}} = [2(m+1)^2]^{1/m} \tag{7.31}$$

For typical commercial ceramics, m values of 10–12 are common, but great efforts are being made to improve processing to achieve m values of 20. For m values of 10, 12, and 20, the ratio of strength at constant survival probability in bending to that in tension is predicted by the Weibull theory to be 1.73, 1.62, and 1.40, respectively. Higher values of strength in bending than in tension are usually observed in practice.

The quantity usually compared between tests is not the stress for a particular survival probability, but the mean strength. Following Jayatilaka (1979), the Weibull distributions and mean strength equations are given in Table 7.1 for tension and several bending tests.

Here the gamma function is defined by

$$\Gamma(z) = \int_0^{\infty} t^{z-1} \exp(-t)\, dt \tag{7.32}$$

Table 7.2 gives values of the gamma function corresponding to the range of m values normally found (Abramowitz and Stegun, 1964). For m values typically encountered $\Gamma(1 + 1/m)$ is nearly equal to one; for example, for $m = 10$, $\Gamma(1 + 1/m) = \Gamma(1.1) = 0.95135$.

The expression for the mean strength, $\bar{\sigma}$, in terms of σ_0 can be derived by evaluating the following integral:

$$\bar{\sigma} = \int_0^{\infty} \sigma\, p_f\, d\sigma \tag{7.33}$$

where the probability of failure in the interval σ to $\sigma + d\sigma$ is given by

$$p_f = \frac{dP_f}{d\sigma} = \frac{d}{d\sigma}(1 - P_s) \tag{7.34}$$

For the case of uniform tension

$$p_f = \frac{d}{d\sigma}\left(1 - \exp\left(-V\left(\frac{\sigma}{\sigma_0}\right)^m\right)\right) = mV\left(\frac{\sigma^{m-1}}{\sigma_0^m}\right)\exp\left(-V\left(\frac{\sigma}{\sigma_0}\right)^m\right) \tag{7.35}$$

so that

$$\bar{\sigma} = \int_0^{\infty} mV\left(\frac{\sigma}{\sigma_0}\right)^m \exp\left(-V\left(\frac{\sigma}{\sigma_0}\right)^M\right) d\sigma \tag{7.36}$$

TABLE 7.1 The Two-Parameter Weibull Distribution and Mean Strength for Several Test Methods for a Volume Distribution of Flaws

Tension
$P_s = \exp\left(-\left(\frac{\sigma_t}{\sigma_0}\right)^m V\right)$
$\bar{\sigma}_t = \frac{\sigma_0}{V^{1/m}} \Gamma\left(1 + \frac{1}{m}\right)$
Three-Point Bending
$P_s = \exp\left(-\left(\frac{\sigma_{3b,max}}{\sigma_0}\right)^m \frac{V}{2(m+1)^2}\right)$
$\bar{\sigma}_{3b,max} = \frac{\sigma_0}{V^{1/m}} \left[2(m+1)^2\right]^{1/m} \Gamma\left(1+\frac{1}{m}\right)$
Pure Bending
$P_s = \exp\left(-\left(\frac{\sigma_{b,\max}}{\sigma_0}\right)^m \frac{V}{2(m+1)}\right)$
$\bar{\sigma}_{b,max} = \frac{\sigma_0}{V^{1/m}} \left[2(m+1)\right]^{1/m} \Gamma\left(1 + \frac{1}{m}\right)$
Four-Point Bending with Load at the Quarter Points
$P_s = \exp\left(-\left(\frac{\sigma_{4b,1/4,\max}}{\sigma_0}\right)^m \frac{V(m+2)}{4(m+1)^2}\right)$
$\bar{\sigma}_{4b,1/4,\max} = \frac{\sigma_0}{V^{1/m}} \left[\frac{4(m+1)^2}{m+2}\right]^{1/m} \Gamma\left(1 + \frac{1}{m}\right)$
Four-Point Bending with Loads at the Third Points
$P_s = \exp\left(-\left(\frac{\sigma_{4b,1/3,max}}{\sigma_0}\right)^m \frac{V(m+3)}{6(m+1)^2}\right)$
$\bar{\sigma}_{4b,1/3,max} = \frac{\sigma_0}{V^{1/m}} \left[\frac{6(m+1)^2}{m+3}\right]^{1/m} \Gamma\left(1 + \frac{1}{m}\right)$

TABLE 7.2 Values of the Gamma Function

z	$\Gamma(z)$	z	$\Gamma(z)$	z	$\Gamma(z)$
1.00	1.00000	1.10	0.95135	1.20	0.91817
1.01	0.99432	1.11	0.94739	1.21	0.91558
1.02	0.98884	1.12	0.94359	1.22	0.91311
1.03	0.98354	1.13	0.93993	1.23	0.91075
1.04	0.97843	1.14	0.93641	1.24	0.90852
1.05	0.97350	1.15	0.93304	1.25	0.90640
1.06	0.96874	1.16	0.92980	1.26	0.90440
1.07	0.96415	1.17	0.92669	1.27	0.90250
1.08	0.95972	1.18	0.92372	1.28	0.90072
1.09	0.95545	1.19	0.92088	1.29	0.89904

Defining $y = V(\sigma/\sigma_0)^m$ and substituting gives

$$\bar{\sigma} = \frac{\sigma_0}{V^{1/m}} \int_0^\infty y^{1/m} \exp(-y)\, dy = \frac{\sigma_0\, \Gamma(1 + 1/m)}{V^{1/m}} \quad \text{(for uniform tension)} \tag{7.37}$$

Using the equations of Table 7.1, the value of the mean strength can be related between tests for specimens of the same volume and between volumes for the same type of test. Consider first relationships at constant specimen volume.

Let $\bar{\sigma}_t$ and $\bar{\sigma}_{3b,\max}$ denote the mean strengths in tension and in three-point bending, respectively. Their ratio is given by

$$\frac{\bar{\sigma}_{3b,\max}}{\bar{\sigma}_t} = [2(m+1)^2]^{\frac{1}{m}} \tag{7.38}$$

Let $\bar{\sigma}_{4b,1/4,\max}$ represent the mean strength in four-point bending with the load at the quarter points. Then

$$\frac{\bar{\sigma}_{4b,1/4,\max}}{\bar{\sigma}_t} = \left[\frac{4(m+1)^2}{m+2}\right]^{\frac{1}{m}} \tag{7.39}$$

Let $\bar{\sigma}_{4b,1/3,\max}$ represent the mean strength in four-point bending with the load at the third points. Then

$$\frac{\bar{\sigma}_{4b,1/3,\max}}{\bar{\sigma}_t} = \left[\frac{6(m+1)^2}{m+3}\right]^{\frac{1}{m}} \tag{7.40}$$

Consider next the effect on mean strength of changing from specimen volume V_1 to V_2 while continuing to use the same type of test. For any of the four types of test considered above

$$\frac{\bar{\sigma}_1}{\bar{\sigma}_2} = \left(\frac{V_2}{V_1}\right)^{\frac{1}{m}} \tag{7.41}$$

Use of these equations in combination allows data taken with one type of test to be used to predict the behavior of specimens of different volume in a different type of test. An example is given in a later section. At this point we give an example of the least-squares method of determining the Weibull parameters using strength data on silicon nitride determined in bending as part of an international round robin and supplied by the Oak Ridge National Laboratory (Brüchner-Foit and Munz, 1989). The analysis is given in Table 7.3 and a Weibull plot with the line determined by a least-squares best fit is shown in Figure 7.1. Here the strength values of column 1 have been put in ascending order in column 2 and assigned an order number in column 4. The probability of failure in column 5 was then calculated from equation (7.23) and the probability of survival from the first equality in (7.17). A linear regression analysis of a plot of ln $\ln(1/P_s)$, given in column 7, as a function of ln (strength), given in column 3, yielded $m = 12.88 \pm 0.29$ and $\sigma_0 = 670.74 \pm 0.21$.

The above procedure assumes that failure is caused by flaws distributed throughout the volume of the sample. This is not necessarily the case. Failure may occur only from flaws located at the surface, as is usually the case with glass. In this case a Weibull distribution may still be fitted to the strength values measured in a common test and a constant size of specimen. The conversion to a different type of stress distribution or a different size specimen must, however, be done by modifying the integration of equation (7.27) for different test methods and different specimen sizes to include only the surface region in which the flaws responsible for failure occur. A further complexity can occur when two types of flaw distributions both contribute to failure (e.g., volume and surface flaws) giving a bimodal distribution of strength values. In this situation, when fractography can be used to determine the type of flaw responsible for each failure, the strength data can be separated into two groups and can then be analyzed for volume and surface values of the Weibull parameters (see Section 7.14).

7.7 SAFETY FACTORS

Another important concept is that of the safety factor, which is the mean strength divided by the value of the maximum allowable design stress that has been chosen. Within the bounds of a statistical treatment of strength, it is not possible to have absolute safety. One can at best choose the working stress as a fraction of the mean stress to provide a calculated low probability of failure.

The safety factor treatment can be developed by expressing the Weibull distribution in terms of the mean strength rather than σ_0. For any of the test methods considered in Table 7.3, the mean strength expression can be solved for σ_0 and the result substituted into the corresponding Weibull distribution to give the following expression in all cases:

$$P_s = \exp - \left[\left(\frac{\sigma}{\bar{\sigma}} \right) \Gamma \left(1 + \frac{1}{m} \right) \right]^m \tag{7.42}$$

One can now choose some value of P_s considered adequate for safety and solve for the the safety factor, sf, defined as the ratio of the mean strength to the working stress ($\sigma_{\text{allowable}}$) necessary to achieve this value of P_s. The result is

$$sf = \frac{\bar{\sigma}}{\sigma_{\text{allowable}}} = \frac{\Gamma\left(1 + \frac{1}{m}\right)}{\left(\ln \frac{1}{P_s}\right)^{\frac{1}{m}}} \tag{7.43}$$

Typical allowable failure probabilities range from, say, 10^{-3} for cases where there is no serious consequence of failure to perhaps 10^{-7} where loss of life or other very serious damage might result from failure. For the former case and m values of 10 and 15, the required safety factors predicted by the Weibull theory are 1.90 and 1.53, respectively. For the latter case the predicted safety factors are 4.77 and 2.83, respectively. A larger value of m permits the use of working stress much nearer to the mean strength. The reason for striving to achieve high m values through good processing is evident.

7.8 AN EXAMPLE OF A SAFE STRESS CALCULATION

The usable stress in a particular design depends on the volume and stress distribution as well as the Weibull parameters. A common situation is one in which the Weibull parameters are determined in a three-point bending test and then applied to a sample of larger volume under some other stress distribution. A simple example that provides for both change of stress distribution and change of volume is given here. It is assumed that the mean strength in three-point bending, $\bar{\sigma}_{3b}$, and the Weibull m parameter of a two-parameter Weibull distribution have been determined from a series of measurements of breaking strength in a given type of test (taken as three-point bending in the present example) of bars of gage-length volume V_{3b}. It is desired to determine the stress level that can be used to give a survival probability P_{s1} in uniform tension for a piece of the same material of volume V.

TABLE 7.3 Determination of Weibull Parameters for Silicon Nitride from Strength Values Measured in Bending

Strength	Ordered Strength	ln (Ordered Strength)	Order Number, i	P_f = Probability of Failure	P_s = Probability of Survival	ln ln $(1/P_s)$	Regression Value of ln ln $(1/P_s)$
616.6	474.0	6.161	1	0.0083	0.9917	-4.783	-4.475
672.0	535.4	6.283	2	0.0250	0.9750	-3.676	-2.904
656.5	548.6	6.307	3	0.0417	0.9583	-3.157	-2.591
636.3	552.2	6.314	4	0.0583	0.9417	-2.812	-2.505
608.2	553.0	6.315	5	0.0750	0.9250	-2.552	-2.488
561.6	561.6	6.331	6	0.0917	0.9083	-2.342	-2.288
693.2	561.7	6.331	7	0.1083	0.8917	-2.166	-2.285
632.7	578.0	6.360	8	0.1250	0.8750	-2.013	-1.918
708.8	580.6	6.364	9	0.1417	0.8583	-1.879	-1.860
716.8	583.7	6.369	10	0.1583	0.8417	-1.758	-1.792
707.2	584.7	6.371	11	0.1750	0.8250	-1.648	-1.769
669.0	588.1	6.377	12	0.1917	0.8083	-1.547	-1.695
588.1	589.1	6.379	13	0.2083	0.7917	-1.454	-1.673
601.2	590.3	6.381	14	0.2250	0.7750	-1.367	-1.647
618.8	595.5	6.389	15	0.2417	0.7583	-1.285	-1.533
626.2	601.2	6.399	16	0.2583	0.7417	-1.208	-1.410
584.7	607.7	6.410	17	0.2750	0.7250	-1.134	-1.272
674.2	608.2	6.411	18	0.2917	0.7083	-1.065	-1.261
548.6	616.1	6.423	19	0.3083	0.6917	-0.998	-1.095
735.2	616.6	6.424	20	0.3250	0.6750	-0.934	-1.085
552.2	618.8	6.428	21	0.3417	0.6583	-0.872	-1.039
668.0	619.6	6.429	22	0.3583	0.6417	-0.813	-1.022
535.4	621.0	6.431	23	0.3750	0.6250	-0.755	-0.994
699.7	624.8	6.437	24	0.3917	0.6083	-0.699	-0.915
624.8	626.2	6.440	25	0.4083	0.5917	-0.645	-0.887
744.1	626.3	6.440	26	0.4250	0.5750	-0.592	-0.883
619.6	631.3	6.448	27	0.4417	0.5583	-0.540	-0.781
621.0	632.7	6.450	28	0.4583	0.5417	-0.489	-0.752
655.1	636.3	6.456	29	0.4750	0.5250	-0.440	-0.679
663.2	640.7	6.463	30	0.4917	0.5083	-0.391	-0.591
699.4	642.2	6.465	31	0.5083	0.4917	-0.343	-0.561
578.0	655.1	6.485	32	0.5250	0.4750	-0.295	-0.304
616.1	656.5	6.487	33	0.5417	0.4583	-0.248	-0.276
731.4	657.7	6.489	34	0.5583	0.4417	-0.202	-0.254
721.8	663.2	6.497	35	0.5750	0.4250	-0.156	-0.145
712.8	667.1	6.503	36	0.5917	0.4083	-0.110	-0.071
626.3	668.0	6.504	37	0.6083	0.3917	-0.065	-0.054
631.3	669.0	6.506	38	0.6250	0.3750	-0.019	-0.034
607.7	672.0	6.510	39	0.6417	0.3583	0.026	0.024
640.7	674.2	6.513	40	0.6583	0.3417	0.071	0.065
590.3	683.5	6.527	41	0.6750	0.3250	0.117	0.243
657.7	684.0	6.528	42	0.6917	0.3083	0.163	0.252
474.0	693.2	6.541	43	0.7083	0.2917	0.209	0.423

TABLE 7.3 *(continued)*

Strength	Ordered strength	ln ordered strength	order number, i	P_f = Probability of failure	P_s = Probability of survival	ln ln $1/P_s$	Regression value of ln ln $1/P_s$
561.7	699.4	6.550	44	0.7250	0.2750	0.255	0.539
580.6	699.7	6.551	45	0.7417	0.2583	0.303	0.545
710.0	699.7	6.551	46	0.7583	0.2417	0.351	0.545
684.0	706.9	6.561	47	0.7750	0.2250	0.400	0.676
595.5	707.2	6.561	48	0.7917	0.2083	0.450	0.682
730.3	708.8	6.564	49	0.8083	0.1917	0.502	0.711
706.9	710.0	6.565	50	0.8250	0.1750	0.556	0.732
718.8	712.7	6.569	51	0.8417	0.1583	0.611	0.782
712.7	712.8	6.569	52	0.8583	0.1417	0.670	0.784
699.7	716.8	6.575	53	0.8750	0.1250	0.732	0.855
667.1	718.8	6.578	54	0.8917	0.1083	0.799	0.892
683.5	721.8	6.582	55	0.9083	0.0917	0.871	0.946
553.0	726.0	6.587	56	0.9250	0.0750	0.952	1.019
583.7	730.3	6.593	57	0.9417	0.0583	1.044	1.096
642.2	731.4	6.595	58	0.9583	0.0417	1.156	1.116
726.0	735.2	6.600	59	0.9750	0.0250	1.305	1.183
589.1	744.1	6.612	60	0.9917	0.0083	1.566	1.337

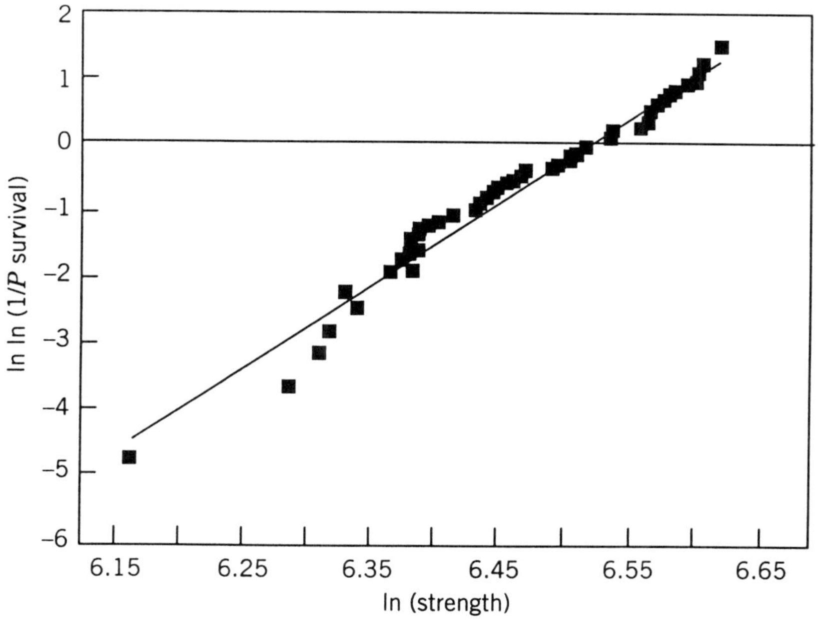

Figure 7.1 Weibull plot and fitted Weibull function for silicon nitride strength values determined in bending.

The problem can be solved by any of several equivalent procedures. A three-step procedure is used here to illustrate the ideas involved. These steps are: (1) determine the stress in three-point bending, σ_{3b}, s_1, to give the survival probability P_{s1} for the original test method and volume; (2) determine the volume of a hypothetical tensile specimen, V_t, eq to give the same survival probability under a tensile stress $\sigma_t = \sigma_{3b}, s_1$; (3) determine for the real tensile specimen of volume V_t and survival probability P_{s1} the allowable tensile stress σ_{t}, s_1 and express this as a safety factor in terms of the measured mean strength in bending. The notation for these successive steps is summarized in Table 7.4.

The first step can be done by writing the survival probability in terms of the mean strength in three-point bending, $\bar{\sigma}_{3b}$, and solving for the allowable bending stress. This survival probability expression is

$$P_{s1} = \exp\left(-\left[\left(\frac{\sigma_{3b,s1}}{\bar{\sigma}_{3b}}\right)\Gamma(1+\frac{1}{m})\right]^m \right) \tag{7.44}$$

This gives

$$\sigma_{3b,s1} = \frac{\bar{\sigma}_{3b}}{\Gamma(1+\frac{1}{m})}(-\ln P_{s1})^{1/m} \tag{7.45}$$

The second step can be done by writing P_{sb} (the survival probability in bending) $= P_{s1} = P_{st}$ and solving for the volume ratio required:

$$P_{sb} = \exp[-\left(\frac{\sigma_{3b,s1}}{\sigma_0}\right)^m \frac{V_b}{2(m+1)^2}] = \exp[-\left(\frac{\sigma_{t,s1}}{\sigma_0}\right)^m V_{t,\text{eq}}] = P_{st} \tag{7.46}$$

TABLE 7.4 Steps in Calculating Allowable Stress

Volume of Specimen	Survival Probability	Allowable Stress for This Survival Probability
Three-Point Bending Specimen Used to Determine Weibull Parameters		
V_{3b}	P_{s1}	$\sigma_{3b,s1}$
Hypothetical Tensile Specimen with Same Survival Probability Under Tensile Stress equal to $\sigma_{3b,s1}$		
$V_{t,eq}$	P_{s1}	$\sigma_{3b,s1}$
Real Tensile Specimen		
V_t	P_{s1}	$\sigma_{t,s1}$

This gives

$$\frac{V_{t,\,\mathrm{eq}}}{V_{3b}} = \frac{1}{2(m+1)^2} \tag{7.47}$$

for the ratio of the volumes of the equivalent tensile and the initial bend specimen at the same survival probability under the condition that the stress in the hypothetical tensile specimen equals the outer fiber stress in the bend specimen.

The third step can be done in an analogous manner using the expression for survival probability in tension:

$$\exp[-\left(\frac{\sigma_{3b,s1}}{\sigma 0}\right)^m V_{t,\,\mathrm{eq}} = \exp[-\left(\frac{\sigma_{t,s1}}{\sigma 0}\right)^m V_t \tag{7.48}$$

This gives

$$\sigma_{t,s1} = \sigma_{3b,s1}\left(\frac{V_{t,\,\mathrm{eq}}}{V_t}\right)^{1/m} = \frac{\bar{\sigma}_{3b}}{\Gamma(1+\frac{1}{m})}\left(\frac{V_b}{V_t}\frac{\ln(1/P_s)}{2(m+1)^2}\right)^{1/m} \tag{7.49}$$

Consider the example of the Weibull data discussed earlier for which m = 12.88. Suppose it is desired to have a probability of failure of 1×10^{-7} (i.e., $P_{s1} = 1 - 1 \times 10^{-7}$) in uniform tension for a piece of the same material having ten times the volume of the gage length of the bend specimen. Step 1 gives $\sigma_{3,\,s1}/\sigma_{3b,\,\mathrm{ave}} = 0.3177$. Step 2 gives $V_{t,\,\mathrm{eq}}/V_{3b} = 0.002595$. Step 3 gives $\sigma_{t,\,s1}/\sigma_{3b,\,s1} = 0.5267$. Combining these steps gives $\sigma_{t,\,s1}/\sigma_{3b,\,\mathrm{ave}} = 0.1673$ or a safety factor of 5.976.

It is interesting to recompute the same problem changing to $m = 20$. Step 1 gives $\sigma_{3b,\,s1}/\sigma_{3b,\,\mathrm{ave}} = 0.4646$. Step 2 gives $V_{t,\,\mathrm{eq}}/V_{3b} = 0.001134$. Step 3 gives $\sigma_{t,\,s1}/\sigma_{3b,\,s1} = 0.6349$. Combining these steps gives $\sigma_{t,\,s1}/\sigma_{3b,\,\mathrm{ave}} = 0.3077$ or a safety factor of 3.250. Increasing m from 12.8 to 20 nearly doubles the allowable stress for the same failure probability of 1×10^{-7}.

It is also interesting to keep $m = 20$ and allow the failure probability to increase by a factor of ten to 1×10^{-6}. In this case step 1 gives $\sigma_{b,\,s1}/\sigma_{3b,\,\mathrm{ave}} = 0.5438$. Step 2 gives $V_{t,\,\mathrm{eq}}/V_{3b} = 0.001134$. Step 3 gives $\sigma_{t,\,s1}/\sigma_{3b,\,s1} = 0.6349$. Combining these steps gives $\sigma_{t,\,s1}/\sigma_{3b,\,\mathrm{ave}} = 0.3453$ or a safety factor of 2.896.

7.9 COMPARISON OF GAUSSIAN AND WEIBULL DISTRIBUTIONS

The $p(\sigma)$ form of the Gaussian and Weibull distributions, (7.11) and (7.36), are compared in Figure 7.2. The Weibull distribution is plotted for a mean strength

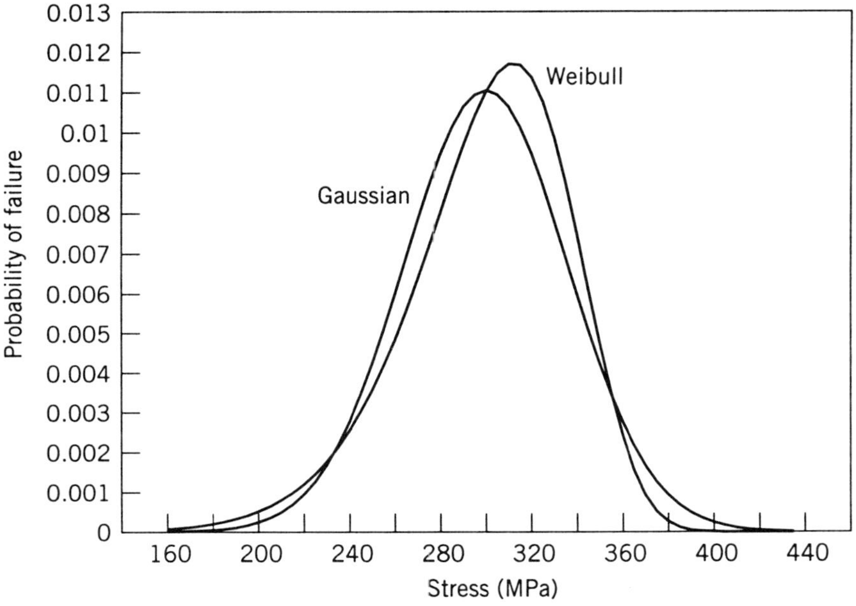

Figure 7.2 Gaussian and Weibull distributions for mean strength of 300 MPa and variance of 36 mPa^2.

of 300 MPa and an m value of 10. The Gaussian distribution is plotted for the same mean strength and the same variance as the Weibull distribution. The variance of the Weibull distribution expressed in terms of m is given by

$$s_W^2 = \int_0^\infty (\sigma - \bar{\sigma})^2 \, p(\sigma) \, d\sigma = (\bar{\sigma})^2 \left(\frac{\Gamma(1+\frac{2}{m})}{[\,\Gamma(1+\frac{1}{m})\,]^2} - 1 \right) \tag{7.50}$$

For $m = 10$ this gives an s value of 36 MPa and the Gaussian distribution of Figure 7.2 is plotted accordingly. The Weibull distribution is seen to be nonsymmetric and to have a mean strength that does not correspond to the maximum failure probability.

7.10 USE OF POOLED FRACTURE DATA IN LINEAR REGRESSION DETERMINATION OF WEIBULL PARAMETERS

The Weibull parameters for a given material are usually determined from a series of strength measurements, all of which were determined using the same test method and specimens of the same size. It is possible to pool data determined from two or more types of strength test with different specimen sizes used in the different test types. Consider two types of tests A and B (perhaps a tensile test and a three-point bend test) for which the specimens have volumes V_A and V_B, respectively, and values k_A and k_B where k was defined in Section 7.5 as

$$k = \frac{1}{V} \int_V \left(\frac{\sigma}{\sigma_{max}} \right)^m dV \tag{7.51}$$

Using the expression given there for survival probability in terms of k and taking the logarithm gives

$$\ln (P_{sA}) = -k_A V_A \left(\frac{\sigma}{\sigma_0} \right)^m \tag{7.52}$$

$$\ln (P_{sB}) = -k_B V_B \left(\frac{\sigma}{\sigma_0} \right)^m \tag{7.53}$$

Dividing at constant stress gives

$$\frac{\ln P_{sA}}{\ln P_{sB}} = \frac{k_A V_A}{k_B V_B} \tag{7.54}$$

One of the specimen sizes and corresponding test methods (say, A) can be chosen as the standard. For each strength value in set B this equation can then be used to calculate the appropriate type A survival probability. This requires assuming a trial value $\hat{m}$ of the Weibull parameter m because k depends on m. The entire resulting data set (set A plus the transformed set B) can then be treated as a single set of type A and the linear regression method of determining the Weibull parameters can then be used. The resulting m value can then be used to again transform the original survival probabilities for set A to those appropriate to set B and the linear regression can be repeated. The linear regression method thus requires iterative calculations when data from different test methods are pooled. Alternate iterative procedures are discussed by Johnson and Tucker (1993) and Tucker and Johnson (1993), who also treat parameter estimation for the case of surface-dependent rather than volume-dependent Weibull statistics.

7.11 METHOD OF MAXIMUM LIKELIHOOD IN WEIBULL PARAMETER ESTIMATION

As noted previously, the Weibull parameters can not only be estimated by the least-squares procedure but also by the method of maximum likelihood.

Johnson and Tucker (1993) have applied both methods not only to the relatively simple case in which all the experimental data are taken by one measurement method using specimens of one size but also to situations in which data must be pooled from more than one size of specimen and/or more than one measurement technique. For the former case the method of least squares involves only a single regression but the method of maximum likelihood involves solving a nonlinear equation by a series of iterations. In the case of pooled data, both the least-squares method and the maximum likelihood method involve a series of iterations to obtain the best estimate of the Weibull parameters for a two-parameter Weibull distribution. Johnson and Tucker recommend the method of maximum likelihood. We content ourselves with presenting the basic equations for this method, which has also been treated by Jayatilaka (1979).

In the method of maximum likelihood, the probability of obtaining the strength value for each experimental measurement is written in terms of the unknown Weibull parameters m and σ_0. For an experimental value of strength σ_i, the probability of obtaining this value is the probability density p_i given by

$$p_i = \left(\frac{dP}{d\sigma}\right)_{\sigma=\sigma_i} = \frac{mkV}{(\sigma_0)^m}\,\sigma^{m-1} \exp\left[-kV\left(\frac{\sigma}{\sigma_0}\right)^m\right] \tag{7.55}$$

where k is the dimensionless structure factor that contains the effect of the stress variation over the volume of the specimen and so differs for different types of test as discussed earlier.

The likelihood of obtaining all the measured strength values is then the product of all the p_i values. It is convenient to deal with l, the logarithm of the likelihood, which is given by

$$l = \sum_{i=1}^{n} \ln(p_i) \tag{7.56}$$

The maximum likelihood of obtaining the observed values will occur for the values of the Weibull parameters that maximize l. Setting the derivatives of l with respect to m and σ_0 equal to zero to obtain the maximum gives two equations that can be combined to eliminate σ_0 and give the equation for m:

$$\frac{n}{m} + \sum_{i=1}^{n} \frac{dk_i/dm}{k_i} + \sum_{i=1}^{n} \ln(\sigma_i) - \frac{n \sum_{i=1}^{n} \left[k_i V_i \sigma_i^m \ln(\sigma_i) + V_i \sigma_i^m \, dk_i/dm\right]}{\sum_{i=1}^{n} V_i k_i \sigma_i^m} = 0 \tag{7.57}$$

where V_i and k_i refer to the respective values for the specimen size and type of test used to obtain the strength value. This equation for m must be solved by iteration and is done with the aid of a computer. Once the m value is obtained the σ_0 value is calculated from

$$n\,\sigma_0^m - \sum_{i=1}^{n} k_i V_i \sigma_i^m = 0 \tag{7.58}$$

Thoman, Bain, and Antle (1969) have given a technique for using the method of maximum likelihood including the use of unbiasing factors.

7.12 STATISTICS OF FAILURE UNDER MULTIAXIAL STRESS

The above statistical treatment of failure has tacitly assumed parallel uniaxial tension in each volume element of the specimen, although the magnitude may vary with position as in bending tests. The statistics of failure under multiaxial stress have been treated by a method due to Batdorf (Batdorf and Crose, 1974; Batdorf and Heinisch, 1978; Batdorf and Chang, 1979) and a multiaxial elemental strength model (Evans, 1978a; Lamon, 1988). The Batdorf and Evans theories have been shown to be equivalent (Chao and Shetty, 1990; Andreasen, 1993) provided that the same fracture criterion is used.

Chao and Shetty (1991) used Batdorf's statistical fracture theory to correlate uniaxial strength results (determined in four-point and three-point bending) with biaxial strength results (determined with uniform pressure on disk). They considered two failure criteria: the critical normal stress criterion and the noncoplanar strain energy release rate criterion. The former takes the simple form

$$K_{\mathrm{I}} = K_{\mathrm{I}C} \tag{7.59}$$

and thus considered failure to be determined solely by mode I loading of a crack. The latter leads to an approximate equation

$$\left(\frac{K_{\mathrm{I}}}{K_{\mathrm{I}C}}\right) + \left(\frac{K_{\mathrm{II}}}{CK_{\mathrm{I}C}}\right)^2 = 1 \tag{7.60}$$

originally suggested by Palaniswamy and Knauss (1978) with $C = (2/3)^{1/2} = 0.82$. Measurements were made on two commercial ceramics, an alumina and a silicon nitride, having the properties listed in Table 7.5. The elastic moduli measured by the three flexure methods and by a compression method are

TABLE 7.5 Properties of Two Commercial Ceramics Used in Uniaxial and Biaxial Comparison				
Ceramic	E (GPa)	ν	ρ (g/cm^3)	K_{IC} $(MPa\, m^{1/2})$
Alumina (Coors AD-94)	297.2	0.23	3.70	3.1
Silicon nitride (GTE SN W-1000)	285.0	0.23	3/28	7.2

listed in Table 7.6 and are seen to be in reasonably good agreement. The Weibull parameters determined in the three flexural tests are given in Table 7.7. All data have been rounded to four places.

For the alumina specimens the data obtained in four-point bending were fitted to the Weibull function using the method of maximum likelihood. The data are shown as a linearized Weibull plot in Figure 7.3 and fit well with a Weibull parameter $m = 23.77$ as shown. The Weibull plots for three-point bending and for biaxial flexure were then predicted from these four-point bending results for the normal stress criterion and for the noncoplanar strain energy release rate criterion. The former does not fit but the latter fits the data well using $C = 1$.

For silicon nitride specimens the method of Batdorf and Sines (1980) was used to obtain $m = 11.51$ and $\sigma_0 = 918.4$ MPa. Figure 7.4 shows a linearized Weibull plot on which the four-point bend strengths have been converted to equivalent three-point bend strengths using effective stressed volumes for each. The fracture probabilities for the biaxial flexure test were then predicted. Neither the critical normal stress nor the critical strain energy release rate criteria gave correct predictions; both overestimated the fracture probabilities. Assuming all flaws perpendicular to the stress and using the critical normal stress criterion gave predictions closer to the observed biaxial strength data.

TABLE 7.6 Elastic Moduli Determined in Different Tests (GPa)				
Ceramic	Four-point bend	Three-Point Bend	Biaxial Flexure	Compression
Alumina	290.1	296.6	301.8	297.2
Silicon nitride	279.9	283.7	291.7	285.0

TABLE 7.7 Weibull Parameters

Ceramic	Four-Point Bend		Three-Point Bend		Biaxial Flexure	
	m	σ_0 (*MPa*)	m	σ_0 (*MPa*)	m	σ_0 (*MPa*)
Alumina	23.77	353.4	25.43	385.9	22.25	338.8
Silicon nitride	10.12	768.4	15.67	908.1	12.02	717.1

7.13 EFFECTS OF SLOW CRACK PROPAGATION AND R-CURVE BEHAVIOR ON STATISTICAL DISTRIBUTIONS OF STRENGTH

The above discussion of statistical effects on strength has tacitly assumed that strength is not time dependent and that the critical stress intensity for crack propagation does not depend on crack length. Both assumptions may be incorrect under certain circumstances.

Subcritical crack propagation is treated in Chapter 8 and leads to a dependence of strength on the rate of loading. We show in Section 8.7 that the Weibull

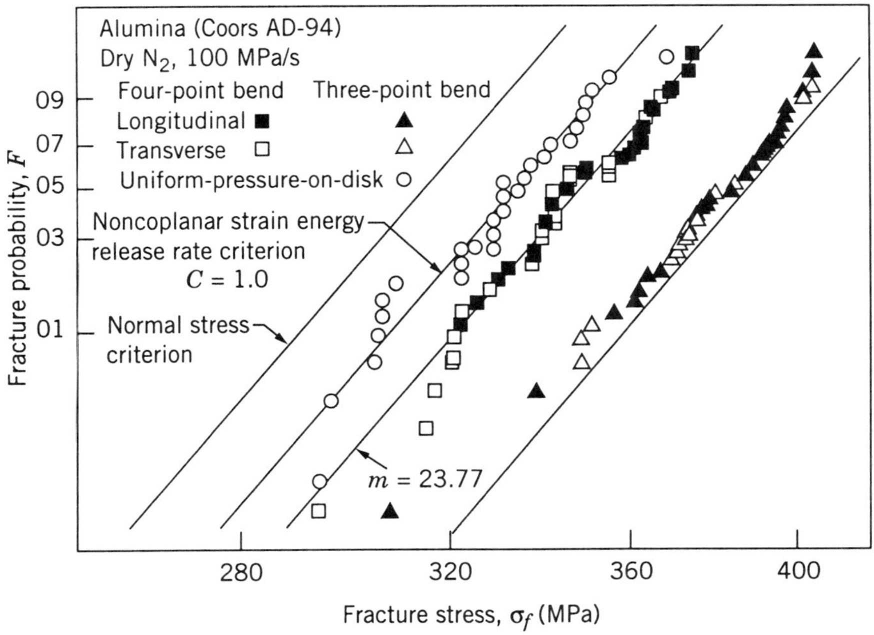

Figure 7.3 Weibull plots for alumina. (From L. Y. Chao and D. K. Shetty, "Reliability Analysis of Structural Ceramics Subjected to Biaxial Flexure," *J. Am. Ceram. Soc.*, Vol. 74, 333–344, 1991. Reprinted by permission of the American Ceramic Society.)

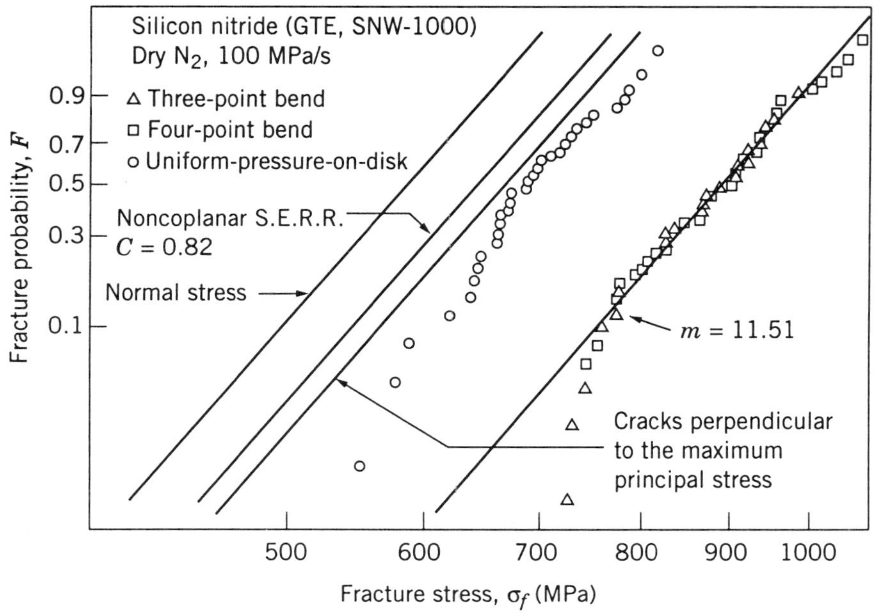

Figure 7.4 Weibull plots for silicon nitride. (From L. Y. Chao and D. K. Shetty, "Reliability Analysis of Structural Ceramics Subjected to Biaxial Flexure," *J. Am. Ceram. Soc.*, Vol. 74, 333–344, 1991. Reprinted by permission of the American Ceramic Society.)

strength distribution can still be used but that the value of the Weibull m parameter as measured in constant strain rate experiments may be different. In Section 8.10 the effect of localized stresses in producing a type of subcritical crack growth that leads to an altered value of the Weibull m is described.

The effect of crack resistance that is dependent on crack length (R-curve behavior) is treated in Chapter 9. In Section 9.4 the resulting effect of statistical distribution of strength is briefly discussed, including the possibility that the Weibull distribution may not accurately describe the actual strength distribution.

7.14 SURFACE FLAW DISTRIBUTIONS AND MULTIPLE FLAW DISTRIBUTIONS

Flaws responsible for fracture have so far in this chapter been assumed to be randomly distributed throughout the volume of material. The treatment for predicting failure probabilities for different volumes was based on this assumption. However, as noted in Chapter 6, some types of flaws of the same size should cause failure at a lower stress when they are located on the surface rather than within the volume. It is therefore possible that, for some types of specimens, failure occurs primarily or even entirely from surface flaws as appears to be the case with glass. For a purely surface distribution of flaws, the Weibull distribution can still be used as before and no change in the equations is needed as long as all specimens have the same shape, type of loading, and volume. However, to deal with different types of loading and different volumes for a surface distribution of flaws, the k integral of equation (7.27) must be taken over the volume within a fixed distance from the surface rather than over the whole volume. This results in different expressions of the type of (7.32) (see below) for stresses to give equal failure probabilities for specimens of different test types. For the case of the same specimen shape and same test method but different specimen sizes, equation (7.42) must be used with the appropriate volume, which is that of a surface layer of constant depth. For the case of cylindrical tensile samples of radii r_1, r_2 and lengths L_1, L_2, equation (7.42) becomes

$$\frac{\bar{\sigma}_1}{\bar{\sigma}_2} = \left(\frac{r_2 L_2}{r_1 L_1}\right)^{\frac{1}{m}} \tag{7.61}$$

When only surface flaws are considered to be the sources of fracture, as is usually the case for glass, the Weibull distribution can be rewritten in terms of a surface rather than a volume dependence. A useful way to approach this is to consider the Weibull distribution for a circular cylinder of length L and radius r under tension along the axis. If only the flaws in a layer of thickness t at the surface cause failure, then the appropriate volume for the first equation in Table 7.1 is $V_s = 2\,\pi r t L = At$, where A is the area of the cylindrical surface and the Weibull distribution becomes

$$P_s = \exp\left(-\left(\frac{\sigma}{\sigma_0}\right)^m At\right) = \exp\left(-\left(\frac{\sigma}{\sigma_{0,\mathrm{surf}}}\right)^m A\right) \tag{7.62}$$

The right-hand expression gives the surface form of the Weibull distribution for uniform tension. This derivation shows that the second Weibull parameter must have different dimensions in the surface and volume cases; we have used $\sigma_{0,\mathrm{surf}}$ here to emphasize this distinction. Equation (7.62) suggests that if t were known, then $\sigma_{0,\mathrm{surf}}$ could be calculated from σ_0. Actually, the distribution of surface flaws may be different from the volume flaws so that the Weibull parameters for surface flaws should be determined directly from experiment. For that purpose the expressions given in Table 7.1 for volume distributions of flaws have been modified by M. J. Matthewson (personal communication, 1995) for the case of surface distributions and are given in Table 7.8.

A further complication sometimes arises in which two distributions of flaws both contribute to failure. For example, failure may result from both volume flaws and surface flaws. Another example that sometimes actually occurs in bending tests with bars of rectangular cross section is that failure occurs from

TABLE 7.8 The Two-Parameter Weibull Distribution and Mean Strength for Several Test Methods for a Surface Distribution of Flaws

A for tension is the surface area under tension; for example, for a cylinder of length L and radius r, $A = 2\pi r L$, while for a rectangular bar of length L, width w, and thickness h, $A = 2(w+h)L$. A for bending of a bar of length L and width w with bending in the direction of the height, h, is expressed as $A_b = wL[1 + \frac{h/w}{m+1}]$
Tension
$P_s = \exp\left(-\left(\frac{\sigma_t}{\sigma_{0,surf}}\right)^m A_b\right)$
$\bar{\sigma}_t = \frac{\sigma_{0,surf}}{A_b^{1/m}} \Gamma(1 + \frac{1}{m})$
Three-Point Bending
$P_s = \exp\left(-\left(\frac{\sigma_{3b,max}}{\sigma_{0,surf}}\right)^m \frac{A_b}{(m+1)}\right)$
$\bar{\sigma}_{3b,max} = \frac{\sigma_{0,surf}}{A_b^{1/m}} (m+1)^{1/m} \Gamma(1 + \frac{1}{m})$
Pure Bending
$P_s = \exp\left(-\left(\frac{\sigma_{b,\max}}{\sigma_{0,surf}}\right)^m A_b\right)$
$\bar{\sigma}_{b,max} = \frac{\sigma_{0,surf}}{A_b^{1/m}} \Gamma(1 + \frac{1}{m})$
Four-Point Bending with Load at the Quarter Points
$P_s = \exp\left(-\left(\frac{\sigma_{4b,1/4,\max}}{\sigma_{0,surf}}\right)^m \frac{A_b(m+2)}{2(m+1)}\right)$
$\bar{\sigma}_{4b,1/4,\max} = \frac{\sigma_{0,surf}}{A_b^{1/m}} \left[\frac{2(m+1)}{m+2}\right]^{1/m} \Gamma(1 + \frac{1}{m})$
Four-Point Bending with Loads at the Third Points
$P_s = \exp\left(-\left(\frac{\sigma_{4b,1/3,max}}{\sigma_{0,surf}}\right)^m \frac{A_b(m+3)}{3(m+1)}\right)$
$\bar{\sigma}_{4b,1/3,max} = \frac{\sigma_{0,surf}}{A_b^{1/m}} \left[\frac{3(m+1)}{m+3}\right]^{1/m} \Gamma(1 + \frac{1}{m})$

both surface and edge flaws. In these cases of multiple flaw distributions the failure probability is no longer described by a single Weibull distribution but requires the combination of two distributions. Johnson was apparently the first to apply this idea to failure in ceramics in an unpublished report that is the basis for later treatments (Jakus, Ritter, Service, and Sonderman, 1981; Johnson, 1983). For the case of volume and surface distributions (designated by subscripts v and s), the overall survival probability can be written as

$$P_S = 1 - (1 - P_{S,v})(1 - P_{S,s}) \tag{7.63}$$

Taking each of the survival probabilities to be a Weibull distribution gives

$$P_S = 1 - \exp\left(-k_v V_v \left(\frac{\sigma}{\sigma_{0,v}}\right) - k_s V_s \left(\frac{\sigma}{\sigma_{0,s}}\right)\right) \tag{7.64}$$

When the origin of failure can be assigned for each specimen to either a surface or a volume flaw by subsequent fractographic examination (see Chapter 16), it is possible to analyze the strength data and determine the four Weibull parameters. The analysis, however, is not straightforward. One cannot simply separate the two distributions and analyze each separately. The proper procedure is discussed by Jakus, Ritter, Service, and Sonderman (1981), Service, Ritter, Jakus, and Sonderman (1985), Sonderman, Jakus, Ritter, Yuhaski, and Service (1985), and in ASTM standard C 1239.

CHAPTER 8

SUBCRITICAL CRACK PROPAGATION

8.1 INTRODUCTION

The discussion of crack behavior so far has assumed essentially instantaneous response of cracks to an applied stress. The general assumption has been that a crack does not extend as the applied stress intensity value is increased until the critical stress intensity value is reached, at which point the crack propagates very rapidly. However, in many ceramics cracks slowly grow under applied stress intensity values well below the critical stress intensity value.

A situation of engineering importance is the behavior of cracks under a service load that causes the crack to grow slowly so that the applied stress intensity factor increases until the critical value is reached and failure occurs. The basis for safe-life design for this situation is presented in this chapter.

The principal subcritical crack propagation mechanism in ceramics at room temperature involves stress-assisted reaction with the environment—especially reaction with water. This mechanism is discussed first in the present chapter. Several other mechanisms have been proposed, including thermal activation of lattice-trapped cracks and mass transport (diffusion). The latter does have practical importance at elevated temperatures and is discussed in the last section of this chapter.

8.2 OBSERVED SUBCRITICAL CRACK PROPAGATION

Slow crack propagation occurs in many ceramics and is especially prevalent in ceramics containing a silica phase and in the presence of an appreciable partial pressure of water vapor (Wiederhorn, 1974). For a given material, temperature, and relative humidity, the controlling variable is the applied stress intensity factor. A typical plot of crack velocity as a function of stress intensity factor for soda-lime–silica glass is shown in Figure 8.1 from work by Wiederhorn (1967) (Wiederhorn and Bolz, 1970; Wiederhorn, 1974). This figure shows the logarithm of crack velocity as a function of stress intensity factor. Several important points are illustrated (Wachtman, 1974). (1) The stress intensity factor K_I is the controlling factor; all data from many cracks with various combinations of crack length and applied stress fall on the same

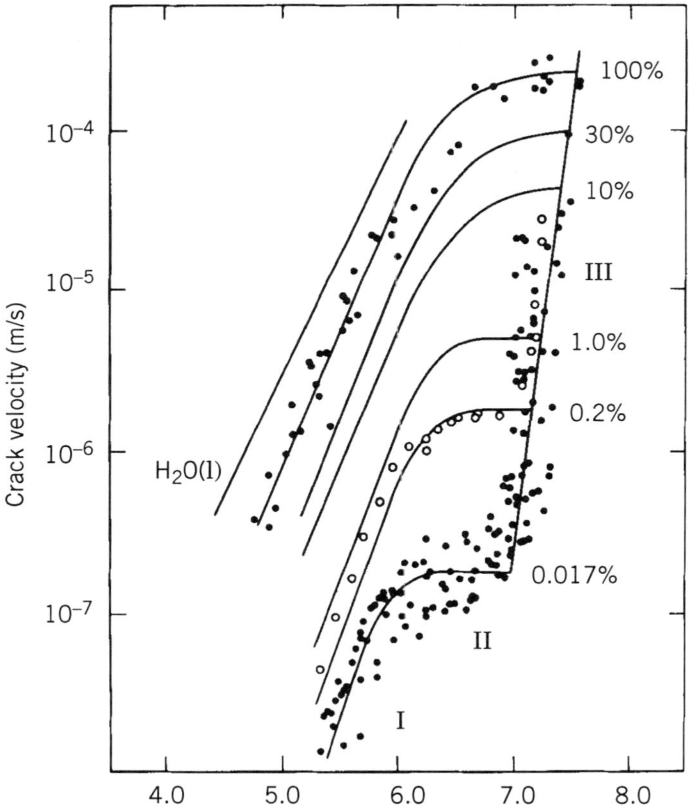

Figure 8.1 Crack velocity in soda-lime–silica glass as a function of stress intensity factor in nitrogen at various values of relative humidity (From S. M. Wiederhorn, "Influence of Water Vapor on Crack Propagation in Soda-Lime Glass," *J. Am. Ceram. Soc.*, Vol. 50, 407–414, 1967. Reprinted by permission of the American Ceramic Society.)

master curves when expressed in terms of K_I. (2) At low values of K_I the logarithmic plot is linear; such linear behavior is predicted by the theory of Charles and Hillig (1962) (Hillig and Charles, 1965), described later. This range of behavior is termed region I. (3) Within region I the crack velocity for a constant K_I value is greater for greater values of relative humidity. Crack growth in region I is attributable to a stress-enhanced chemical reaction between water and the glass. The rate of crack growth is reaction-rate-controlled so that increasing the concentration of reactant should increase the rate of growth. This prediction is borne out by the results shown in Figure 8.1. Further evidence of chemical control in region I is provided by Figure 8.2 (Wiederhorn and Johnson, 1973), which shows that increasing the pH increases the rate of crack propagation at constant K_I for silica glass in water. The Charles–Hillig theory predicts a temperature dependence of crack velocity; this dependence is observed as shown in Figure 8.3 (Wiederhorn and Bolz, 1970). (4) Figure 8.1 also shows the existence at higher K_I values of region II, in which the crack velocity is still dependent on the partial pressure of water, but independent of K_I. Here the rate of crack growth is still controlled by a stress-enhanced chemical reaction between water and glass, but now the rate of growth is determined by the rate of transport to the crack tip. (5) At still higher K_I values, region III occurs, in which crack propagation becomes independent of water in the environment. Here the growth rate at constant K_I depends on temperature and compo-

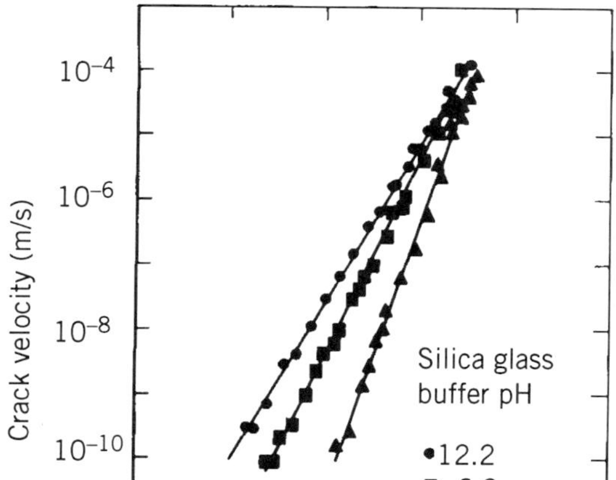

Figure 8.2 Effect of pH on crack velocity in silica glass in aqueous environments. (From S. M. Wiederhorn and H. Johnson, "Effect of Electrolyte pH on Crack Propagation in Glass," *J. Am. Ceram. Soc.*, Vol. 56, 192–197, 1973. Reprinted by permission of the American Ceramic Society.)

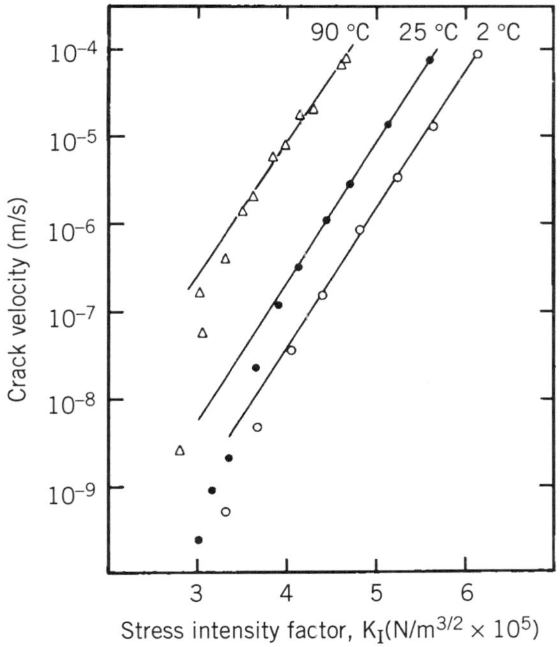

Figure 8.3 Effect of temperature on crack velocity in soda-lime–silica glass in water. (From S. M. Wiederhorn and L. H. Bolz, "Stress Corrosion and Static Fatigue of Glass," *J. Am. Ceram. Soc.*, Vol. 53, 543–548, 1970. Reprinted by permission of the American Ceramic Society.)

sition of the glass, as shown in Figure 8.4 (Wiederhorn, Johnson, Diness, and Heuer, 1974).

Results similar to those found for glass at room temperature have been found for a variety of ceramic materials including porcelains, high alumina ceramics, silicon nitride, vitreous carbon, portland cement, PZT, and barium titanate. At room temperature both porcelain and polycrystalline aluminum oxide show crack growth rate dependent on environment; the three regions found for glass are also found in these cases.

Although the data in region I are well described by a semilog plot as shown in Figure 8.1, they can be equally well described by a straight line on a plot of ln v as a function of ln K_I. The latter plot leads to the empirical law of crack velocity,

$$v = A\, K_I^{\,n} \tag{8.1}$$

This law is convenient mathematically and is usually used in lifetime calculations as described later. The effect on lifetime predictions of the type of velocity law that is used has been discussed by Jakus, Ritter, and Sullivan (1981). Tomozawa (1996) argues that crack blunting by water may invalidate such lifetime calculations for glass annealed in water vapor.

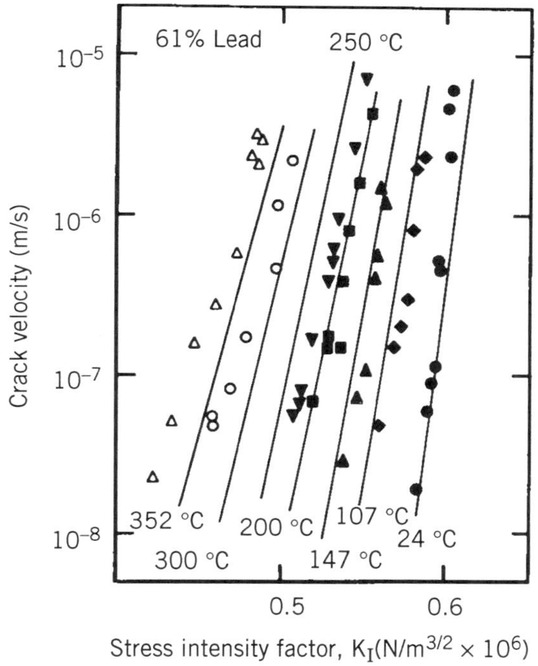

Figure 8.4 Effect of temperature on crack velocity of high-lead glass in vacuum. (From S. M. Wiederhorn, H. Johnson, A. M. Diness, and A. H. Heuer, "Fracture of Glass in Vacuum," *J. Am. Ceram. Soc.*, Vol. 57, 8, 1974. Reprinted by permission of the American Ceramic Society.)

8.3 CRACK VELOCITY THEORY AND A MOLECULAR MECHANISM

According to the Griffith theory, cracks do not propagate below the critical stress required for rapid propagation to complete failure. Charles and Hillig (1962) assumed that the rate of subcritical crack propagation is controlled by a chemical reaction between the glass and the water in the environment. Because chemical reactions are activated processes, the crack velocity should depend on an activation energy, which should in turn depend on stress. The Charles–Hillig theory gives the velocity of chemically assisted crack growth as

$$v = v_0 \exp\left[(-E^+ + V^+\sigma_t - V_M\gamma/\rho)/RT\right] \tag{8.2}$$

Here E^+ is the stress-free activation energy, V^+ is the activation volume, σ_t is the stress at the crack tip, V_M is the molar volume of the glass, γ is the interfacial surface energy between the glass and the reaction products, and ρ is the radius of the crack tip. Wiederhorn (1970) expressed the stress in terms of the stress intensity factor using equation (5.20) as

$$\sigma = \frac{2K_I}{(\pi\rho)^{1/2}} \tag{8.3}$$

and defined

$$E^* = E^+ + \frac{\gamma V_M}{\rho} \tag{8.4}$$

leading to

$$v = V_0 \exp\left(\left\{-E^* + \frac{2V^+}{(\pi\rho)^{1/2}} K_I\right\}/RT\right) \tag{8.5}$$

This gives

$$\ln v = \left(\frac{2V^+}{(\pi\rho)^{1/2}}\right)\frac{K_I}{RT} - \frac{E^*}{RT} + \ln V_0 \tag{8.6}$$

This provides a basis for the linear dependence of ln v upon K_I at constant T for environmentally sensitive subcritical crack growth in region I. It is perhaps sufficient to argue qualitatively that the stress intensity factor is the appropriate variable to describe the driving force not only for sudden crack propagation but for any stress-controlled process at a crack tip including stress corrosion. The experimental data on crack velocity in samples of a given glass with different crack lengths do fall on a single curve when plotted as a function of the stress intensity factor. As noted below, the functional dependence is as well described by a power of the stress intensity factor as by an exponential; the former is more convenient in lifetime calculations.

The above treatment did not depend on a specific molecular mechanism; a detailed qualitative molecular model has been proposed by Michalske and Freiman (1983) for silicates. For the case of water interacting with stressed silica they suggested a three-step process as follows:

> Step 1. A water molecule from the environment attaches to a bridging Si—O—Si bond at the crack tip. The water molecule is aligned by: (1) formation of the hydrogen bond with the $O_{(bridging)}$ atom and (2) interaction of the lone-pair orbitals from $O_{(water)}$ with the Si atom. The lone-pair orbital interaction may involve either Van der Waals attraction or some covalent bonding with unoccupied orbitals of Si.
>
> Step 2. A concerted reaction occurs in which proton transfer to the $O_{(br)}$ is accomplished simultaneously with electron transfer from the $O_{(w)}$ to the Si atom. As a result of this reaction, two new bonds are formed, one between $O_{(w)}$ and Si, and one between hydrogen and $O_{(br)}$; the original bridging bond between $O_{(br)}$ and Si is destroyed.
>
> Step 3. Rupture of the hydrogen bond between $O_{(w)}$ and transferred hydrogen occurs to yield surface Si—O—H groups on each fracture surface. Since the

hydrogen bond is weak, this step is expected to occur immediately after proton transfer.

They note that there is no need for prior dissociation of the water molecule nor do any reaction products need to be removed from the surface. They studied several nonaqueous environments also and found several that enhanced stress-corrosion crack growth. Those that enhanced crack growth contained active groups with electron donor sites on one end and proton donor sites at the other. Other nonaqueous environments had no effect provided care was taken to exclude trace amounts of water. They suggest that earlier results by other workers indicating a dependence of the effectiveness in assisting slow crack growth upon chain length were more likely due to systematic variations in the affinity for water and that trace amounts of water were probably the cause of the earlier results.

8.4 TIME TO FAILURE UNDER CONSTANT STRESS

The effect of subcritical crack propagation on the failure stress under different conditions of loading has been treated by several investigators including Davidge, McLaren, and Tappin (1973), Evans and Wiederhorn (1974), and Ritter (1978). The general fracture mechanics relation for the stress intensity factor K_I for a crack of length c under stress σ is

$$K_I = Y \sigma c^{1/2} \tag{8.7}$$

Consider the situation when the stress distant from the crack remains constant but the crack slowly grows longer. The stress intensity factor will increase at a rate given by

$$\frac{dK_I}{dt} = \frac{\sigma Y}{2c^{1/2}} \frac{dc}{dt} + \sigma c^{1/2} \frac{dY}{dt} \tag{8.8}$$

Using $v = dc/dt$ gives

$$dt = \frac{2}{Y^2 \sigma^2} \frac{K_I}{v} \left(dK_I - \frac{K_I}{Y} d \ln Y\right) \tag{8.9}$$

For cracks small compared to the specimen dimensions, the change in Y as the crack grows is small so that the last term can be dropped, giving

$$dt = \frac{2}{Y^2 \sigma^2} \frac{K_I}{v} dK_I \tag{8.10}$$

The crack will grow from its initial length at which $K_I = K_{Ii}$ until it reaches a length at which $K_I = K_{IC}$, at which time the specimen will fail. The time interval $t_{f,\sigma}$ required for crack growth to failure under constant stress σ is then given by

$$t_{f,\sigma} = \frac{2}{\sigma^2 Y^2} \int_{K_{Ii}}^{K_{IC}} \left(\frac{K_I}{v} \right) dK_I \tag{8.11}$$

As mentioned earlier, the empirical crack growth law commonly used for region I is $v = A\, K_I^n$, equation (8.1). An exact treatment would require using this crack velocity expression only for region I and using a constant value of v for region II in the integration. Because the crack is traveling relatively fast in region II, the time spent in traversing this region is small compared with the time spent in region I if the experiment begins well down in region I. For simplicity, the above crack growth law is used for the complete integration. This procedure slightly underestimates the time to failure; that is, it gives a conservative estimate of the failure time. Substituting this relation into the integral gives

$$t_{f,\sigma} = \frac{2}{\sigma^2 Y^2 A(n-2)} \left(\frac{1}{K_{Ii}^{n-2}} - \frac{1}{K_{IC}^{n-2}} \right) \tag{8.12}$$

If the constant stress σ is well below the short time failure stress $\sigma_{f,t=0}$ and the n value is large, the second term can be ignored. For example, for $\sigma = 0.6\sigma_{f,t=0}$ and $n = 12$, dropping the second term introduces only 0.6% error. Accordingly, the equation commonly used to estimate failure time under constant stress is

$$t_f = \frac{2}{\sigma^2 Y^2 A(n-2)} \left(\frac{1}{K_{Ii}^{n-2}} \right) \tag{8.13}$$

The parameter n can be evaluated by determining the failure time in a series of experiments at different values of constant applied stress, all well below the short time fracture stress. The stress dependence concealed in K_{Ii} can be made explicit by writing

$$K_{Ii} = \left(\frac{\sigma}{\sigma_{f,t=0}} \right) K_{IC} \tag{8.14}$$

which gives

$$t_{f,\sigma} = \frac{2}{\sigma^n Y^2 A(n-2)} \left(\frac{\sigma_{f,t=0}}{K_{IC}} \right)^{n-2} \tag{8.15}$$

The value of n is then obtained from a plot of ln $t_{f,\sigma}$ as a function of ln σ, which has a slope of $-n$. Values of $\sigma_{f,t=0}$ and K_{IC} can be obtained from short time or inert atmosphere strength and fracture mechanics experiments and A can then be calculated from the above equation.

This equation can be used to express the ratio of the times to failure, $t_{f,\sigma 1}$ and $t_{f,\sigma 2}$, for two specimens with the same short time strengths under two corresponding subcritical stresses, σ_1 and σ_2, as

$$\frac{t_{f,\sigma_1}}{t_{f,\sigma_2}} = \left(\frac{\sigma_2}{\sigma_1}\right)^n \quad \text{(for the same short time strength).} \tag{8.16}$$

One might wonder about the usefulness of this equation considering that strengths of specimens are statistically distributed. We shall subsequently (Section 8.7) show that this equation applies to mean strengths measured in tension so that its use is justified for this case. However, in bend tests the existence of a flaw distribution combined with a stress that is not constant over the specimen introduces more complex behavior. The reader is referred to Hu, Mai, and Cotterell (1988) for a theory for this case. In the present chapter we shall not introduce this complication and instead continue the conventional treatment.

8.5 FAILURE UNDER CONSTANT STRAIN RATE

The effect of subcritical crack propagation on failure at constant strain rate was considered by Evans (1973). If a specimen is loaded in a testing machine with constant rate of head motion the strain rate $\dot{\varepsilon}$ is constant as is the stress rate $\dot{\sigma} = E\,\dot{\varepsilon}$. The rate of change of stress with crack length can be expressed in terms of the crack velocity as

$$\frac{d\sigma}{dc} = \frac{d\sigma}{dt}\frac{dt}{dc} = \frac{\dot{\sigma}}{v} \tag{8.17}$$

Using $v = AK_I{}^n$, $K_I = Y\sigma c^{1/2}$, and $d\sigma/dc = (d\sigma/dt)(dt/dc) = \dot{\sigma}/v$ gives

$$\frac{d\sigma}{dc} = \frac{\dot{\sigma}}{AK_I^n} = \frac{\dot{\sigma}}{AY^n\sigma^n c^{n/2}} \tag{8.18}$$

Separating variables and integrating from the initial stress $\sigma = 0$ and initial crack length c_i to the fracture stress $\sigma_{f,\dot{\varepsilon}}$ and the crack length at failure c_f gives

$$\int_0^{\sigma_{f,\dot{\varepsilon}}} \sigma^n d\sigma = \int_{c_i}^{c_f} \frac{\dot{\sigma}}{AY^n} \frac{dc}{c^{n/2}} \tag{8.19}$$

so that

$$\frac{\sigma^{n+1}_{f,\dot{\varepsilon}}}{n+1} = \frac{2\dot{\sigma}}{(n-2)AY^n}\left[-\frac{1}{c_f^{\frac{n-2}{2}}} + \frac{1}{c_i^{\frac{n-2}{2}}}\right] \tag{8.20}$$

Taking the first term in brackets as small compared to the second term and using

$$c_i = \left(\frac{K_{IC}}{Y\sigma_{f,t=0}}\right)^2 \tag{8.21}$$

where as before $\sigma_{f,t=0}$ is the fracture stress in an inert environment or at very high strain rate (i.e., with no subcritical crack growth), leads to

$$(\sigma_{f,\dot{\varepsilon}})^{n+1} = \frac{2(n+1)E\dot{\varepsilon}\,\sigma_{f,t=0}^{n-2}}{AY^2K_{IC}^{n-2}\,(n-2)} = \frac{2(n+1)\dot{\sigma}}{AY^2(n-2)}\left(\frac{\sigma_{f,t=0}}{K_{IC}}\right)^{n-2} \tag{8.22}$$

Taking logarithms gives

$$Ln\,\sigma_{f,\dot{\varepsilon}} = \frac{1}{n+1}\,Ln\,\dot{\varepsilon} + \frac{1}{n+1}\,Ln\left[\frac{2(n+1)E\sigma_{f,t=0}^{n-2}}{(n-2)AY^2K^{n-2}}\right] \tag{8.23}$$

so that n can be determined from a plot of the log fracture stress as a function of the log strain rate. Another useful result is the comparison of strength values $\sigma_{f,\dot{\varepsilon}_1}$ and $\sigma_{f,\dot{\varepsilon}_2}$ of the specimens with the same short time strengths at two corresponding constant strain rates $\dot{\varepsilon}_1$ and $\dot{\varepsilon}_2$. From above

$$\left(\frac{\sigma_{f,\varepsilon_1}}{\sigma_{f,\varepsilon_2}}\right)^{n+1} = \frac{\dot{\varepsilon}_1}{\dot{\varepsilon}_2} \quad \text{(for the specimens with the same short time strength).} \tag{8.24}$$

Some precautions concerning the use of this equation are appropriate. First, this holds for specimens with the same short time strength, but this cannot be guaranteed for individual specimens. We show in the next section that equation

(8.23) holds if the strengths of individual specimens are replaced with the average strengths at two strain rates. Second, the equations (8.22) through (8.24) are based on dropping the first term in brackets in (8.20). As the strain rate is increased the measured strength at constant strain rate will increase according to (8.24) until the above condition is violated. With further increase in strain rate the strength will level off and approach the strength in the absence of subcritical crack propagation. Requiring that the first term in (8.20) be only 0.01 or less, the magnitude of the second term leads to the condition $(c_i/c_f)^{n-1} < 0.01$. For $n = 20$ the condition is $c_i/c_f < 0.80$. One could check this by measuring the strength without crack propagation, but this requires an additional experiment of a different type. Alternatively, one can depend entirely on the measurements of strength at two different strain rates, make a consistency check by using (8.22) in the form $(\sigma_{f,\dot{\varepsilon}})^{n+1} = C\dot{\varepsilon}$, and calculating C for both strain rates. If the C values agree the use of equation (8.24) is justified.

Lifetime predictions are very sensitive to the values of the crack growth parameters used. This sensitivity and precautions in determination of crack growth parameters have been discussed by Jakus, Coyne, and Ritter (1978) and Ritter, Bandyopadhyay, and Jakus (1979, 1981).

8.6 COMPARISON OF TIMES TO FAILURE UNDER CONSTANT STRESS AND CONSTANT STRAIN RATE

In a constant strain rate experiment with stress rate $\dot{\sigma} = E\dot{\varepsilon}$, the time to failure $t_{f,\dot{\varepsilon}}$ and the corresponding stress at failure $\sigma_{f,\dot{\varepsilon}} = \dot{\sigma} t_{f,\dot{\varepsilon}}$ are determined. If the same specimen had instead been subjected to the constant stress $\sigma = \dot{\sigma} t_{f,\dot{\varepsilon}}$, the time to failure $t_{f,\sigma}$ would be observed. These times to failure can be related using equations from the previous two sections. From equation (8.22) for stress to failure under constant strain rate

$$(\sigma_{f,\dot{\varepsilon}})^{n+1} = (\dot{\sigma}\, t_{f,\dot{\varepsilon}})^{n+1} = \frac{2(n+1)\dot{\sigma}\, \sigma_{f,t=0}^{n-2}}{AY^2 K_{IC}^{n-2}\,(n-2)} \tag{8.25}$$

or

$$\dot{\sigma}^{\,n}\,(t_{f,\dot{\varepsilon}})^{n+1} = \frac{2(n+1)}{AY^2(n-2)}\left(\frac{\sigma_{f,t=0}}{K_{IC}}\right)^{n-2} \tag{8.26}$$

The equation for time to failure at constant stress is

$$t_{f,\sigma} = \frac{2}{\sigma^n A\, Y^2(n-2)}\left(\frac{\sigma_{f,t=0}}{K_{IC}}\right)^{n-2} \tag{8.27}$$

Substituting

$$\sigma = \dot{\sigma}\, t_{f,\dot{\varepsilon}} \tag{8.28}$$

in this equation gives

$$\dot{\sigma}^{\,n}(t_{f,\dot{\varepsilon}})^n\, t_{f,\sigma} = \frac{2}{AY^2(n-2)}\left(\frac{\sigma_{f,t=0}}{K_{IC}}\right)^{n-2} \tag{8.29}$$

Comparing equations gives

$$t_{f,\dot{\varepsilon}} = (n+1)t_{f,\sigma} \quad \text{for the same short term strength when} \quad \sigma = E\,\dot{\varepsilon}\, t_{f,\dot{\varepsilon}} \tag{8.30}$$

Ceramics in service may be subject to stress that varies with time in a more complicated manner than we have discussed so far. Evans and Fuller (1974) have derived lifetime expressions for subcritical crack propagation under cyclic loading while Jakus and Ritter (1981) have treated time to failure for subcritical crack propagation under arbitrary application of stress as a function of time.

8.7 RELATION OF WEIBULL STATISTICAL PARAMETERS WITH AND WITHOUT SUBCRITICAL CRACK GROWTH

So far in this chapter the discussion has tacitly assumed a single specimen or at least a single initial flaw size. In real situations one must consider the combined effect of a statistical distribution of flaws in each specimen (with resulting statistical distribution of strengths in a set of specimens) and of subcritical crack growth. If the two-parameter Weibull distribution is assumed for strengths and the power law crack growth equation (8.1) is assumed, then some simple results for the effects of slow crack growth rate on the Weibull m can be obtained, as in done in the following paragraphs. We caution, however, that neither of these assumptions is necessarily correct in real materials; the situation has been considered by Singpurwalla, Wilson, and Fuller (1995) and a detailed example of combining fracture mechanics and moisture-assisted crack growth to develop a statistical treatment of failure of glass has been given by Fuller, Freiman, Quinn, Quinn, and Carter (1994). Despite this warning the relatively simple treatment based on the above assumptions is a valuable first approximation and so is presented next.

We consider a population of samples, some of which are measured first under conditions with no slow crack growth to give instantaneous strength values $\sigma_{f,\,t\,=\,0}$. Some of the population are then measured under constant strain rate $\dot{\varepsilon}$ to give strength values $\sigma_{f,\,\dot{\varepsilon}}$ and a failure time $t_{f,\,\dot{\varepsilon}}$ with $\sigma_{f,\,\dot{\varepsilon}}$ =

$E\dot{\varepsilon}t_{f,\dot{\varepsilon}}$ where E is Young's modulus. Other samples are then subjected to the constant stress $\sigma_{f,\dot{\varepsilon}}$ to give a failure time $t_{f,\sigma}$. In all tests we use samples with the same volume V and the same test method. We assume that the instantaneous strengths are distributed according to a Weibull distribution and inquire what distribution the strengths at constant strain rate follow.

The instantaneous strength values are taken to obey the Weibull distribution

$$P_s = \exp\left(-V\left(\frac{\sigma_{f,t=0}}{\sigma_0}\right)^m\right) \tag{8.31}$$

Solving for the instantaneous strength gives

$$\ln \sigma_{f,t=0} = \frac{1}{m}\ln\ln\frac{1}{P_s} + \frac{1}{m}\ln V + \ln\sigma_0 \tag{8.32}$$

We now express $\sigma_{f,t=0}$ in terms of $\sigma_{f,\dot{\varepsilon}}$, substitute in the above equation, and solve for $\sigma_{f,\dot{\varepsilon}}$ in terms of P_s to determine what distribution $\sigma_{f,\dot{\varepsilon}}$ follows. We have

$$\sigma_{f,\dot{\varepsilon}} = E\,\dot{\varepsilon}\,t_{f,\dot{\varepsilon}} \tag{8.33}$$

This failure time under constant strain rate is related to the failure time, $t_{f,\sigma}$, under constant stress $\sigma_{f,\sigma}$ equal to $\sigma_{f,\dot{\varepsilon}}$ by

$$t_{f,\dot{\varepsilon}} = (n+1)\,t_{f,\sigma} \tag{8.34}$$

This gives

$$t_{f,\sigma} = \frac{\sigma_{f,\dot{\varepsilon}}}{(n+1)\,E\,\dot{\varepsilon}} \tag{8.35}$$

We now use the expression for failure time under constant stress

$$t_{f,\sigma} = \frac{2}{\sigma_{f,\dot{\varepsilon}}^{n}\,Y^2 A\,(n-2)}\left(\frac{\sigma_{f,t=0}}{K_{IC}}\right)^{n-2} \tag{8.36}$$

Equating the two expressions for $t_{f,\sigma}$ gives

$$\frac{\sigma_{f,\dot{\varepsilon}}}{E\dot{\varepsilon}\,(n+1)} = \frac{2}{\sigma_{f,\dot{\varepsilon}}^{n}\,Y^2 A\,(n-2)}\left(\frac{\sigma_{f,t=0}}{K_{IC}}\right)^{n-2} \tag{8.37}$$

Solving for the instantaneous failure stress gives

$$\sigma_{f,t=0} = \left(\sigma_{f,\dot{\varepsilon}}\right)^{\frac{n+1}{n-2}} K_{IC}\left(\frac{Y^2 A(n-2)}{E\dot{\varepsilon}\,(n+1)}\right)^{\frac{1}{n+1}} \tag{8.38}$$

or

$$\ln \sigma_{f,t=0} = \frac{n+1}{n-2} \ln \sigma_{f,\dot{\varepsilon}} + \ln K_{IC} + \frac{1}{n-2} \ln \frac{Y^2 A(n-2)}{E\dot{\varepsilon}\,(n+1)} \tag{8.39}$$

Equating this expression and the Weibull expression for $\ln \sigma_{f,\,t\,=\,0}$ and solving the resulting equation for $\ln \sigma_{f,\,\dot{\varepsilon}}$ gives

$$\ln \sigma_{f,\dot{\varepsilon}} = \left(\frac{n-2}{n+1}\right)\frac{1}{m}\ln \ln \frac{1}{P_s} + \left(\frac{n-2}{n+1}\right)\frac{1}{m} \ln V + \left(\frac{n-2}{n+1}\right)\ln \sigma_0 + \frac{1}{n+1} \ln \frac{Y^2 A(n-2)}{E\dot{\varepsilon}\,(n+1)} \tag{8.40}$$

Comparing equations, we see that the strength values measured under constant strain rate do follow a Weibull distribution but that the Weibull parameters determined from a plot of ln σ as a function of ln ln $(1/P_s)$ will be different. We now introduce the notation $m_{t\,=\,0}$ and $\sigma_{0,\,t\,=\,0}$ for values of the Weibull parameters determined in the absence of subcritical crack growth and $m_{\dot{\varepsilon}}$ and $\sigma_{0,\,\dot{\varepsilon}}$ for Weibull parameters determined from strength values measured at constant strain rate under conditions allowing subcritical crack growth and using the usual Weibull distribution. From equations (8.39) and (8.40) we have

$$m_{\dot{\varepsilon}} = \left(\frac{n+1}{n-2}\right) m_{t=0} \tag{8.41}$$

and

$$\ln \sigma_{0,\dot{\varepsilon}} = \left(\frac{n-2}{n+1}\right)\ln \sigma_{0,t=0} + \frac{1}{n+1} \ln \frac{Y^2 A(n-2)}{E\dot{\varepsilon}\,(n+1)} \tag{8.42a}$$

The limit of negligible slow crack growth before failure is obtained by letting $n \to \infty$. The above equations then give the same value for the Weibull parameters determined with or without slow crack growth, as they should. For slow crack growth with $n = 10$, the m value determined from plots using constant strain rate data should be 1.375 times the m value determined in the absence of slow crack growth according to the above treatment. No comparative data to test this prediction are known to the author.

In Section 8.6 the equation

$$\left(\frac{\sigma_{f,\dot{\varepsilon}_1}}{\sigma_{f,\dot{\varepsilon}_2}}\right)^{n+1} = \frac{\dot{\varepsilon}_1}{\dot{\varepsilon}_2} \quad \text{(for the same short time strength).} \tag{8.42b}$$

was derived. It is, of course, not possible to use the same specimen for two failure experiments. We now show that the same equation holds for the predicted failure stress at the same survival probability with different specimens from a common Weibull distribution. Writing the above ln $\sigma_{f,\dot{\varepsilon}}$ equation twice for two different strain rates and subtracting gives

$$\ln \sigma_{f,\dot{\varepsilon}\,1} - \ln \sigma_{f,\dot{\varepsilon}\,2} = \frac{1}{n+1}\,(-ln\,\dot{\varepsilon}_1 + \ln \dot{\varepsilon}_1) \quad \text{(for constant } P_s\text{)} \tag{8.43}$$

or

$$\left(\frac{\sigma_{f,\dot{\varepsilon}_1}}{\sigma_{f,\dot{\varepsilon}_2}}\right)^{n+1} = \frac{\dot{\varepsilon}_1}{\dot{\varepsilon}_2} \quad \text{(for constant } P_s\text{)} \tag{8.44}$$

Since this equation holds for all P_s values, it follows that the average strengths at two strain rates also obey it:

$$\left(\frac{\overline{\sigma}_{f,\dot{\varepsilon}_1}}{\overline{\sigma}_{f,\dot{\varepsilon}_2}}\right)^{n+1} = \frac{\dot{\varepsilon}_1}{\dot{\varepsilon}_2} \tag{8.45}$$

This equation provides a useful way to determine the subcritical crack propagation parameter n from

$$n = \frac{\ln(\dot{\varepsilon}_1/\dot{\varepsilon}_2)}{\ln(\overline{\sigma}_{f,\dot{\varepsilon}\,1}/\overline{\sigma}_{f,\dot{\varepsilon}\,2})} - 1 \tag{8.46}$$

Another relation of interest in predicting lifetime under conditions of subcritical crack growth is the distribution of times to failure under constant load (Paluszny, 1977; Fett and Munz, 1994; Singpurwalla, Wilson, and Fuller, 1995). The times to failure under constant load can be seen to follow a Weibull distribution by showing that this distribution leads to a Weibull distribution in terms of short time failure stress. To do this one can write

$$P_s(t_{f,\sigma}) = \exp\left(-\left(\frac{t_{f,\sigma}}{t_{f,\sigma,0}}\right)^{m_\sigma}\right) \tag{8.47}$$

and substitute the failure time under constant load from equation (8.15) to obtain a Weibull distribution in terms of stress. The resulting equation shows that

$$m_\sigma = \frac{m}{n-2} \tag{8.48}$$

and

$$t_{f,\sigma,0} = \frac{2(\sigma_{f,t=0})^{n-2}}{\sigma^n \, Y^2 \, A \, (n-2) \, (K_{IC})^{n-2}} \tag{8.49}$$

Fett and Munz also develop failure time equations for subcritical crack conditions when R-curve behavior is also present and for cyclic stress as well as constant stress.

8.8 CONSTRUCTION OF STRENGTH–PROBABILITY–TIME DIAGRAMS

Davidge, McLaren and Tappin (1973) devised a method of constructing strength–probability–time (STP) diagrams from data on the time to failure of a number of specimens. Probability considerations were also introduced by Evans and Wiederhorn (1974), Fuller, Wiederhorn, Ritter and Oates (1980), and Wiederhorn and Fuller (1985). In the method of Davidge et al., a series of specimens must be measured at each of at least two strain rates. A series of specimens is run at one strain rate and the failure times and corresponding failure stresses are determined. The data are then arranged in order by failure stress and the failure probability is calculated in the same manner as is normally done with Weibull data. A plot on Weibull paper then gives the slope $m_{\dot\varepsilon}$ = the Weibull constant for constant strain rate. A series of specimens is then run at a second strain rate and the mean ratio of strengths at the two strain rates is used to determine the subcritical crack growth parameter *n*. This *n* value is used to calculate a static lifetime associated with each strength value using

$$t_{f,\sigma} = \frac{t_{f,\dot\varepsilon}}{(n+1)} \tag{8.50}$$

For each static failure time, the constant load strength at a chosen constant failure time t_1, say, 1 second, is then calculated using

$$\frac{t_{f,\sigma_1}}{t_{f,\sigma_2}} = \left(\frac{\sigma_2}{\sigma_1}\right)^n \qquad (8.51)$$

These strength values are used with the original survival probability values to make a plot on Weibull paper for 1-second life. A straight line is drawn through the data. The time to failure under constant stress equation is

$$t_{f,\sigma} = \frac{2}{\sigma^n Y^2 A(n-2)}\left(\frac{\sigma_{f,t=0}}{K_{IC}}\right)^{n-2} \qquad (8.52)$$

For two service stresses σ_{s1} with failure time $t_{f,\sigma s1}$ and σ_{s2} with failure time $t_{f,\sigma s2}$, this gives

$$\frac{t_{f,\sigma_{s1}}\,\sigma_{s1}{}^n}{t_{f,\sigma_{s2}}\,\sigma_{s2}{}^n} = 1 \qquad (8.53)$$

Accordingly, parallel straight lines are shifted along the ln (strength) axis for a different survival time according to

$$\ln \sigma_{s2} - \ln \sigma_{s1} = \frac{1}{n+1} Ln\left(\frac{t_{f,\sigma_{s1}}}{t_{f,\sigma_{s2}}}\right) \qquad (8.54)$$

This family of curves can be used to produce plots of ln t as a function of ln σ at constant survival probability by drawing lines at constant failure probability on the plot just constructed and reading off the stress corresponding to each failure time.

An example of this procedure is shown in Table 8.1 and Figure 8.5. The table shows ten values of strength for a strain rate of 0.0001/s and ten values for a rate of 0.000001/s. These are actually simulated data obtained by assuming an average strength of 350 MPa at the slower rate, a Weibull $m_{\dot{\varepsilon}}$ value of 12, and a slow crack growth parameter n of 15. The expected strength values were calculated and random error was then introduced. The resulting first two columns of Table 8.1 are treated as experimental data for this illustration.

The average strengths at the two strain rates, $\sigma_{f,\dot{\varepsilon}1}$ and $\sigma_{f,\dot{\varepsilon}2}$, were used to calculate the slow crack propagation parameter n from

$$n = \frac{\dot{\varepsilon}_1 - \dot{\varepsilon}_2}{\overline{\sigma}_{f,\dot{\varepsilon}1} - \overline{\sigma}_{f,\dot{\varepsilon}2}} - 1 \qquad (8.55)$$

This procedure gave $n = 15.2$ for the strength data with random errors included. The time to failure for each specimen is listed in the third and fourth columns and was calculated from the strength and strain rate using

TABLE 8.1 Construction of Survival Probability as a Function of Time

Strength at 0.0001/s	Strength at 0.000001/s	Time to Fail at 0.0001/s	Time to Fail at 0.000001/s	Strength for 1 s from 0.001/s Data	Strength for 1 s from 0.000001/s Data
258.2	196.2	0.3980	30.2448	243.0	245.5
292.6	219.7	0.4510	33.8606	277.6	276.9
300.1	226.8	0.4626	34.9645	285.3	286.5
314.8	238.5	0.4853	36.7687	300.2	302.3
317.7	240.0	0.4897	36.9973	303.1	304.3
327.3	246.9	0.5045	38.0595	312.9	313.6
336.6	255.3	0.5188	39.3591	322.3	325.0
345.4	259.6	0.5324	40.0151	331.4	330.8
355.2	266.5	0.5476	41.0883	341.4	340.3
370.7	273.3	0.5714	42.1286	357.3	349.4
ln ln $(1/P_s)$	ln 1 s Strength from 0.001/s	ln 1 s Strength from 0.00000/s	Fitted ln 1 s Strength	Calculated ln 100 s Strength	Calculated in 10,000 s Strength
−2.970	5.493	5.503	5.507	5.204	4.901
−1.817	5.626	5.624	5.610	5.307	5.005
−1.246	5.653	5.658	5.661	5.358	5.056
−0.842	5.704	5.711	5.697	5.394	5.092
−0.514	5.714	5.718	5.726	5.424	5.121
−0.225	5.746	5.748	5.752	5.450	5.147
0.049	5.776	5.784	5.777	5.474	5.171
0.327	5.803	5.801	5.801	5.499	5.196
0.640	5.833	5.830	5.829	5.527	5.224
1.097	5.878	5.856	5.870	5.568	5.265

$$t_{f,\dot{\varepsilon}} = \sigma_f / (E\,\dot{\varepsilon}) \tag{8.56}$$

where σ_f is the fracture stress, E is Young's modulus, and $\dot{\varepsilon}$ is the constant strain rate used. The equivalent time to failure under constant stress equal to the failure stress in constant strain rate loading was then calculated from

$$t_{f,\sigma} = t_{f,\dot{\varepsilon}} / (n+1) \tag{8.57}$$

For each specimen the stress σ_1 required to cause failure in 1 second under constant stress was calculated from

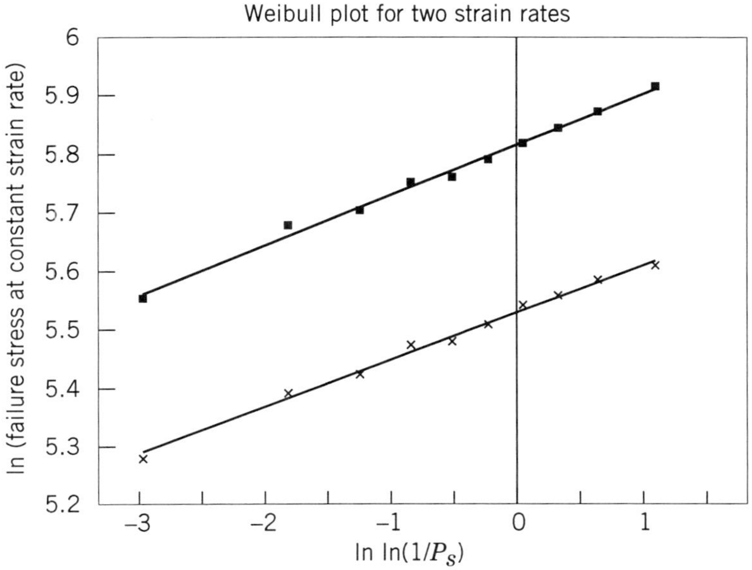

Figure 8.5 Failure stress as a function of survival probability, P_s, at 0.0001/s and 0.000001/s strain rates.

$$\sigma_1 = \sigma_f \left(\frac{t_f}{1 \text{ second}} \right)^{1/n} \tag{8.58}$$

and the results are listed in columns 5 and 6. The probability of survival was calculated from

$$P_s = 1 - \frac{i - 1/2}{10} \tag{8.59}$$

The Weibull plot of Figure 8.5 was than made from ln ln$(1/P_s)$ and ln σ_1 as tabulated in the first three columns of the second row of the table. A least-squares straight line fit gave $m = 12.4$. The fitted values are given in the fourth column of the second row of the table and plotted as the line through the data. Finally, the predicted stresses for 100 s and 10,000 s survival were calculated from

$$\sigma_2 = \sigma_1 \left(\frac{t_{f1}}{t_{f2}} \right)^{1/n} \tag{8.60}$$

and are listed in the last two columns and plotted in Figure 8.6. This figure can be used to read off the predicted maximum stress that can be used for this material for a given failure probability and a given time.

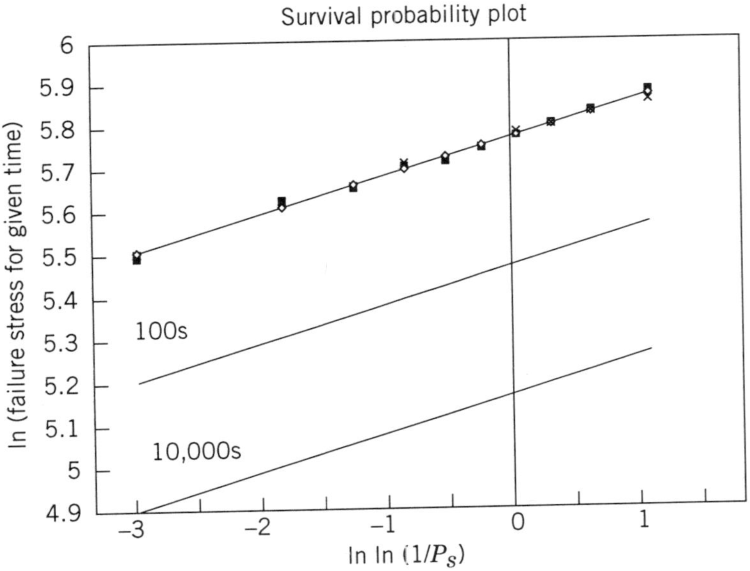

Figure 8.6 Pooled data for strength at 1 s and calculated strength values at 100 s and 100,000 s as a function of survival probability, P_s.

8.9 PROOF TESTING TO GUARANTEE MINIMUM LIFE

If a proof stress, σ_p, is applied to each specimen, those that survive will have

$$K_{Ip} = \sigma_p Y (c_0)^{1/2} < K_{IC} \tag{8.61}$$

In subsequent service under stress σ_s the initial stress intensity factor will be

$$K_{Ii} = \sigma_s Y (c_0)^{1/2} \tag{8.62}$$

A minimum estimate of failure time can be obtained by taking $K_{Ip} = K_{IC}$. Substituting in the time to failure under constant stress equation gives

$$t_s = \frac{2(K_{IC})^{2-n}}{A\sigma_s^2 Y^2 (n-2)} \left(\frac{\sigma_s}{\sigma_p}\right)^{2-n} \tag{8.63}$$

This equation gives a straight line with slope −2 on a plot of ln t as a function of ln σ_s for each value of σ_p/σ_s. Plots can be constructed as described in the preceding section (failure time as a function of stress for each of a series of survival probabilities). The desired time of survival can then be chosen and the intersection found with the chosen level of survival probability to establish a point on the ln (survival time) versus ln (stress) plot. The lines of constant proof stress can then be used to select the minimum proof stress to ensure this level of survival probability.

An important caution is that this procedure assumes that the proof test does not cause crack growth during unloading so that fast unloading or inert conditions must be used. A further difficulty with proof testing is that the stress state imposed in the proof test should duplicate that in service. This is difficult to do for complex shapes and loadings. For example, a turbocharger rotor can be proof tested by spinning at an angular velocity exceeding that to be used in service. This procedure gives a good simulation of the service stresses caused by centrifugal force but simple spinning does not duplicate the additional stresses caused by gas pressure on the rotor blades in service and so is not a fully faithful proof test. Despite this shortcoming, spinning has been used successfully as a proof test for turbocharger rotors (Matsui, Ishida, Soma, and Oda, 1986).

Jakus (1986) has reviewed proof testing and given the following summary: "Proof testing is a widely used, viable tool for increasing product reliability of many glass and ceramic components. It is, however, a technique whose practical benefits can be diminished by fatigue, unfortunately a rather common strength degradation phenomenon in glass and ceramic materials. To realize the greatest benefits from proof testing, good "proof-test control" must be maintained: Fatigue during the test must be minimized, either by providing a dry, inert environment or by using fast loading and very fast unloading rates. If fatigue is eliminated, the after-proof distribution becomes truncated; hence, a minimum strength is ensured, and furthermore the strength distribution also substantially improves. In fatigueless cases, the proof-test theory predicts the experimental results very well. When fatigue becomes significant, one can no longer ensure a minimum strength for the survivors, although the strength of the population as a whole still may be substantially improved by proof testing. When very severe fatigue conditions exist, proof testing may not be beneficial, and proof-test theory is ill-equipped to predict the outcome. Although eliminating fatigue during proof testing may not be a trivial task in many industrial environments, it may nevertheless be necessary in critical applications, since it is perhaps the only way of ensuring ceramic-component reliability."

8.10 SUBCRITICAL CRACK GROWTH AND FAILURE FROM FLAWS ORIGINATING FROM RESIDUAL STRESS CONCENTRATIONS

Fuller, Lawn, and Cook (1983) have considered subcritical crack growth under circumstances where local residual stress modifies crack growth. In particular, they considered failure from flaws caused by hardness indentations. Their treatment is an extension of the earlier work of Lawn and his collaborators on hardness indentations summarized in Sections 6.5 and 9.3. They wrote equation (8.15) for time to failure under constant stress of a Griffith crack without residual stress in the form

$$t_f \sigma_f^n = \frac{2}{n-2} (\sigma_i^0)^n \frac{c_i}{A} \tag{8.64}$$

For the case of residual stress caused by an indentation, they obtained the following approximate result for large n:

$$t_f \sigma_f^{n'} = \left(\frac{8\pi}{(r+1)n'} \right)^{1/2} (\sigma_i)^{n'} \frac{c_m}{A} \tag{8.65}$$

where

$$n' = \frac{rn+2}{r+1} \tag{8.66}$$

and where $r = 1$ for line flaws and $r = 3$ for point flaws. A general conclusion from this theory is that experiment on delayed failure from subcritical crack growth will show the same characteristics for cracks subject to residual stress and cracks free of such stress. Thus the tests themselves do not distinguish between the types of cracks despite the fact that a different interpretation of the parameters obtained is appropriate for the two cases. These authors discuss the use of tests at constant strain rate to obtain slow crack propagation parameters for use in lifetime prediction. The reader considering the use of data obtained in dynamic tests for the purpose of lifetime prediction should consult their discussion, which is too long to summarize here.

8.11 SLOW CRACK PROPAGATION AT HIGH TEMPERATURE

Subcritical crack propagation at and near room temperature was discussed earlier in Chapter 8. At such temperatures environmentally assisted subcritical crack propagation appears to be the major mechanism. Other mechanisms are possible and a variety of theories have been developed (Evans and Dalgleish, 1987). It appears that some of the other mechanisms do have practical importance in creep rupture at high temperatures. Chuang, Tressler, and Minford (1986) have examined kinetic theories of these mechanisms and have compared their predictions to experiments on alpha-silicon carbide. They classify the mechanisms underlying these theories into three types: bond breaking, chemical corrosion, and matter transport. The Charles–Hillig theory, discussed in Chapter 8, is of the second type. Kinetic theories of crack growth typically lead to the possibility of crack healing below a threshold value of the applied stress intensity factor and so predict a threshold value, K_{th}, below which static fatigue should not take place. Several theories of diffusive crack growth exist; that of Chuang (1982) appears to be in reasonable agreement with experiment.

His theory gives a threshold stress intensity factor for diffusive crack growth greater than the stress intensity factor for a crack not modified by this process. Table 8.2 gives the theoretical lower limit values and the experimental values from Chuang, Tressler, and Minford (1986). Column 2 gives the predicted values of K_{th}/K_G, where

$$K_G = \left(\frac{E(2\gamma_s - \gamma_{gb})}{1 - \nu^2} \right)^{1/2} \tag{8.67}$$

is the Griffith value for a grain boundary crack. Chemical corrosion, as expected, gives a threshold value below the Griffith value. The observed threshold is well above the Griffith value, leading the authors to the conclusion that matter transport (i.e., diffusive growth of the crack) is the controlling mechanism for the materials and conditions studied. The fact that the experimental values are considerably greater than the predicted values is rationalized by the authors as being due to effects of crack deflection and shielding.

It should be noted that at values below K_{th} an infinite life is not to be expected. Instead, the process of cavity formation (not considered in the above kinetic theories) would presumably come into play at long times and lead eventually to failure for K values above a cavitation threshold.

Yavuz and Tressler (1993) used static fatigue tests to determine K_{th} for a fully dense α-SiC hot isostatically pressed (HIPed) without additives and for a pressureless-sintered α-SiC, which contained aluminum as a sintering aid. CVD SiC also exhibited delayed failure at 1300 and 1400 °C but a reliable K_{th} could not be determined for this material because of scatter in the data. Table 8.3 shows the comparison of observed values with predictions of Chuang's theory. The agreement is good at 1300 and 1400 °C for the sintered material. There was evidence of a damage process concurrent with slow crack growth at

TABLE 8.2 Predicted and Experimental Values of K_{th} in MPa · $m^{1/2}$ for SiC at 1200 °C

Mechanism	K_{th}/K_G	Hot-Pressed α-SiC	Oxidized Sintered α-SiC	As-machined Sintered α-SiC
Bond breaking	1.0	0.752	0.711	0.564
Chemical corrosion	0.61	0.459	0.433	0.344
Matter transport	1.69	1.1271	1.201	0.954
Experimental results		2.25	1.75	2.25

Source: Chuang, Tressler, and Minford (1986).

TABLE 8.3 Comparison of Observed K_{th} in MPa · $m^{1/2}$ with Values Predicted by Chuang's Model

	Sintered		HIPed	
	Predicted	Observed	Predicted	Observed
1200	2.00	3.3	2.13	3.1
1300	1.97	2.2	2.10	2.7
1400	1.94	1.7	2.03	2.2
Source: Yavuz and Tressler (1993).				

1400 °C. For the hot isostatically pressed material, static loading below the threshold stress intensity strengthened the material at 1300 and 1400 °C probably by a stress relaxation mechanism at the crack tip in the viscous silica phase present along the grain boundaries and some of the triple-grain junctions.

CHAPTER 9

STABLE CRACK PROPAGATION AND R-CURVE BEHAVIOR

9.1 INTRODUCTION

As discussed in previous chapters, the Griffith theory and its extension by Irwin and others lead to the conclusion that a crack in a body under uniform stress far from the crack will propagate catastrophically for values of this stress slightly exceeding a critical value given by

$$\sigma_f = \frac{K_C}{Yc^{1/2}} \tag{9.1}$$

where K_C is the critical stress intensity factor, c is the appropriate crack size, and Y is a geometrical factor. In the Griffith theory the crack will be stable under σ_f and will heal for smaller stress values. The predicted healing effect is usually ignored on the grounds that various effects such as crack surface contamination prevent it in practice although a freshly extended crack can show some degree of healing when the stress is removed. The point of interest here is that the Griffith theory as discussed to this point does not allow stable crack propagation; once the critical stress is exceeded the crack extends indefinitely. There is, of course, the phenomenon of subcritical crack propagation behavior as discussed in Chapter 8. This subcritical crack propagation is associated with transport effects such as diffusion of water and stress-assisted chemical attack. This category of subcritical crack growth introduces concepts outside the basic Griffith theory and is not the subject of the present chapter. Without introducing such effects it is still possible to have stable crack propagation within the conceptual framework of the Griffith theory under appropriate circumstances. There are two fundamentally different sets of such circumstances: (1) combinations of specimen geometry and details of load application that lead to a decreasing applied stress intensity factor as the crack elongates and (2) in materials that exhibit a critical stress intensity factor that is not constant but increases faster than the applied stress intensity factor increases as the crack elongates (i.e., in materials that show R-curve behavior). In practice, both effects can act together, but it is convenient to separate them for discussion.

The first category of changing stress in material of constant critical stress intensity factor is certainly important and is the basis for methods of creating a controlled crack such as forcing a wedge into a slot. It is, however, the second category that has become of great importance in designing microstructures to produce high toughness in ceramics and that is the subject of this chapter. Discussion of the microstructural effects producing R-curve behavior is left primarily to subsequent chapters. The present chapter concentrates on formulating the concept of R-curve behavior, reviewing some of the early evidence for the existence of the effect, and describing an ingenious extension of indented-strength measurements by Lawn and co-workers that has been used to study R-curve behavior for short cracks.

The R-curve concept arose from studies of fracture in metals and alloys in the 1960s as described by Broek (1986) and was later applied to ceramics (Huebner and Jillek, 1977; Knehans and Steinbrech, 1982; Marshall, 1986). Particularly striking effects occur in transformation-toughened ceramics (Swain and Hannink, 1983; Heuer, 1987; Lutz, Claussen, and Swain, 1991) and in fiber-reinforced ceramic–ceramic matrix composites and will be discussed in later chapters. There is some difference in the notation used by various authors. In particular, Lawn (1993) has advocated the use of the term T-curve (i.e., toughness curve) to emphasize the distinction between applied stress intensity factor and the critical stress intensity factor of the material. Butler and Fuller (1993) summarize what is apparently becoming standard usage in speaking of R-curve or T-curve behavior: to use R-curve when the effect is discussed in terms of strain energy release rate and T-curve when the discussion is in terms of stress intensity factor. The R-curve notation is used here for the general effect and the R versus T distinction is used for the corresponding specific descriptions. It should be noted, however, that the R versus T distinction has not yet become universal in the literature on ceramics; the term R-curve is still used for either description of crack-length-dependent toughening behavior by some authors.

9.2 THE R-CURVE (T-CURVE) CONCEPT

We begin the discussion in terms of the crack extension force, G, the crack resistance, R, the stored elastic energy, U_e, the energy associated with the crack,

U_s, and the crack length, c (Broek, 1986; Butler and Fuller, 1993). As discussed in Section 5.6 these are defined by

$$G = -\left(\frac{dW}{dc} + \frac{dU_e}{dc}\right) \text{ and } R = \frac{dU_s}{dc} \tag{9.2}$$

Here we have taken the crack area as $A = ct$ and taken the thickness, t, as unity. The energy condition for crack propagation is

$$G \geq R \tag{9.3}$$

where the equality is a condition of balance and slight excess causes catastrophic propagation. To guarantee that an excess of driving force persists as the crack extends, the additional condition

$$\frac{dG}{dc} \geq \frac{dR}{dc} \tag{9.4}$$

is required. We examine the meaning of these conditions in graphical form beginning with the case of a constant crack resistance as shown in Figure 9.1. The upper portion of this figure shows G and R plotted as a function of crack length. G is given by

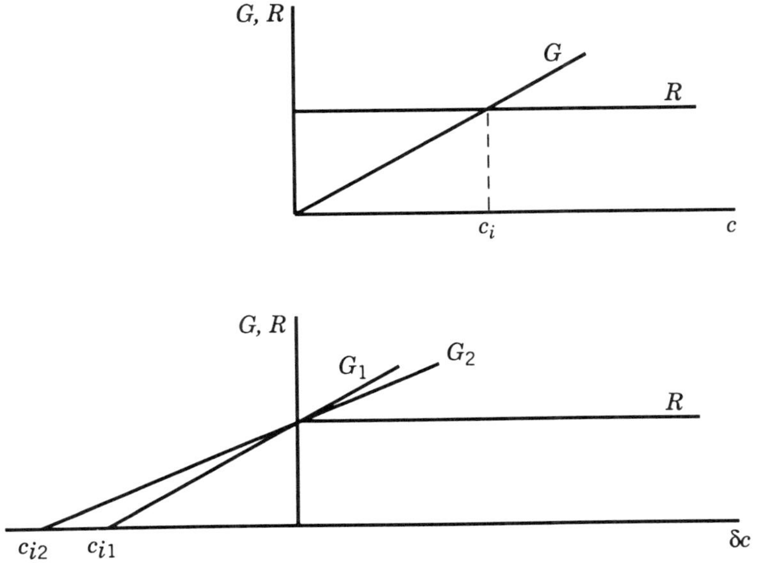

Figure 9.1 Representation of energy balance with constant resistance.

$$G = \frac{2\sigma^2 \pi c}{E} \tag{9.5}$$

for plane stress σ. G thus increases linearly with crack length. The G line has been drawn for a stress value just large enough to give $G = R$ at the initial crack length c_i. The crack will be unstable and will propagate under the slightest increase in crack length or stress because the other condition, $dG/dc > dR/dc$, is satisfied. It is convenient to draw this diagram with the origin taken at the initial crack tip, as in the lower portion of Figure 9.1. In this case G curves for two different initial crack lengths, c_{i1} and c_{i2} are shown. Different values of stress have been taken so that either crack is just in balance and would propagate catastrophically for any slight disturbance.

We can use a second drawing to illustrate behavior for the case in which the resistance increases as a crack grows from its initial length. This is a common and very important case in polycrystalline ceramics and ceramic–ceramic composites and can arise from several types of crack shielding and crack bridging discussed in Chapter 10. For the moment we omit discussion of the causes of R-curve behavior and focus on the effects on crack propagation. Figure 9.2 shows a rising R-curve. The G_1 curve is drawn for a stress such that $G_1(c_i) = R(c_i)$. However, as is evident from the drawing, $dG_1/dc < dR/dc$ and so the crack will not extend. For a somewhat higher value of stress the curve G_2 exceeds the initial crack resistance so that the crack will extend an amount δc, such that $G_2(c_i + \delta c) = R(c_i + \delta c)$. The crack will not extend further because $dG_2/dc < dR/dc$ at this point. Thus the existence of a rising R-curve can lead to a region of stable crack propagation under stress.

Further increase in stress leads eventually to the G_3 curve, which is tangent to the R-curve. The crack will extend from its initial length as before but will continue extending catastrophically because at the point of tangency $dG/dc = dR/dc$. There is thus a region of stable crack propagation for stresses between the lines and G_1 and G_3.

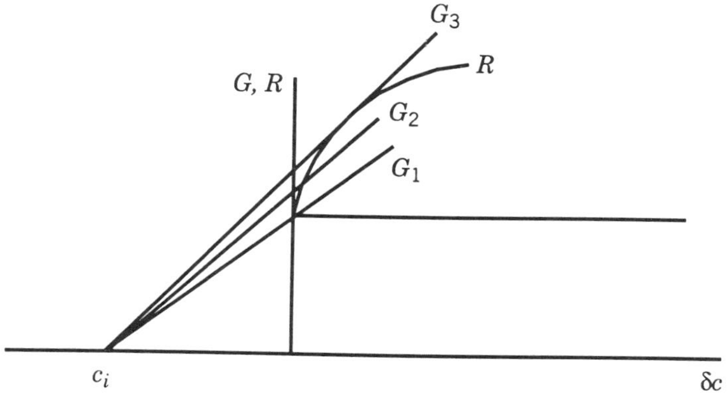

Figure 9.2 Representations of energy balance with increasing resistance.

We now turn to a description of the same behavior in terms of the applied stress intensity factor K; that is, to T-curve behavior. In a material without microstructural features (such as an ideal glass) the value of the stress intensity factor required for crack propagation is a constant independent of crack length and can be written as

$$T = K_{IC} = K_0 \tag{9.6}$$

In ceramics with microstructural features an advancing crack interacts with the microstructure and one can take two views as to how to describe what is happening. In the first view, the stress intensity factor required to propagate a planar crack remains the same but the local stress intensity acting at the crack tip is no longer that calculated from a sample equation of the type

$$K = Y \sigma c^{1/2} \tag{9.7}$$

where σ is the uniform tensile stress at a long distance from the crack caused by external forces on the ceramic specimen. Instead, the applied stress intensity factor at the crack tip is also affected by internal stresses caused by the interaction of the crack with the microstructure as the crack advances. The crack tip is said to be shielded by the interaction processes and the resulting internal stresses. In other words, the toughness is regarded as unchanged in this first view and the effect of microstructural interaction is taken as a change in the local applied stress intensity factor.

It is found to be useful to consider the same effects in terms of a different description. In this second view, the applied stress intensity factor due to the external loads alone is considered to be the effective driving force for crack propagation and the internal stresses arising from the interaction of the advancing crack with microstructure are regarded as changing the toughness of the material by an amount K_μ. The stress intensity factor required for crack propagation changes as the crack length increases and can be written as

$$T = K_0 + K_\mu \equiv T(c) \tag{9.8}$$

The conditions for fracture are analogous to those involving G and R and are

$$K \geq T \quad \text{and} \quad \frac{dK_a}{dc} = \frac{dT}{dc} \tag{9.9}$$

A plot analogous to Figure 9.2 but in terms of K and T is shown in Figure 9.3. Following a suggestion by Evans, the arc tangent function has been used to represent the T-curve. Although experimental data sometimes indicate a flatter T-curve, the arc tangent provides a useful qualitative illustration (Heuer, 1987). The T-curves of Figures 9.3 through 9.5 are given by

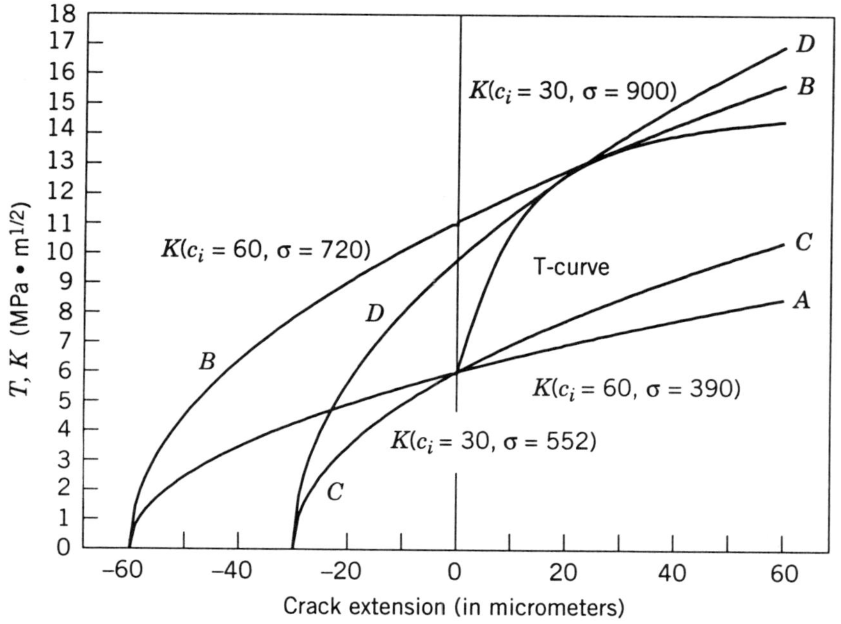

Figure 9.3 T-curve in K linear format with linear crack length scale.

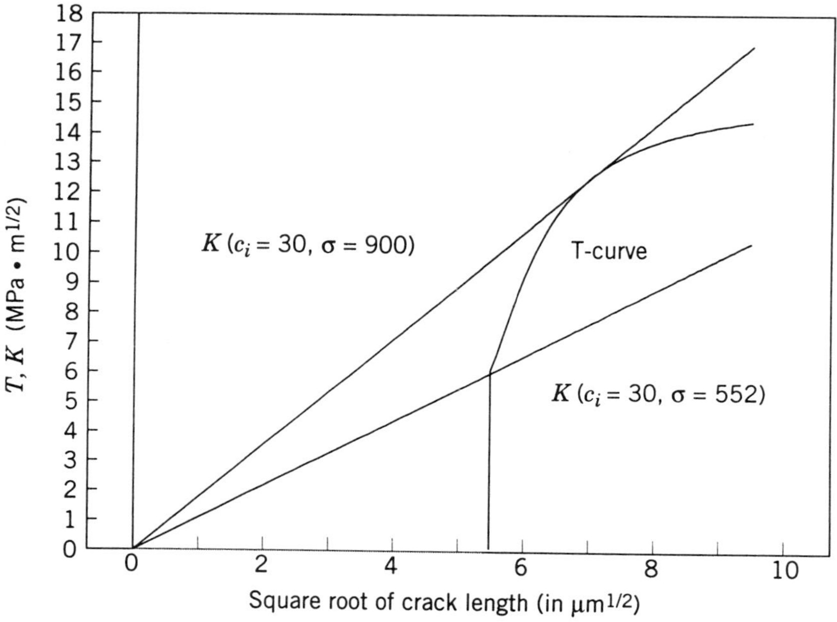

Figure 9.4 T-curve in K linear format with square-root crack length scale.

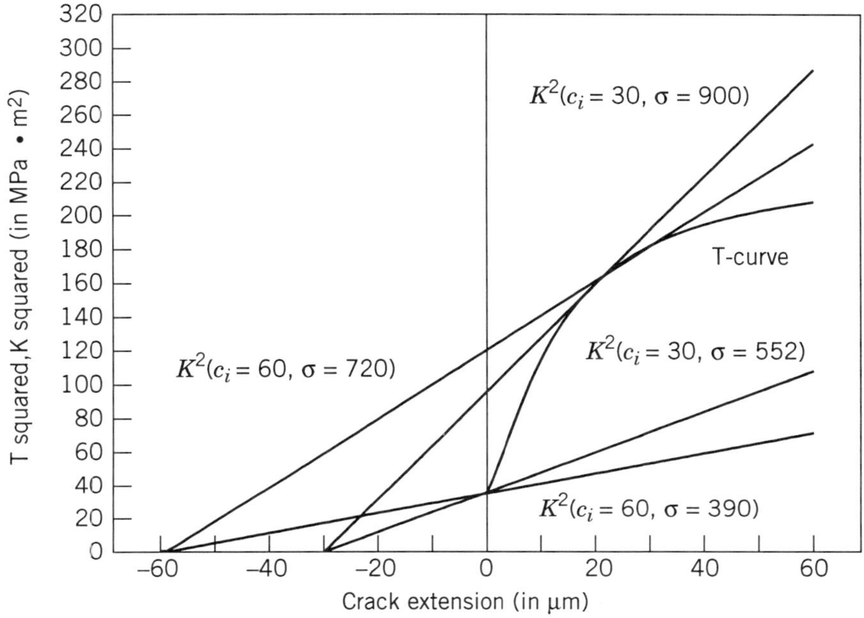

Figure 9.5 T-curve in K-squared format with linear crack length scale.

$$T = 6 + 6\tan^{-1}\left(\frac{\delta c}{10}\right) \tag{9.10}$$

where T is in MPa · $m^{1/2}$ and δc is crack extension in micrometers (μm).

Two K curves in Figure 9.3 (A and C) for stress values (390 and 552 MPa, respectively) chosen so that K passes through the initial T value are given for two values of initial crack length (60 and 30 μm, respectively). Under these conditions the crack will not propagate because $dK/dc < dT/dc$. Two K curves (B and D) for the same initial crack lengths but for higher stress values (720 and 900 MPa, respectively) such that the K curves are tangent to the T curve are also shown. For the latter conditions the crack will propagate catastrophically. For intermediate stress values (390–720 MPa for the 60-μm initial crack length and 552–900 MPa for the 30μm initial crack length) some stable crack propagation would occur.

The plot of Figure 9.3 is not the best format because the K curves are not linear. Two ways to linearize the K curves are shown in Figures 9.4 and 9.5. In Figure 9.4 the abscissa is taken as the square root of the crack length and the resulting linear K curve is shown for the values of stress bounding the range of stable crack propagation. In Figure 9.5 the ordinate is taken as T squared and K squared and the abscissa as crack length so that the curve of K^2 is linear. The representation of Figure 9.5 is perhaps most useful because the origin can be taken at the initial crack tip so that curves for two initial crack lengths can be placed on the same plot as shown. Again the K-squared curves are drawn for the values of stress bounding the region of stable crack propagation.

It should be noted that more complex R- and T-curves may exist leading to more complex behavior. In particular, cracks below a certain initial length may be unstable under the action of internal stresses arising from thermal expansion differences in a polycrystalline ceramic. In this case the crack in believed to "pop-in" to a certain length corresponding to a short initial portion of decreasing R-curve and T-curve before the subsequent increase. Butler and Fuller (1993) give examples of such curves.

9.3 EFFECT OF RESIDUAL STRESS ON R-CURVE BEHAVIOR

An important question concerning R-curve behavior is what the toughness values for small cracks responsible for failure in high-strength specimens are and how to measure these values. Small cracks can be produced by hardness indentations and used to determine toughness by the direct, indirect, or modified indentation methods as described in Chapter 6. In the indirect indentation method, for example, the strength is measured after indentation to determine the toughness. By varying the load the toughness can be studied as a function of crack size for relatively small cracks. However, as the indentation load, P, is decreased the strength is often found to depart from the predicted variation with $P^{-1/3}$ for small loads. An example from Lawn, Freiman, Baker, Cobb, and Gonzalez (1984) for polycrystalline alumina is shown in Figure 9.6. Here the strength is seen to approach a constant value as the load is decreased. This is not surprising: it is reasonable that for sufficiently small loads the strength should be controlled by inherent flaws rather than by the flaws induced by the indentation. Figure 9.6 also shows that the low-indentation load strength decreases with increasing grain size for this alumina. Again, this is not surprising because the inherent flaws are generally expected to be larger for alumina with larger grain size. Lawn and his collaborators have proposed an internal stress model to account for this behavior and have had considerable success in fitting this model to experimental data (Lawn, Freiman, Baker, Cobb, and Gonzalez, 1984; Cook, Lawn, and Fairbanks, 1985a, 1985b).

This model begins by setting the applied stress intensity factor, K, caused by indentation load, P, and applied stress, σ, equal to an effective critical stress intensity factor, K_C^{eff}, that is a function of the half-length, c, of the crack

$$K = \frac{\chi P}{c^{3/2}} + Y \sigma c^{1/2} = K_C^{\text{eff}} \tag{9.11}$$

It is reasonable to assume that K_C^{eff} depends on the ratio of the crack size to the grain size. Following Evans (1980), Lawn, Freiman, Baker, Cobb, and Gonzalez (1984) wrote the effective critical stress intensity factor as

$$K_C^{\text{eff}} = K_C^P - \mu K_C^P g(c/D) \tag{9.12}$$

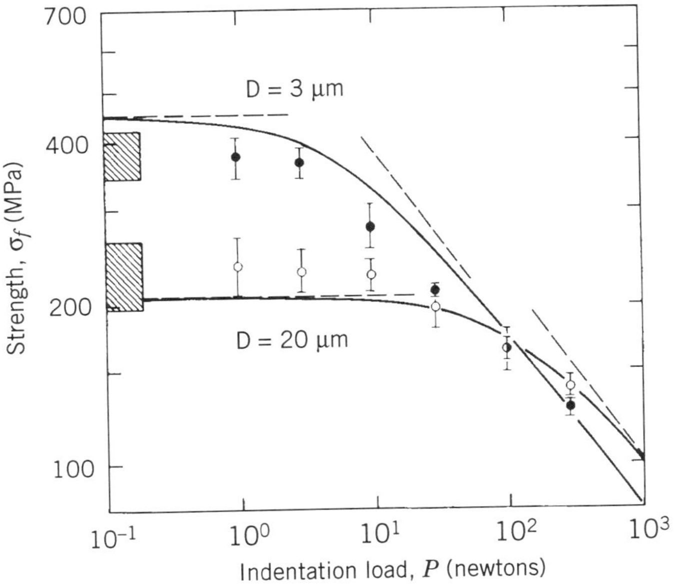

Figure 9.6 Inert strength of 3-μm and 20-μm aluminas as a function of Vickers hardness indentation load. (From B. R. Lawn, S. W. Freiman, T. L. Baker, D. D. Cobb, and A. C. Gonzalez, "Study of Microstructural Effects in the Strength of Alumina Using Controlled Flaws," *J. Am. Ceram. Soc.*, Vol. 67, 67–69, 1984. Reprinted by permission of the American Ceramic Society.)

where μ is a dimensionless quantity, $K_C{}^P$ is the critical stress intensity factor for a long crack in polycrystalline material, and D is the average grain diameter. The function g decreases monotonically from $g = 1$ for $c/D = 1$ to $g = 0$ for $c/D = \infty$.

Lawn and his collaborators took

$$g\,(c/D) = (D/c)^{3/2} \tag{9.13}$$

They argue for this form of equation by analogy to the form obtained for the residual stresses produced by an indentation. That is, residual stresses from thermal expansion mismatch of a grain with the average thermal expansion of the matrix will be localized similarly to the stresses from the plastic deformation zone caused by indentation. They also argue that the results are not strongly dependent on the exact functional form assumed and that the present assumption is convenient because it leads to an explicit solution.

The crack length, c_0, after indentation and before stressing is determined by

$$K_a = \frac{\chi\, P}{c^{3/2}} = K_C^{\text{eff}} = K_C^P - \mu\, K_C^P\,(D/c)^{3/2} \tag{9.14}$$

giving

$$c_0 = \left(\frac{\chi P}{K_C^P} + \mu D^{3/2}\right)^{2/3} \tag{9.15}$$

When stress is subsequently applied at a value less than that for failure, the crack will extend stably to a length given by

$$K = \frac{\chi P}{c^{3/2}} + Y\sigma c^{1/2} = K_C^{\text{eff}} = K_C^P - \mu K_C^P (D/c)^{3/2} \tag{9.16}$$

or

$$\frac{\chi P + \mu K_C^P D^{3/2}}{c^{3/2}} + Y\sigma c^{1/2} = K_C^P \tag{9.17}$$

The solution for the maximum stable crack size for the case $\mu = 0$ was previously obtained in Section 6.5 and is

$$c_{m,\sigma=0} = \left(\frac{4\chi P}{K_C}\right)^{2/3} \tag{9.18}$$

The solution for the present case can be obtained by replacing χP by $\chi P + \mu K_C{}^P D^{3/2}$ and K_C by $K_C{}^P$, giving

$$c_m = \left[4\left(\frac{\chi P}{K_C^P} + \mu D^{3/2}\right)\right]^{2/3} \tag{9.19}$$

The stress at failure, σ_f, is obtained by substituting the maximum crack length in the stress intensity equation (9.17) and is

$$\sigma_f = \frac{3K_C^P}{4^{4/3}\, Y\,(\chi P/K_C^P + \mu D^{3/2})^{1/3}} \tag{9.20}$$

In the limit of large indentation loads this expression becomes the usual relation for the indented strength of materials with no R-curve effect:

$$\sigma_f^P = \frac{3(K_C^P)^{4/3}}{4^{4/3}\, Y\chi^{1/3} P^{1/3}} \tag{9.21}$$

At sufficiently low loads the P term can be neglected and the equation becomes

$$\sigma_f^P = \frac{3K_C^P}{4^{4/3}\,Y\,\mu^{1/3}D^{1/2}} \tag{9.22}$$

which is consistent with the usual reciprocal square-root grain size dependence of strength of brittle polycrystalline ceramics.

In a subsequent investigation (Cook, Lawn, and Fairbanks, 1985a, 1985b) the same model was used but with slightly modified notation. Because the subsequent results are discussed in terms of the later notation, our discussion uses the later notation from this point on. The crack-size-dependent critical stress intensity factor K_C^{eff} was replaced by K_C' and the term $\mu K_C^P(D/c)^{3/2}$ was replaced by $\mu Q/c^{3/2}$. The above equations for stress intensity factor become

$$K = Y\sigma c^{1/2} + \frac{\chi P + \mu Q}{c^{3/2}} = K_C^\infty \tag{9.23}$$

The maximum stress for large injected cracks ($\chi P > \mu Q$) in this notation is

$$\sigma_m^P = \frac{3(K_C^\infty)^{4/3}}{4^{4/3}\,Y\,(\chi P)^{1/3}} \tag{9.24}$$

and the maximum stress for small injected cracks ($\chi P < \mu Q$) in this notation is

$$\sigma_m^Q = \frac{3(K_C^\infty)^{4/3}}{4^{4/3}\,Y\,(\mu Q)^{1/3}} \tag{9.25}$$

A value of P denoted P^* is defined by setting $\sigma_m^P(P^*) = \sigma_m^Q$, which leads to

$$\chi P^* = \mu Q \tag{9.26}$$

The maximum crack size before failure in this notation is

$$c_m = \left[\frac{3\chi(P+P^*)}{Y\sigma_m}\right]^{1/2} = \left[\frac{4\chi(P+P^*)}{K_C^\infty}\right]^{2/3} \tag{9.27}$$

and the initial crack length after indentation but before stressing is

$$c_0 = \left(\frac{\chi (P + P^*)}{K_C^\infty} \right)^{2/3} \tag{9.28}$$

The crack length before indentation is

$$c_0^Q = \left(\frac{\chi P^*}{K_C^\infty} \right)^{2/3} \tag{9.29}$$

Thus the implication of this model is that any flaw nucleus will expand to a crack of size $c_0{}^Q$ and that under the action of stress this crack will grow stably until it reaches a size giving $\frac{c_m}{c_0} = 4^{2/3} = 2.5$.

The critical stress can be expressed in several equivalent forms:

$$\sigma_m = \frac{3(K_C^\infty)^{4/3}}{4^{4/3} \, Y \chi^{1/3} \, (P + P^*)^{1/3}} \tag{9.30}$$

$$\sigma_m = \frac{\sigma_m^Q (P^*)^{1/3}}{(P + P^*)^{1/3}} \tag{9.31}$$

$$\sigma_m = \frac{\sigma_m^P \, P^{1/3}}{(P + P^*)^{1/3}} \tag{9.32}$$

Figure 9.7 (Cook, Lawn, and Fairbanks, 1985a) shows a plot of σ_m as a function of P with the latter taken in units of P^*. In this model the strength of an indented material is characterized by two parameters, which can be taken as any pair of the set $\sigma_m{}^Q$, $\sigma_m{}^P P^{1/3}$, and P^*.

Cook, Lawn, and Fairbanks defined an apparent toughness, $K_C{}^{app}$, in terms of this model. The toughness at large loads (i.e., $P > \mu Q/\chi$) is given in terms of the failure stress, $\sigma_m{}^P$, for this condition by

$$K_C^\infty = \left(\frac{256 \, Y^3 \chi}{27} \right)^{1/4} \left(\sigma_m^P \, P^{1/3} \right)^{3/4} \tag{9.33}$$

They define an apparent toughness for a smaller load, P, in terms of the failure stress, σ_m, for this condition by

$$K_C^{app} = \left(\frac{256 \, Y^3 \chi}{27} \right)^{1/4} \left(\sigma_m \, P^{1/3} \right)^{3/4} \tag{9.34}$$

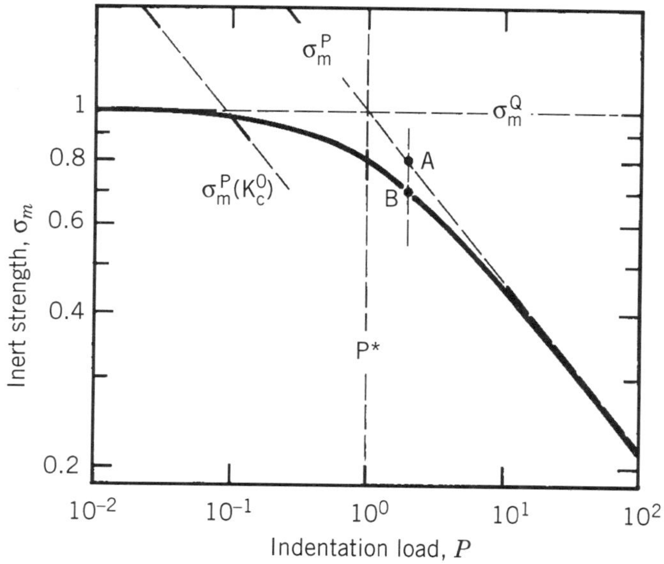

Figure 9.7 Normalized plots of strength as a function of indentation load. Solid curve is general solution and dashed lines asymptotic to this curve are limiting solutions. (From R. F. Cook, B. R. Lawn, and C. J. Fairbanks, "Microstructure–Strength Properties in Ceramics: I, Effect of Crack Size on Toughness," *J. Am. Ceram. Soc.*, Vol. 68, 604–615, 1985. Reprinted by permission of the American Ceramic Society.)

This leads to

$$K_C^{app} = K_C^{\infty}\left(\frac{\sigma_m}{\sigma_m^P}\right)^{3/4} \tag{9.35}$$

or

$$K_C^{app} = K_C^{\infty}\left(\frac{P}{(P+P^*)}\right)^{1/4} \tag{9.36}$$

A plot of this function is shown in Figure 9.8 from Cook, Lawn, and Fairbanks (1985a). This theory leads to an unbounded decrease in the apparent toughness as determined from indentation load as the indentation load decreases. Cook et al. suggest that the theory is not valid below some bound shown as K_C^0 in Figure 9.8. This bound may be either the toughness for single crystals or for grain boundary fracture. The existence of such a bound within the experimentally accessible range would lead to a transition in the plot of inert strength as a function of indentation load to the dotted line labeled $\sigma_m^P(K_C^0)$ in Figure 9.7.

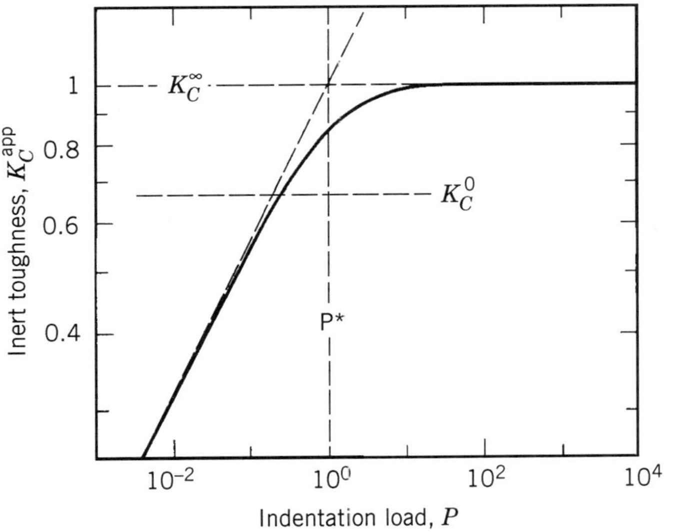

Figure 9.8 Normalized plot of apparent toughness as a function of indentation load showing R-curve behavior. (From R. F. Cook, B. R. Lawn, and C. J. Fairbanks, "Microstructure–Strength Properties in Ceramics: I, Effect of Crack Size on Toughness," *J. Am. Ceram. Soc.*, Vol. 68, 604–615, 1985. Reprinted by permission of the American Ceramic Society.)

Cook, Lawn, and Fairbanks determined the inert strength for a variety of alumina, barium titanate, and glass-ceramic materials and found a good fit to the model described above. A particular success of the model is that for each material all the data could be put on a master plot by expressing the load in terms of P/P^* and the apparent toughness as K_C^{app}/K_C^{∞}. The result in shown in Figure 9.9 (Cook, Lawn, and Fairbanks, 1985a) for alumina materials. Similarly good fits were obtained for barium titanate and glass-ceramic materials. Subsequently, Bennison and Lawn (1989) have more closely identified the R-curve with the bridging mechanism and related it more closely to microstructural variables.

This work has been extended by Cook, Liniger, Steinbrech, and Deuerler (1994) to a series of aluminas using a large range of indenter loads. Three aluminas produced by successively longer sintering times had most-frequent intercept lengths of 10, 15, and 20 μm. The corresponding initial and steady-state T values were 1.69–2.81, 1.36–3.33, and 1.28–3.64 MPa · $m^{1/2}$. In apparent contrast, Bleise and Steinbrech (1994) found for 16-μm grain size alumina that the initial toughness value was 4 MPa · $m^{1/2}$ and remained constant for a substantial range of crack growth before starting to increase with further growth. They did not find an approach to a steady-state value.

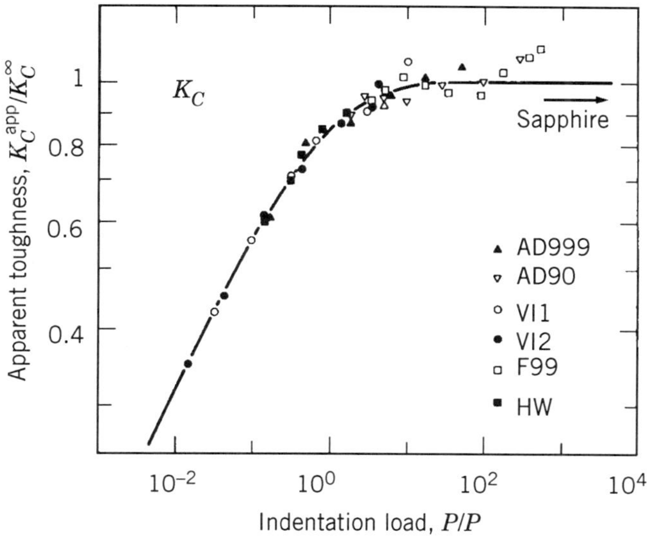

Figure 9.9 Apparent toughness as a function of indentation load. The constant value for sapphire is indicated by the arrow. (From R. F. Cook, B. R. Lawn, and C. J. Fairbanks, "Microstructure–Strength Properties in Ceramics: I, Effect of Crack Size on Toughness," *J. Am. Ceram. Soc.*, Vol. 68, 604–615, 1985. Reprinted by permission of the American Ceramic Society.)

9.4 R-CURVE EFFECTS OF STRENGTH DISTRIBUTIONS

Kendall, Alford, and Birchall (1987) considered the effect of toughness on reliability as measured by the Weibull m; their work was extended by Cook and Clarke (1988) and Shetty and Wang (1989). Kendall et al. concluded that increasing the toughness of a material to a higher value, which was still not dependent on crack length, did not change the m value. Only when there was R-curve behavior did the Weibull modulus increase, as in the cases of transformation toughening or toughening by reinforcement with ceramic fibers. They developed a model for the effect of R-curve behavior on the Weibull modulus by representing the variation in toughness with crack length over a measured range of crack lengths as

$$K_{IC} = A_K (\Delta c)^{n_K} \tag{9.37}$$

where we have introduced subscript K to distinguish from the parameters in the empirical crack velocity law (8.1). According to their model the ratio of the Weibull modulus m_R for material with R-curve behavior to that for the same material without R-curve behavior m is given by

$$m_R = m \frac{0.5}{0.5 - n_K} \tag{9.38}$$

Li and Yamanis (1989) reported on a silicon nitride with R-curve behavior, $m_R = 18.5$ and $n_K = 0.221$, with the latter determined by indented-strength measurements. They compared this to another silicon nitride without significant R-curve behavior and $m \approx 10$. The model of Kendall et al. gives a predicted ratio of $m_R/m = 1.79$ in good agreement with the directly measured ratio of 1.8. This silicon nitride is discussed further in Section 29.3.

Cook and Clarke (1988) extended the treatment of Kendall et al. by adding a term for a localized stress favoring crack growth. Shetty and Wang (1989) reviewed and extended the earlier work. In particular, the latter showed that if the initial strength distribution without R-curve behavior was of Weibull form, the distribution would also have Weibull form with a modified Weibull m for uniform loading as given by Kendall et al. They showed that the modified Weibull distribution would have a modified σ_0 value, $\sigma_0 R$, given by

$$\sigma_{0R} = \frac{A_R \, \pi^{n_K} \, K_C^{2n_K - 1}}{4^{n_K} \, \sigma_0^{2n_K - 1}} \tag{9.39}$$

They also concluded that in comparing distributions in specimens where flaws have been generated by point contact loads the effect of R-curve behavior is to give a modified Weibull modulus:

$$m_R = \frac{m\,(2n_K + 3)}{3(1 - 2n_K)} \tag{9.40}$$

These equations, based on a power law representation of R-curve behavior, are illuminating, but Shetty and Wang have found R-curve behavior that is not well described by a power law in MgO–PSZ and have used other functional representations.

The above discussion of the effect of a rising R-curve on increasing the Weibull modulus needs to be taken with caution. Tandon, Green, and Cook (1993) considered the effect of strength variability of both (1) subjecting flaws to a residual surface stress and (2) T-curve behavior. In the first case they considered uniform compressive stress extending a depth δ into the specimen. In the latter case they considered a rising T-curve. In both cases they considered a distribution of flaws from a minimum length c_0 to a maximum c_N. They point out that under these circumstances the strength distribution may no longer be described by a Weibull distribution and they take the standard deviation as a more general and appropriate descriptor. The resulting strength distributions depend on the relative sizes of the parameters describing the residual stress and the T-curve in comparison to the crack size distribution so that the authors' work must be consulted for details. The only generalization that we can give here is that for some conditions either residual stress or a rising T-curve does give an increase in mean strength but can also give an increase in the standard deviation of the strength rather than a decrease.

Another caution in regard to attempting to improve reliability through an R-curve effect is provided by the work of Kovar and Readey (1994). Grain bridging has been identified as a mechanism for toughening polycrystalline alumina. These authors studied aluminas with grain sizes of 5, 16, and 27 μm. Increasing the grain size was found to given enhanced flaw tolerance caused by a rising R-curve. However, no reduction in strength variability was found with increasing grain size.

9.5 EFFECT OF R-CURVE ON SUBCRITICAL CRACK GROWTH

Fett and Munz (1992) have studied subcritical crack growth in polycrystalline alumina using an edge-notched beam with constant load and have provided a summary of references to earlier work showing R-curve effects on subcritical crack growth. They found that despite a monotonically increasing value of the externally applied stress intensity factor with time due to crack growth, the crack velocity actually decreased with time during a portion of the experiment. They attribute this behavior to crack shielding in these materials that exhibit strong R-curves. Evidently the life prediction techniques given in Chapter 8 for subcritical crack growth cannot be applied without modification to materials with strong R-curve behavior.

CHAPTER 10

AN OVERVIEW OF TOUGHENING MECHANISMS IN CERAMICS

10.1 INTRODUCTION

An issue of the greatest importance to the load-bearing use of ceramics is that of increasing toughness. From the fundamental fracture mechanics equation

$$\sigma_f = K_{IC}/Yc^{1/2} \tag{10.1}$$

one can see that increasing strength, σ_f, requires some combination of increasing the toughness, K_{IC}, and reducing the size, c, of the largest crack. Becher (1991) summarizes the fracture toughness of glasses and single crystal ceramics as lying in the range 0.5–2 MPa · $m^{1/2}$ and states that in polycrystalline ceramics not employing special microstructural design toughening techniques the toughness usually does not exceed 5 MPa · $m^{1/2}$. With this toughness, to achieve a strength of 800 MPa for a penny-shaped crack far from an external surface (Y = 1.128) requires a maximum crack radius of about 30 μm. Extreme efforts in control of processing from the synthesis of raw materials through forming and sintering to the final finishing are constantly made to reduce the maximum crack size and the continuing development of very fine powders and improved processing techniques offers promise of further progress (Lange, 1989). However, there are limits to how small cracks can be made in practical ceramics. Also, a ceramic with a small crack size used near its fracture strength is in danger of sudden failure from a larger crack introduced by contact with a hard material. Designing and producing microstructures to increase toughness have thus become perhaps the main goals of research on structural ceramics. Evidently a theory of how microstructure can influence toughness is crucial to guide development of tougher ceramics. A general classification of toughening mechanisms in ceramics is given in Table 10.1. These mechanisms will be taken up in later sections of this chapter.

The practical achievements in toughening ceramics are illustrated in Table 10.2 from Becher (1991).

The behavior of real cracks in ceramics with complex microstructures is difficult to model quantitatively but has been discussed extensively by R. Rice (1981, 1984, 1985). There are many theories of the effects of microstructure on crack propagation in ceramics including toughening. These have been discussed in several review papers (Evans, 1990; Becher, 1991; Rodel, 1992; Steinbrech, 1992). There does not appear to be a single master theory that includes all toughening effects. This is because each effect becomes very complex when treated mathematically in full generality. Indeed, it is generally necessary in treating a single toughening effect to introduce simplifying assumptions with the goal of making the treatment mathematically practicable while attempting to preserve the essential physics of the toughening behavior. The situation is thus that many theories of individual toughening effects exist. It is recognized that several toughening effects can take place together and that the combined effect is not necessarily a simple linear combination of the individual effects. Furthermore, some of the effects considered in early

TABLE 10.1 A Classification of Toughening Mechanisms in Ceramics

General Mechanism	Detailed Mechanisms
Crack deflection	1. Tilt and twist out of crack plane around grains and second phase additions
Crack bowing	1. Bowing in crack plane between second-phase crack-pinning points
Crack branching	1. Crack may subdivide into two or more roughly parallel cracks
Crack tip shielding by process zone activity	1. Microcracking
	2. Transformation toughening
	3. Ductile yielding in process zone
Crack tip shielding by crack bridging	1. Second-phase brittle fibers with partial debonding
	2. Frictional and ligamentary grain bridges
	3. Second-phase ductile ligament bridging

TABLE 10.2 Fracture Toughness Values for Various Ceramic Materials

Material	Grain Size (μm)	Fracture Toughness ($MPa \cdot m^{1/2}$)
Alumina	1–2 10–12	2.5–3 4.5
Alumina + 20 vol. % SiC whiskers	1–2	8–10
Alumina + 20 vol. % TZP (1 mol % yttria)[a]	2	8
Alumina + 40 vol. % TZP (12 mol % ceria)[a]	2	13
Polycrystalline cubic zirconia	50	3
TZP (12 mol % ceria)[a]	4–6	15–18
TZP (2 mol % yttria)[a]	1.5	12
TZP (2 mol % yttria)[a]	0.5	7
PSZ (9 mol % magnesia) with increasing precipitate size[a]	50	8–16
Silicon nitride, equiaxed grains	2–3	4
Silicon nitride, elongated grains	~4[b]	10
Silicon carbide, densification additive—alumina	2	3.5–4
Silicon carbide, densification additive—boron and carbon	5–7	2.5–3
Silicon carbide + 25 vol. % titanium carbide	2.5	6

Values determined at room temperature using precracked (3- to 5-mm precrack length) applied moment double cantilever beam specimen. TZP is polycrystalline tetragonal zirconia PSZ is partially stabilized zirconia. Grain size of equiaxed polycrystalline matrix phase.
[a]Toughness values are quite sensitive to test temperature.
[b]Diameter of grain.

Source: P. F. Becher "Microstructural Design of Toughened Ceramics," *J. Am. Ceram. Soc.* Vol. 73, 255–269, 1991. Reprinted by permission of the American Ceramic Society.

work, although significant in some special ceramic microstructures, now appear to be less important in most high-strength and high-toughness ceramics than other effects treated later. Thus attention to toughening polycrystalline ceramics at the time of this writing centers on crack tip shielding processes often acting in conjunction with crack deflection. Other toughening mechanisms discussed in earlier literature, such as crack bowing, now appear to be of more specialized interest. The present chapter presents a brief summary of the main categories of Table 10.1. Subsequent chapters treat transformation toughening and toughening with fibers in more detail because of their importance and complexity. There appears to be general consensus in the literature concerning the qualitative aspects of the major toughening mechanisms in polycrystalline ceramics, but the detailed descriptions vary somewhat, as does the way in which the mechanisms are grouped for discussion. For example, Lutz, Claussen, and Swain (1991) state that "the toughening mechanisms can be subdivided into (1) crack deflection and crack branching, (2) contact shielding processes (wedging caused by broken-out grains or rough crack surfaces, and crack bridging), and (3) stress-induced zone-shielding processes (transformation and microcracking, and residual stress fields)."

10.2 TOUGHENING BY CRACK DEFLECTION

Crack deflection can take place when there are local areas in a ceramic that have lower resistance to crack propagation than an average plane cutting through at right angles to the tensile stress (Wiederhorn, 1984). An important case is that of grain boundaries, which typically require about half the fracture surface energy of a single crystal. A crack advancing through a polycrystalline ceramic should thus attempt to follow along the grain boundaries. However, as shown in Chapter 5, the stresses acting in inclined planes near a crack tip depend on the angles involved. For a straight crack that tilts about a direction perpendicular to the crack advance through an angle θ, the stress intensity factor required for crack propagation is given by

$$K(\theta) = K(\theta = 0) \sec^2 (\theta/2) \tag{10.2}$$

A crack that twists about the direction of crack advance through an angle φ will experience a stress intensity factor given by

$$K(\varphi) = K(\varphi = 0) \sec^2(\varphi) \tag{10.3}$$

Thus a given twist angle is more effective than the same angle of tilt in increasing the stress intensity factor in the original crack plane necessary to drive the crack out of the plane.

A crack advancing in a polycrystalline body and following grain boundaries will be made up of many sections at various angles. Wiederhorn (1984) states that K_{IC} for a polycrystalline body should be approximately twice that for a corresponding single crystal based on the geometrical effect of deflection alone.

Faber and Evans (1983a, 1983b) considered cracks deflected by spheres, rods, and disks in an otherwise uniform body. They reached several important conclusions:

1. The increase in toughness due to crack deflection depends on the particle shape and volume fraction of particles but not on the particle size.

2. The most effective morphology for increased toughening is the rod shape followed in turn by the disk and sphere. For rods the toughening effect increases with increasing aspect ratio up to a ratio of about 12.
3. The toughness increases with increasing volume fraction of deflecting particles up to a volume fraction of about 20% and increases very little with further increases in volume fraction.
4. The maximum toughness caused by rods (i.e., with aspect ratio 12 and volume fraction 20%) is about four times the toughness without deflecting particles when expressed in terms of R (i.e., in terms of fracture surface energy) or about two times when expressed in terms of T (i.e., in terms of critical stress intensity factor).

The above remarks about polycrystalline materials and deflecting particles relate only to the geometrical effects of deflection. Crack deflection can also cause partial bridging by grains or second-phase particles with additional important toughening effects to be treated later. Rodel (1992) has reviewed the interaction between crack deflection and crack bridging in whisker-toughened alumina and in silicon nitride containing rod-shaped grains. He concludes that crack bridging is probably the most important mechanism of toughening in these materials (rather than the direct effect of crack deflection) but that crack deflection is required for the development of crack bridging.

Crack deflection appears to be the dominant toughening mechanism in platelet-reinforced alumina, as discussed in Section 13.5.

10.3 TOUGHENING BY CRACK BOWING

It has long been recognized (Hasselman and Fulrath, 1966) that the introduction of second-phase inhomogeneities into a brittle material (such as the introduction of small nickel spheres into glass) can lead to an increase in strength. It is now also recognized that the toughness can also be increased and theories of crack bowing have been introduced in an attempt to explain these effects. Crack bowing is the development of a nonlinear crack front for a planar crack and must be distinguished from crack deflection, which is the development of out-of-plane portions of the crack. The concept of crack bowing was introduced by Lange (1970, 1971) and studied further by Evans (1972) and by Green, Nicholson, and Embury (1979).

Lange (1970) proposed that strong second phase particles could act as barriers to crack advance. He visualized a crack advancing in a plane, but pinned by barriers with an average spacing of d and bowing out between the pinning points under the action of a stress. He assumed that the crack acts as if it had a line tension, T, and concluded that the critical value of the strain energy release rate required to propagate the crack is

$$G_C = 2(\gamma_0 + \frac{T}{d}) \tag{10.4}$$

where γ_0 is the fracture surface energy for the glass without the inclusions. He noted that the strength, S, is then given from the Griffith condition by

$$S \propto (G_C)^{1/2} \tag{10.5}$$

Lange related the obstacle spacing d to the volume fraction of particulates, V_f, and the particle diameter, D, by taking the spacing to be equal to the mean free path as calculated by Fullman (1953):

$$d = 2D\left(\frac{1 - V_f}{3V_f}\right) \tag{10.6}$$

Lange noted that his theory was consistent with the strength data of Hasselman and Fulrath (1966) and went on (Lange, 1971) to obtain both strength and fracture energy data on borosilicate glass containing dispersed alumina particles. His results showed increases in both strength and toughness with increasing volume fraction of alumina and supported his theory.

10.4 GENERAL REMARKS ON CRACK TIP SHIELDING MECHANISMS

A simplified view of toughening by crack tip shielding that illustrates the basic physical processes is useful (Marshall, Drory, and Evans, 1983; summarized by Steinbrech, 1992). Crack tip shielding is divided, as in Table 10.1, into two basic categories: process zone effects and bridging zone effects.

Process zone effects depend on volume elements of the material near the crack undergoing a nonlinear stress–strain effect. This is idealized as elastic behavior up to σ_c followed by a permanent tensile strain of ϵ^T upon a small increase in stress. Further increases in stress are assumed to produce elastic behavior, perhaps with a different elastic constant. If this permanent strain occurs in a volume fraction V_f over a distance h on either side of an advancing crack, the increase in the critical strain energy release rate is

$$\Delta G_C = 2V_f \sigma_c \varepsilon^T h \tag{10.7}$$

as discussed in more detail in Chapter 11. Alternatively, a stress intensity factor approach can be taken, leading to an increase in toughness due to transformation toughening of

$$\Delta K_C^T = 0.30\, V_f \varepsilon^T E\, h^{1/2} \tag{10.8}$$

The equivalence of these two approaches is shown in Chapter 11. We discuss the relationship between the energy approach and the stress intensity factor approach in general later in the present section after discussing bridging zone effects.

Bridging zone effects depend on a closure stress $p(u)$, where u is the crack opening. This stress acts up to a maximum value of crack opening u^* at which the bridging element fails. If a bridging zone exists for a distance from the crack tip, and attention is focused on the end of the bridging zone away from the crack tip, as shown in Section 5.7 by the J-integral method (J. Rice, 1968b; Evans and McMeeking, 1986; Lawn, 1993), the corresponding increase in the critical strain energy release rate is

$$\Delta G_C = 2\int_0^{u^*} p(u)\, du \tag{10.9}$$

If $\sigma(u)$ is the stress in a bridging element (fiber, particulate, grain, etc.) and a volume fraction V_b are bridging, $p(u) = V_b\sigma(u)$ so that

$$\Delta G_C = 2\, V_b \int_0^{u^*} \sigma(u)\, du \tag{10.10}$$

As in the case of process zone effects, bridging zone effects can also be treated by a stress intensity factor approach. This treatment leads to a set of coupled equations, which are discussed in Chapter 12 but are not given here because of their complexity.

As discussed in the following sections, permanent deformation in a process zone can arise from several different effects, as can the bridging forces. However, all the effects fit into the general form given in the preceding equations for ΔG and the equivalent ΔK formulation. If the permanent deformation process also leads to a reduction in the local elastic modulus, a contribution to ΔK due to shielding by reduction of the elastic modulus in the crack process zone must also be taken into account.

We next consider how these effects appear to a person performing a fracture mechanics test to measure a critical stress intensity factor. Typically, a crack of length c is created and a stress σ that is uniform tension at a distance from the crack is applied. An applied stress intensity factor K_a is calculated from

$$K_a = Y\sigma c^{1/2} \tag{10.11}$$

The value of K_a at the instant of fracture is taken to be the observed critical stress intensity factor K_C:

$$K_{IC} = K_{\text{tip at fracture}} \tag{10.12}$$

When no toughening process is operating, this measured critical stress intensity factor is equal to the critical stress intensity factor required at the crack tip, which we here designate $K_0 = K_{\text{tip at fracture}}$. That is, for no toughening mechanism operating

$$K_{IC} = K_0 \tag{10.13}$$

However, when some toughening mechanism such as transformation toughening or fiber bridging is operating, the stress intensity factor characterizing the stress near the crack tip is

$$K_{\text{tip}} = K_a - \Delta K \tag{10.14}$$

where ΔK is the reduction in the stress intensity at the crack tip as a result of some toughening process such as transformation toughening or crack bridging. We have taken ΔK as a positive number and recognized that the action of fiber bridging or transformation is to produce a force tending to close the crack. The crack tip is said to be shielded from the full effect of the applied stress. The applied stress intensity factor is now

$$K_a = K_{\text{tip}} + \Delta K \tag{10.15}$$

At the instant of fracture the applied stress intensity factor is again taken to be the measured critical stress intensity factor and the value of K at the tip must again be K_0 so that

$$K_C = K_0 + \Delta K \tag{10.16}$$

From the point of the observer determining the measured critical stress intensity factor, the material has been toughened by ΔK despite the fact that the required local stress intensity factor for fracture is unchanged. In other words, the effect of the toughening process in partially shielding the crack tip is to cause a larger applied stress to be required for crack propagation and so a larger macroscopic toughness is observed.

The same effect can be described in terms of required mechanical energy release rate. The observer performing the fracture mechanics test will determine the critical mechanical energy release rate, G_C, required of the applied stress at the instant of fracture, which is

$$G_C = G_{a\text{ at fracture}} = G_{\text{tip at fracture}} + \Delta G = G_0 + \Delta G \tag{10.17}$$

where G_0 is the critical mechanical energy release rate in the absence of any toughening mechanism.

An important question is the relationship between the critical stress intensity factor and the critical mechanical energy release rate. In Section 5.7 the Irwin relation between G and K was established in the absence of toughening by carrying out the integration in equation (5.39). This integration was performed along a length Δx from the crack tip and the limit was taken as $\Delta x \to 0$. If the same calculation is now performed when a toughening mechanism is operating over a process zone near the crack, the result is that the stresses and strains entering the integral are those characterized by a stress intensity factor K_0 and the corresponding energy release rate is G_0. The result of the integration in the notation of the present section is

$$G_0 = \frac{K_0^2}{E} \text{ (for plane stress)} \tag{10.18}$$

and

$$G_0 = \frac{K_0^2}{E}(1-\nu^2) \text{ (for plane strain)} \tag{10.19}$$

The same integration can be done from the point of view of the external observer, leading to

$$G_C = \frac{K_C^2}{E} \text{ (for plane stress)} \tag{10.20}$$

and

$$G_C = \frac{K_C^2}{E}(1-\nu^2) \text{ (for plane strain)} \tag{10.21}$$

Combining (10.16) and (10.17) gives, for plane stress,

$$E(G_0 + \Delta G) = (K_0 + \Delta K)^2 \tag{10.22}$$

or

$$\Delta K = (K_0^2 + E\,\Delta G)^{1/2} - K_0 \tag{10.23}$$

Note especially that $\Delta K \neq (E\Delta G)^{1/2}$. Similarly, for plane strain,

$$\frac{E}{(1-\nu^2)}(G_0 + \Delta G) = (K_0 + \Delta K)^2 \tag{10.24}$$

or

$$\Delta K = \left(K_0^2 + \frac{E}{(1-\nu^2)}\Delta G\right)^{1/2} - K_0 \tag{10.25}$$

The same results were obtained by Lawn (1993) using a J-integral argument. Our discussion of the crack tip and the external observer corresponds, respectively, to his enclave observer and global observer.

The conclusion of this discussion is that, in considering toughening caused by either process zone or bridging zone effects, one may use either a mechanical energy release rate approach (in which the total critical mechanical energy release rate when toughening is present is the sum of that for the untoughened material plus the total resistance effect due to all shielding mechanisms that are operating, with care taken to include energy dissipation effects such as frictional sliding of fibers against the matrix) or a stress intensity factor approach (in which the total critical stress intensity factor when toughening is present is the sum of that for the untoughened material plus the total toughening due to all shielding effects that are operating). We emphasize two points. First, the preceding statement applies to the contributions to energy release rate or stress intensity factor due to the simultaneous effect of all the shielding processes that are operating. Second, there is a nonlinear relationship between ΔK and ΔG as shown in the immediately preceding equation. Examples of the use of equation (10.25) in calculating toughening ΔK from ΔG determined by energy considerations and comparing the results to a direct calculation of ΔK from stress considerations are given in Chapter 11 for transformation toughening and in Chapter 14 for whisker bridging.

10.5 TOUGHENING BY CRACK TIP SHIELDING BY MICROCRACKING

It is possible in some microstructures for a large number of small secondary cracks to be caused near the main crack as a result of the high local stresses around the primary crack. Work on R-curves of polycrystalline alumina (Hübner and Jillek, 1977) led to the development of a model for toughening by crack tip shielding by microcracking (Evans and Faber, 1984; Hutchinson, 1987), which is summarized here. Other work (Knehens and Steinbrech, 1982; Ruhle, Claussen, and Heuer, 1986; Swanson, Fairbanks, Lawn, Mai, and Hockey, 1987; Vekinis, Ashby, and Beaumont, 1990) suggested that the main R-curve effect in polycrystalline alumina is probably caused by a degree of grain bridging and frictional pullout behind the crack tip rather than by secondary microcracking. This bridging toughening mechanism is discussed in a later section. The microcracking model indicates that significant toughening effects could arise when the conditions are right for appreciable local microcracking and so the microcracking model is discussed here.

The fracture resistance curves from Knehens and Steinbrech and from Hübner and Jillek as redrawn by Evans and Faber are shown in Figure 10.1. A significant increase in toughness (about a factor of 2) occurs as the crack extends by 2 mm. Evans and Faber considered the possibility that a zone of microcracks developed both ahead and in the wake (on the sides) of an advancing primary crack. They suggested that such microcracking causes both a permanent strain and a decrease in elastic modulus. To simplify mathematical treatment of the problem, they assumed that microcracking is limited to a zone extending a distance w from the crack and that within this zone only a volume fraction V_f of material undergoes microcracking. They took ϵ^T as the permanent strain associated with microcracking. They found the effect on toughness of the transformation zone ahead of the crack to be negligible. However, the

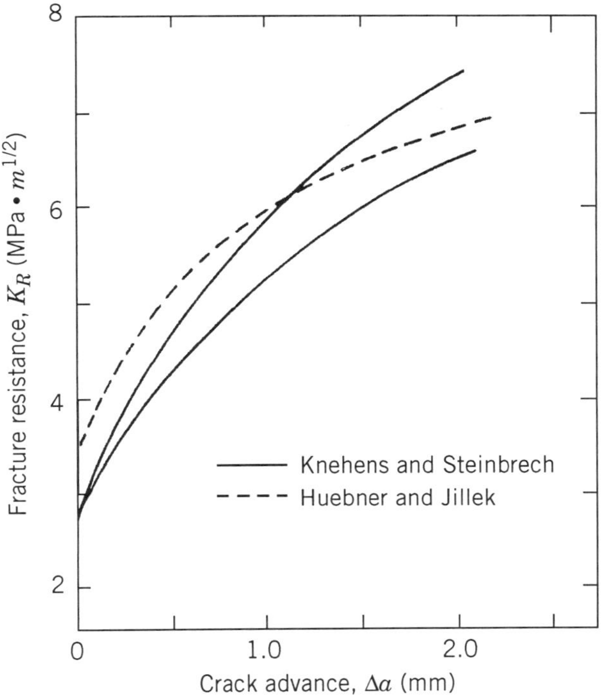

Figure 10.1 R-curve data for alumina. (From A. G. Evans and K. T. Faber, "Crack-Growth Resistance of Microcracking Brittle Materials," *J. Am. Ceram. Soc.*, Vol. 64, 255–260, 194. Reprinted by permission of the American Ceramic Society.)

transformation zone in the wake can cause a large toughening effect, which increases with crack extension and for values of crack extension greater than $5w$ approaches a value of

$$\Delta K_{microcracking} = \frac{0.22\, E\, \varepsilon^T\, V_f\, w^{1/2}}{1 - \nu} \tag{10.26}$$

where E is Young's modulus and ν is Poisson's ratio. However, Evans and Faber also consider the situation in which the zone width increases continuously with increasing crack length. This situation leads to a ΔK contribution that increases indefinitely with the square root of the crack extension. The theory does not appear to be sufficiently developed to allow prediction of the magnitude of toughening effects from the present state of knowledge of factors controlling microcracking. While the effect of microcracking may be important in practical toughening effects in certain polycrystalline ceramics, the weight of opinion appears to favor crack bridging as the controlling effect in polycrystalline alumina, polycrystalline silicon nitride, and perhaps in other polycrystalline ceramics exhibiting pronounced R-curve effects. One case in which microcracking appears to be the predominant mechanism of toughening is that of silicon carbide–titanium diboride composite ceramic (Gu, Faber, and Steinbrech, referenced in Steinbrech, 1992).

10.6 TRANSFORMATION TOUGHENING

Transformation toughening in ceramics containing zirconia in metastable form is a major form of toughening with important practical consequences. As such, it is subsequently treated in a separate chapter. An overview of the main features is given here for comparison with the other toughening mechanisms based on work by McMeeking and Evans (1982), Marshall, Evans and Drory (1983), and Evans and Cannon (1986).

It is found that the stresses around a crack tip cause transformation of a volume fraction V_t of tetragonal zirconia particles to monoclinic form out to a distance w from the crack tip dependent on the applied stress and a critical stress value σ^c required for the transformation. As the crack advances, a wake of transformed material of half-width w is left and the toughness increases until a steady-state value is reached after crack advance of about $5w$. Evans and Cannon give the resulting steady-state increase in toughness as

$$\Delta K_{IC} = 0.22\, E\, V_t\, \varepsilon_{ii}^{T}\, w^{1/2}/(1-\nu) \tag{10.27}$$

where E is Young's modulus, ν is Poisson's ratio, and and $\epsilon_{ii}{}^{T}$ is the volumetric strain change upon transformation, as discussed in Section 11.2. The value of w is given by

$$w = \frac{3^{1/2}(1+\nu)^2}{12\pi}\left(\frac{K_\infty}{\sigma^c}\right)^2 \tag{10.28}$$

Working from an energy balance approach, they obtain for the change in required mechanical energy release rate

$$\Delta G_C = 2V_T\, \varepsilon_{ii}^{T}\, \sigma_c\, w \tag{10.29}$$

They show that the results can be shown to be equivalent by using $G = (K^2/E)(1 - \nu^2)$. Equations (10.27) through (10.29) are discussed more fully in Chapter 11.

10.7 TOUGHENING BY DUCTILE YIELDING IN THE PROCESS ZONE

An important effect in metals is the development of a plastic zone at the crack tip. This effect is not believed to occur in brittle ceramics that are instead believed to fail by propagation of sharp cracks. However, ceramic composites that include a ductile phase may undergo some degree of yielding at points along the crack front. The effect is generally thought to lead to the development of ductile bridging ligaments. This effect is accordingly treated in a later section on ductile bridging.

10.8 TOUGHENING BY FIBER BRIDGING

Fiber bridging may take place in composites made with either continuous, oriented fibers or short, randomly oriented whiskers. The first situation has been treated theoretically in considerable detail. It is developed in Chapter 12 both for its own importance and for the insight into bridging effects in general that it provides. We present only qualitative remarks on the effects of crack bridging by fibers here and refer to Chapter 12 for bridging by aligned, continuous fibers and to Chapter 13 for composites having relatively short fibers with random orientation.

The fiber-reinforced composites of greatest interest are those having fibers much stronger than the matrix that are only weakly bonded to the matrix and that slide against the matrix with a frictional stress τ when a crack passes nearby. Provided a sufficient volume fraction of such fibers are aligned, the stress to propagate a crack in a plane perpendicular to the fibers is increased over the stress necessary to propagate the same size crack in the unreinforced matrix; that is, the composite is tough compared with the matrix. This toughness increase is associated with energy dissipation from τ. Also, when the crack propagates it does so only in the matrix itself, leaving the fibers bridging the matrix. If the fibers are very strong the matrix crack will propagate entirely through the composite, leaving all the fibers unbroken for an indefinite length behind the crack front. If the fibers are not strong enough to carry the entire load alone, a bridged section of finite length develops behind the crack tip. The composite is again tougher and stronger than the matrix alone. The ultimate strength for complete failure of the composite can be much greater than for the unreinforced matrix. The stress–strain curve for the composite is, however, no longer elastic in the sense of being recoverable above the stress necessary to propagate a crack in the matrix.

10.9 TOUGHENING BY FRICTIONAL AND LIGAMENTARY GRAIN BRIDGES

Crack–interface bridging has emerged as important (indeed, perhaps the most important) toughening mechanism in conventional polycrystalline ceramics; that is, in those that do not undergo stress-induced transformation or have bridging effects due to fiber reinforcements or fiber-like elongated grains produced by special processing. Swanson, Fairbanks, Lawn, Mai, and Hockey (1987) present strong evidence for toughening by a grain bridging mechanism in polycrystalline alumina as well as summarizing earlier indications of bridging behavior. The basic phenomenon is similar to toughening by whisker or fiber bridging but is different in detail. They found bridging to occur about every 2–5 grain diameters and to extend about 100 grain diameters or more behind the crack tip. Mai and Lawn (1987) developed a theory and fitted it to data on 16-μm grain size alumina by Swain showing a T-curve behavior in which the critical stress intensity factor rose from about 3.5 MPa · $m^{1/2}$ for an initial crack to about 6.5 MPa · $m^{1/2}$ after 9-mm crack extension.

Their model was based on an assumed form for the stress, $p(u)$, caused by grain bridging to act on the surface of the crack as a function of the crack opening, u. They chose the function

$$p(u) = p^*(1 - u/u^*)^m \tag{10.30}$$

where p^* and u^* are limiting values of the stress and separation, respectively, and m is an exponent. These quantities were not determined directly but were absorbed into other parameters that in turn were determined by fitting the experimental T-curve. This work was extended by Hu and Mai (1992), who used the same function for $p(u)$ and developed a compliance method of determining this function directly.

10.10 TOUGHENING BY SECOND-PHASE DUCTILE LIGAMENT BRIDGING

A theory of toughening due to bridging by elastic, ideally plastic particles has been developed by Budiansky, Amazigo, and Evans (1988). They assume that the crack opening is $2v$ and is elastic up to v_y at stress σ_y and then deforms plastically under this stress until failure occurs at v_p at the last bridging ligament. The stress is given by

$$K_p = K_m(1-V_f)\,\Lambda_p \tag{10.31}$$

where

$$\Lambda_p = \left(1 + \frac{2V_p}{1 - V_p}\frac{E_c\,S\,v_f}{K_m^2\,(1 - \nu^2)}\right)^{1/2} \tag{10.32}$$

The length, L_p, of the crack that is bridged at fracture by the plastic particles is related to Λ_p by

$$\Lambda_p = 1 + \frac{V_p\,S}{K_m}\left(\frac{8L_p}{\pi(1 - V_p)}\right)^{1/2} \tag{10.33}$$

where K_m is the toughness of the matrix. They also discuss several related cases with somewhat different constitutive relations for the bridging ligaments.

Their theory is a small-scale bridging theory in that the assumption is that only a small fraction of the crack length is bridged by ductile particles.

Subsequently, a large-scale bridging theory was developed by Cox (1991) and Cox and Lo (1992) and used by Flinn, Lo, Zok, and Evans (1993) to interpret results on an Al_2O_3/Al composite. Fracture resistance curves (T-curves) were found that were concave upward and gave *K* resistance values as high as 25 MPa · $m^{1/2}$.

10.12 SYNERGISTIC EFFECTS IN TOUGHENING

So far, individual toughening effects have been considered as acting alone. However, a specimen can contain more than one type of toughening mechanism and the effect of the simultaneous action of two or more mechanisms needs to be considered (R. W. Rice, 1985; Becher and Tiegs, 1987). The simplest assumption would be that the toughening effects act independently and would simply add, but the possibility exists of greater toughening through interaction (synergism) than that resulting from simple additivity. Amazigo and Budiansky (1988) and Stump (1991) have considered the effect of dilatant transformation toughening and toughening by ductile particles acting in concert. The treatments are beyond our scope but we quote the results (Amazigo and Budiansky, 1988; Budiansky and Stump, 1991; Stump, 1991) to show that the theory predicts interesting and significant synergistic effects. Denoting the saturation value of $\Lambda_p = K_{ss}/[K_m(1 - V_t)^{1/2}]$ (i.e., the value for a fully developed wake) as $\Lambda_T{}^{ss}$ and defining the analogous quantity for the ductile particulate toughening as

$$\lambda_T^{ss} = \frac{K_T^{ss}}{K_m\,(1-V_t)^{1/2}} \tag{10.34}$$

where V_t is the volume fraction of transforming particles, leads in their theories to the results for overall toughening

$$\Lambda_{ss} = \lambda_T^{ss}\,\Lambda_{p_{ss}} \quad \text{for } \rho \rightarrow \infty \tag{10.35}$$

and

$$\Lambda_{ss} = [(\lambda_T^{ss})^2 + (\Lambda^{p^{ss}})^2 - 1]^{1/2} \quad \text{for } \rho \rightarrow 0 \tag{10.36}$$

where ρ is a coupling parameter given by

$$\rho = \frac{V_p\,S(1+\nu)}{\sigma_c} \tag{10.37}$$

The result for small ρ is simple additivity of crack resistance increments from the two mechanisms expressed in terms of the required strain energy release rate. The result for large ρ is a multiplicative formula showing synergism. Incidentally, values of $\rho \approx 1$ in the full theory give overall toughness ratio values Λ_{ss} almost as large as those for $\rho \to \infty$.

Another case of simultaneous action of different toughening mechanisms is that of toughening by fiber bridging in conjunction with transformation toughening. Becher and Tiegs (1987) have present experimental evidence for synergistic effects. Cui (1992) has presented a theory for the interaction of fiber and transformation toughening analogous to that of Amazigo and Budiansky (1988) for transformation toughening in conjunction with ductile particulate bridging. In Cui's theory the interaction parameter is defined by

$$\rho = \frac{V_f \sigma_y (1 + \nu)}{\sigma_c} \tag{10.38}$$

and he obtains for overall toughening

$$\Lambda = \lambda_T^{ss} \Lambda_f \text{ for } \rho \to \infty \tag{10.39}$$

and

$$\Lambda = [(\lambda_T^{ss})^2 + (\Lambda_f)^2 - 1]^{1/2} \text{ for } \rho \to 0 \tag{10.40}$$

Again a synergistic effect is apparent for large ρ.

CHAPTER 11

TOUGHENING BY TRANSFORMATION

11.1 INTRODUCTION

Transformation-toughened ceramics constitute a major family of tough and flaw-tolerant ceramics. These desirable qualities can be combined with high strength. These good properties are usually optimized at room temperature and decrease with increase in temperature. Thus far, only zirconia (including some typical hafnia content) has been used as the transforming phase in practical transformation toughening, although other possible transforming ceramics have been considered by Kriven (1988). The effect of transformation of some zirconia grains on the thermal expansion of a zirconia ceramic was studied long ago by Geller and Yavorsky (1945), and it was realized that there were advantages to using partially stabilized zirconia rather than fully stabilized zirconia, although the mechanisms were not initially understood. King and Yavorsky (1968) spoke of "a rather remarkable ceramic material which displays a great tolerance to stress concentration" and considered stress relief by plastic deformation as a possible contributing mechanism. Recognition that stress-induced martensitic transformation from the tetragonal to the monoclinic phase can occur in the high stress field near a crack tip and cause significant toughening initiated development of a variety of toughened zirconias and other ceramics (such as alumina) toughened by the addition of zirconia grains (Garvie, Hannink, and Pascoe, 1975; Evans and Heuer, 1980; Claussen, 1983; Heuer, Lange, Swain, and Evans 1986; Evans and Cannon, 1986; Heuer, 1987; W. R. Cannon, 1989; Green, Hannink, and Swain, 1989; Hannink and Swain, 1994).

At least three mechanisms of toughening can operate in zirconia and in dual-phase oxides containing zirconia (Stevens, 1991): (1) toughening by microcracking as discussed in Chapter 10; (2) stress-induced transformation toughening; and (3) the development of compressive surface layers as a result of transformation caused by surface stresses. Hannink and Swain (1994) discuss transformation toughening in terms of three crack tip shielding mechanisms: "1. direct crack-tip shielding: deflection and meandering as a result of precipitates and grain size; 2. crack wake-zone shielding: which for zirconia-bearing ceramics includes transformation toughening and microcrack toughening; 3. contact shielding: which includes tetragonal grain wedging, sliding of grain facets and elongated precipitates, and particle bridging." The present chapter focuses on the phenomenology and theory of stress-induced transformation toughening. Further data on the properties of various zirconias developed for engineering purposes are given in Chapter 26.

11.2 BASIC FACTS OF TRANSFORMATION TOUGHENING

A notation used to designate various types of transformation-toughened ceramics has been developed and is summarized in Table 11.1 (Claussen, 1983; Heuer, 1987).

The basic crystallographic facts about zirconia pertinent to transformation toughening will now be summarized following Heuer (1987). Pure zirconia exhibits a stable, cubic form (fluorite structure) from the melting point (about

TABLE 11.1 Types and Acronyms for Zirconia-Toughened Ceramics (ZTCs)

Precipitation-Toughened Ceramics (PS = Partially Stabilized)	
Mg–PSZ, Ca–PSZ. etc.	Mg–partially stabilzed zirconia, Ca–partially stabilized zirconia
Nearly Single-Phase Ceramics Based on Tetragonal ZrO_2 Solid Solutions (T-ZrO_2)	
Y–TZP	Yttria-containing tetragonal solid solutions
Dispersion-Toughened Ceramics (ZDC or DZC = Dispersed Zirconia Ceramics)	
ZTA, ZTM, etc.	ZrO_2-toughened Al_2O_3, ZrO_2-toughened $3Al_2O_3 \bullet 2SiO_2$ (mullite), etc.

Sources: N. Clausen, "Microstructural Design of Zirconia-Toughened Ceramics," pp. 325–351 in N. Claussen, M. Ruhle, and A. Hever, eds., *Science and Technology of Zirconia II,* Advances in Ceramics Vol 12, American Ceramic Society, Westerville, OH, 1983. A.H. Heuer, "Tranformation Toughening in $Zr0_2$-Containing Ceramics," *J. Am. Ceram. Soc.*, Vol. 70, 689–698, 1987. Reprinted by permission of the American Ceramic Society.

2850 °C) to about 2350 °C. With further decrease in temperature pure zirconia transforms to a tetragonal form that remains stable to about 1240 °C with transformation strains $c \rightarrow$ t of $\epsilon_{11} = \epsilon_{22} = -0.00878$, $\epsilon_{33} = 0.01209$. Further decrease in temperature causes a transformation from tetragonal to monoclinic that is stable at lower temperatures and undergoes a $t \rightarrow$ m transformation strain of $\epsilon_{11} = -0.00149$, $\epsilon_{22} = 0.02442$, $\epsilon_{33} = 0.02386$, $\epsilon_{13} = \epsilon_{31} = 0.08188$. That is, the $t \rightarrow$ m transformation of an isolated crystal is accompanied by both a volume change ($\epsilon_{ii}{}^{T} = \epsilon_{11} + \epsilon_{22} + \epsilon_{33} = 0.04759$) and a shape change. The phase diagram for zirconia–magnesia as determined by Grain (1967) and redrawn by Porter and Heuer (1979) is shown in Figure 11.1. The transformation temperatures can be considerably modified by solid solution and the term stabilization of zirconia is used to designate the lowering of the cubic to tetragonal transformation temperature. The $c \rightarrow$ t transformation has not been studied in pure zirconia but is displacive and nonmartensitic in cubic solid solutions in the Y–ZrO_2 system. The tetragonal to monoclinic transformation is martensitic. A model of the mechanics of the stress-induced phase transformation in zirconia has been given by Budiansky and Truskinovsky (1993).

The strains associated with the tetragonal to martensitic transformation cause microcracking in pure zirconia when sintered materials are cooled through the transformation temperature. Zirconia fully stabilized with an additive such as MgO can be sintered in the cubic phase at moderate temperatures and subsequently cooled to room temperature without cracking because of the sluggish transformation. However, these materials have low thermal

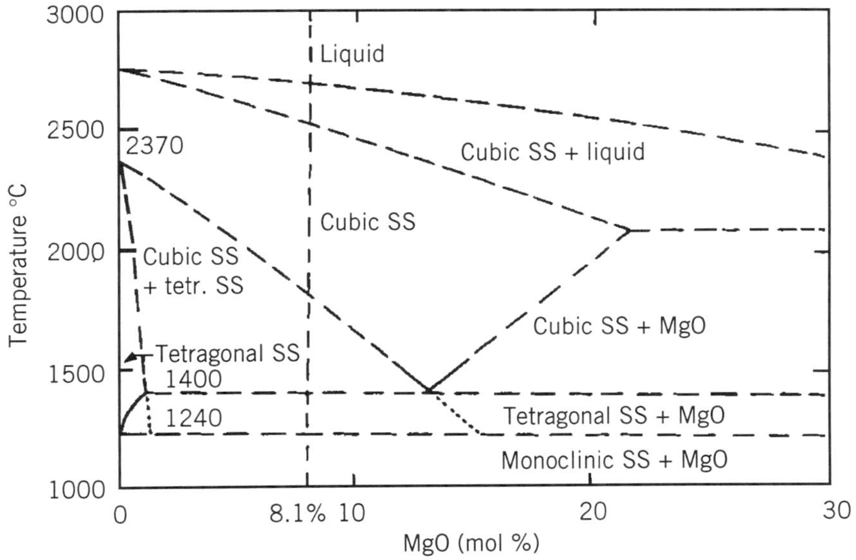

Figure 11.1 Zirconia-rich end of the zirconia–magnesia phase diagram. (From C. F. Grain, "Phase Relations in the ZrO_2–MgO System," *J. Am. Cerm. Soc.*, Vol. 50, 288–290, 1967. Reprinted by permission of the American Ceramic Society.)

conductivity and high thermal expansion, giving poor thermal shock resistance. An early empirical development was that of partially stabilized zirconias containing enough MgO to produce a mixture of cubic and tetragonal precipitates. These materials have lower thermal expansion due to the volume expansion associated with the transformation upon cooling.

Subsequently, it was recognized that the tetragonal to monoclinic transformation temperature of a precipitate constrained by a matrix depends on the precipitate size. The temperature at which some precipitates begin to transform from tetragonal to monoclinc as the temperature is being lowered is termed the martensitic start temperature, M_S. A controlled distribution of tetragonal precipitate sizes is usually achieved by sintering in the single phase cubic field and then aging in the field in which a mixture of cubic and tetragonal phases is stable. This process produces precipitates of a tetragonal solution that are coherent with the cubic phase from which they precipitate. The size of these tetragonal precipitates and hence their martensitic start temperatures are sensitive to the thermal history.

11.3 THEORY OF TRANSFORMATION TOUGHENING

As noted above, the $t \rightarrow$ m transformation of zirconia causes both a volume and shape change so that one might expect the stresses driving the transformation to include both dilatational and shear terms. However, a constrained zirconia grain usually undergoes extensive twining as the $t \rightarrow m$ transformation occurs so that the shape change is minimized. Accordingly, it is generally assumed in a first order theory that only the dilatational component of stress drives the $t \rightarrow m$ transformation. A basic theoretical treatment of this type was given by McMeeking and Evans (1982) followed by Budiansky, Hutchinson, and Lambropoulos (1983). The latter investigators introduced the term supercritical transformation to refer to conditions where the tetragonal particles fully transform to monoclinic when a critical dilatational tensile stress is exceeded as opposed to subcritical transformation when the transformation is not complete even though the critical stress to begin transformation is exceeded. It is fully recognized that shear stresses also play some part as discussed, for example, by I. W. Chen (1991) and Stump (1991). We shall here consider only the effect of the dilatational strain and its coupling with the dilatational component of stress because this is sufficient to account for the main effects. As discussed in Section 1.7, the dilatational component of the stress field (minus the pressure) is given by

$$\sigma = -p = \frac{\sigma_{xx} + \sigma_{yy} + \sigma_{zz}}{3} \tag{11.1}$$

Following Marshall, Evans, and Drory (1983) and Evans and Cannon (1986), $-p$ is written as σ in this section. It is assumed that tetragonal grains constrained in a matrix will transform to monoclinic at any point at which the dilatational stress exceeds a critical value (dependent on grain size) at a given temperature. If σ_c is the critical value of dilatational stress needed to cause transformation, one can solve for the radius from the crack tip at which the dilatational stress becomes equal to this critical value. That is, given the value of the applied stress intensity factor, K_I, one can solve for r_c for which $\sigma > \sigma_c$ as long as $r < r_c$. Recalling the equations for the stresses for plane strain near a crack tip

$$\sigma_{xx,I} = \frac{K_I}{(2\pi r)^{1/2}} \cos\frac{\theta}{2} \left[1 - \sin\frac{\theta}{2} \sin\frac{3\theta}{2}\right] \tag{11.2}$$

$$\sigma_{yy,I} = \frac{K_I}{(2\pi r)^{1/2}} \cos\frac{\theta}{2} \left[1 + \sin\frac{\theta}{2} \sin\frac{3\theta}{2}\right] \tag{11.3}$$

$$\sigma_{zz,I} = \nu\,(\sigma_{xx} + \sigma_{yy}) \tag{11.4}$$

leads to

$$r_c = \frac{2(1+\nu)^2}{9\pi} \left(\frac{K_I}{\sigma^c}\right)^2 \cos^2\frac{\theta}{2} \tag{11.5}.$$

A plot of this stress contour is shown in Figure 11.2. The ordinate is

$$y = r_c \sin\theta \tag{11.6}$$

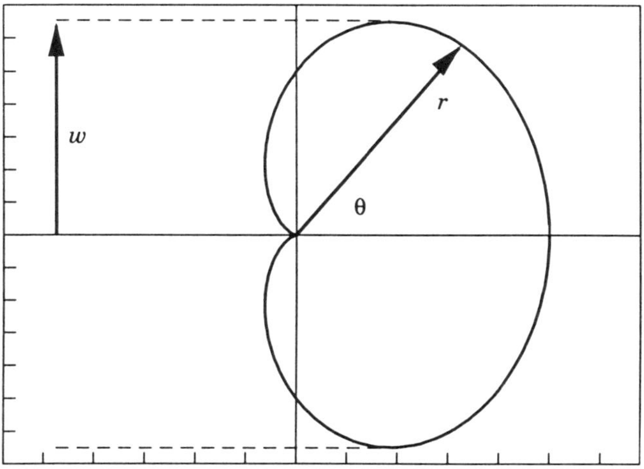

Figure 11.2 Stress contour for transformation and wake.

A crack under no applied stress in a ceramic containing a volume fraction V_f of tetragonal zirconia grains is assumed to cause no transformation. When a stress is applied to produce a stress intensity factor K, it is assumed that all tetragonal grains within the region having $\sigma > \sigma_c$ transform to the monoclinic phase. If the crack propagates as shown in Figure 11.2, the grains are assumed to remain in the monoclinic phase, producing a "wake" of width $w = y_{max}$ on either side of the crack as shown. The ordinate y has its maximum value at $\cos\theta = \frac{1}{2}$ so that [as stated earlier in equation (10.28) without proof]

$$w = \frac{3^{1/2}(1+\nu)^2}{12\pi}\left(\frac{K_I}{\sigma_c}\right)^2 = 0.046\,(1+\nu)^2\left(\frac{K_I}{\sigma_c}\right)^2 \tag{11.7}$$

or

$$K_I = \frac{(4\times 3^{1/2}\,\pi)^{1/2}\sigma_c\, w^{1/2}}{(1+\nu)} \approx \frac{\sigma_c\, w^{1/2}}{0.21\,(1+\nu)} \tag{11.8}$$

The total change in free energy accompanying the transformation from tetragonal to monoclinic is

$$\Delta F = -\Delta F_{chem} + \Delta U_T - \Delta U_{int} + \Delta U_D \tag{11.9}$$

Here ΔF_{chem} is the free energy change for an unconstrained particle and depends on the temperature and the amount of solute such as MgO, CaO, or Y_2O_3 present. The term ΔU_T is the change in strain energy caused by transformation in the absence of an applied stress. The term ΔU_{int} is the change in strain energy caused by the interaction of the strain caused by transformation with the local stress caused by the applied stress. If a transformation strain ϵ_{ij}^T occurs when the stress is σ_{ij}, then $\Delta U_{int} = \epsilon_{ij}^T\sigma_{ij}$, where the repeated index indicates summation. The term ΔU_D is energy dissipation. It is useful to group the nondissipation terms and write

$$\Delta F = \Delta\varphi + \Delta U_D \tag{11.10}$$

Marshall, Evans and Drory (1983) treat the effect of transformation in two ways: by considering the energy balance and by considering the stress intensity factors. The energy balance treatment begins with consideration of an infinitesimal crack advance. As the crack advances a distance dc, the total energy change will be

$$dU = -Gdc + 2\gamma\, dc + d\varphi + dU_D \tag{11.11}$$

and the energy release rate will be $G = G_C^{\infty}$. Here $G\,dc$ is the mechanical energy released, $2\gamma\,dc$ is the energy to form the new surfaces, and $d\varphi$ and dU_D are as above. Applying the Griffith condition $dU/dc = 0$ gives

$$G_C^{\infty} = 2\gamma + \frac{d\varphi}{dc} + \frac{dU_D}{dc} \tag{11.12}$$

Here G_C^{∞} is the toughness for incremental crack advancement in a material in which a crack has already advanced under stress sufficiently to develop a full transformation zone (i.e., a wake). This definition allows effects such as microcracking and crack deflection to be included in the surface energy term $2\gamma = G_C^M$. The increase in toughness directly associated with the transformation is then

$$\Delta G_C^T = G_C^{\infty} - G_C^M = \frac{d\varphi}{dc} + \frac{dU_D}{dc} \tag{11.13}$$

Marshall, Evans, and Drory assume that the strain associated with transformation is primarily dilatational, ϵ^T, and take $\Delta U_{\text{int}} = \epsilon^T \sigma_c$. They consider the changes as the crack advances and show that the result is

$$\Delta G_C^T = 2V_f \varepsilon^T \sigma_c w \tag{11.14}$$

Plots of G_C^T as a function of crack advance for a crack with no initial transformation zone show that an advance of three to four times the half-width, w, of the wake is required to develop this full contribution to the toughness.

Marshall, Evans, and Drory then treat the toughening induced by transformation by direct calculation of the change in the applied stress intensity factor caused by transformation. The applied (remote) stress field causes an applied stress intensity factor K^{∞} for incremental crack advance with additional transformation for a crack that already has a fully developed transformation zone. The transformations occurring within the zone $r < r_c$ cause a reduction to a local value K^L at the crack tip. The propagation condition is then that this local value equal the critical stress intensity value K_C^M for crack propagation in a fully transformed material:

$$K^L = K^{\infty} + \Delta K^T = K_C^M \tag{11.15}$$

where ΔK^T is the change in the stress intensity factor caused directly by the transformation zone. The critical stress intensity factor for crack propagation calculated from the remote applied stresses, K_C^{∞}, is then

$$K_C^{\infty} = K_C^M - \Delta K^T \tag{11.16}$$

and the toughness increase due to transformation is

$$\Delta K_C^T = -\Delta K^T \tag{11.17}$$

Consideration of the stresses exerted by the transformed particles on the crack surface and integration followed by integration with a weight function to give the contribution to the stress intensity factor lead (Marshall, Evans, and Drory, 1983; D. B. Marshall, personal communication, 1994) to the result

$$\Delta K_C^T = \frac{V_f \varepsilon^T E\, w^{1/2}}{2 \times 3^{1/4} \times \pi^{1/2}} \approx 0.30 V_f \varepsilon^T E\, w^{1/2} \tag{11.18}$$

As in the case of the transformation contribution to the toughness in the energy balance approach as indicated by $G_C{}^T$, a plot of $K_C{}^T$ as a function of crack extension shows that an extension of three to four times the wake half-width, w, is required to develop the full transformation contribution to the stress intensity factor.

The energy balance and stress intensity factors should lead to the same results. Marshall, Evans, and Drory (1983) use the basic relation

$$G^\infty = \left(K^\infty\right)^2 \frac{(1-\nu^2)}{E} \tag{11.19}$$

with

$$G_C^M = \frac{(K_C^M)^2 (1-\nu^2)}{E} \tag{11.20}$$

and

$$G^\infty = \frac{(K^\infty)^2 (1-\nu^2)}{E} \tag{11.21}$$

to obtain

$$\Delta G_C^T = G_C^\infty - G_C^M = \frac{1-\nu^2}{E} [(K_C^\infty)^2 - (K_C^\infty + \Delta K_C^T)^2] \tag{11.22}$$

or

$$\Delta G_C^T = 2 K^\infty \Delta K_C^T \left(1 - \frac{\Delta K_C^T}{2 K^\infty}\right) \frac{1-\nu^2}{E} \tag{11.23}$$

Assuming that the stress outside the zone is not modified by the zone allows equation (11.17) to be used for K^{∞}. Combining this with equation (11.18) gives

$$\Delta G_C^T = 2.14 V_f \varepsilon^T \sigma^c w \left(1 - \frac{\Delta K^T{}_C}{2K^{\infty}}\right) \tag{11.24}$$

This agrees with equation (11.14), the result for the direct energy calculation, apart from the multiplying factor. Most importantly, the two methods agree in the influence of various factors on toughening. Specifically, both the energy balance and the direct stress intensity factor calculation methods show the same dependence on the factors V_f, ϵ^T, σ^c, and w. D. B. Marshall (personal communication, 1994) has subsequently done an unpublished calculation that avoids approximations used in the above treatment and concluded that the toughening results from the energy calculation agree exactly with those from the stress calculation. His treatment uses the work of Budiansky, Hutchinson, and Lambropoulos (1983), who showed that the true zone shape was very close to that defined by the undisturbed elastic field (as used by Marshall, Evans and Drory) but that the zone size was different. Thus the true zone width h must be substituted for the width w given by the undisturbed stress field in equation (11.18). Following Marshall by combining equations (11.23) with the first equalities (i.e., with the exact form) of equations (11.8) and (11.18) gives

$$\Delta G_C^T = 2V_f \varepsilon^T \sigma^c h \left[\left(\frac{w}{h}\right)^{1/2} - \frac{3 \times 3^{1/2}}{16} \frac{2(1+\nu)E\, V_f \varepsilon^T}{9\pi(1-\nu)\,\sigma_c}\right] \tag{11.25}$$

The numerical calculations of Budiansky, Hutchinson, and Lambropoulos show that the quantity in square brackets is closely equal to 1 so that the final result is

$$\Delta G_C^T = 2V_f \varepsilon^T \sigma_c h \tag{11.26}$$

Thus the results of the stress intensity change calculation agree exactly with the results of the direct energy calculation within the accuracy of the numerical calculations.

An interesting feature of the above treatment is that the initial transformation zone ahead of the crack does not contribute to the toughening. Consider a crack in a specimen without remote applied stress. There is no transformation zone. If a remote stress is applied to give a stress intensity factor just below that required to propagate the crack, a transformation zone develops ahead of the crack as shown in Figure 11.2. Marshall, Evans and Drory(1983) show that the transformations in the angular sector up to $\theta = 60°$ make a negative contribution and that the contribution from $60° < \theta < 180°$ is positive and just cancels, giving zero net toughening. The direct contribution to toughening from transformation comes from the wake.

In a later, more complete treatment, Evans and Cannon (1986) considered not only the possibility of a purely dilatational effect but several other possibilities, as summarized in Table 11.2.

11.4 SHEAR-DILATANT TRANSFORMATION THEORY

As noted earlier, shear stresses should play a role in causing the tetragonal-to-monoclinic transformation as well as dilatational stress. I. W. Chen (1991) has summarized theoretical work by others taking shear stress into account and presented an extension of the theory whose results we briefly summarize here. For a dilatational stress criterion alone Chen obtains

$$\Delta K = \frac{0.2143}{1-\nu}\, \varepsilon^T E h^{1/2} \tag{12.27}$$

which is the McMeeking–Evans result. When a shear-dilatational criterion is used, he obtains

$$\Delta K = \frac{0.4783}{1-\nu}\, \varepsilon^T E h^{1/2} \tag{12.28}$$

TABLE 11.2 Steady State Toughness Values

Net Strain Coupling	Zone Shape	Toughness	Comment
Dilation	Hydrostatic contour	$\frac{0.22 E V_T \varepsilon_{ii}^T w^{1/2}}{(1-\nu)}$	No reversal
Dilation	Hydrostatic contour	$\frac{0.21 E V_t \varepsilon_{ii}^T w^{1/2}}{(1-\nu)}$	Reverses with maximum hysteresis
Dilation	Shear-band profile	$\frac{0.38 E V_T \varepsilon_{ii}^T w^{1/2}}{(1-\nu)}$	No reversal
Uniaxial dilation	Maximum principal stress contour	$0.55 E\, V_T \varepsilon_{ii}^T h^{1/2}$	No reversal
Dilation and relaxed shear strain	Relaxed shear front	$\frac{0.22 E\, V_T \varepsilon_{ii}^T h^{1/2}}{(1-\nu)}$	

Source: After Evans and Cannon (1986).

The stronger toughening effect appears to be in better agreement with experimental data.

11.5 GRAIN-SIZE-DEPENDENT TRANSFORMATION BEHAVIOR

As noted above, the transformation temperature of a tetragonal zirconia particle constrained in a matrix will be altered from that of a free particle. The temperature at which a constrained particle will spontaneously transform is termed the martensitic start temperature M_s. It is commonly observed that the martensitic start temperature depends on the size of the transforming grains. We now summarize a theory of this effect due to Becher and Swain (1992). Adopting their notation, the free energy change associated with the transformation of a tetragonal particle embedded in a matrix is

$$\Delta F_0 = \Delta F_{CH} + \Delta U_e + \Delta U_s - \Delta U_I \tag{11.29}$$

where ΔF_{CH} is the chemical free energy change, $\Delta U_e = E(\epsilon^T)^2$ is the strain energy change, ΔU_s is the surface energy change (taken to be negligible), and

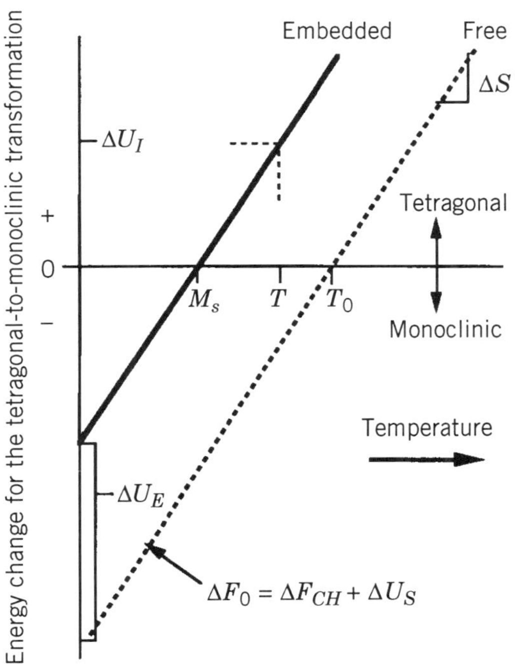

Figure 11.3 Free energy of tetragonal to monoclinic transformation of zirconia. (From P. F. Becher and M. V. Swain, "Grain-Size Dependent Transformation Behavior in Polycrystalline Tetragonal Zirconia," *J. Am. Cerm. Soc.*, Vol. 75, 493–502, 1992. Reprinted by permission of the American Ceramic Society.)

ΔU_I is the interaction term due to the application of a stress. Figure 11.3 shows the free energy change with temperature for a free particle and the curve for an embedded particle shifted vertically by ΔU_e. The chemical free energy term varies with temperature as

$$\Delta F_{CH} = \Delta S^{t \to m}(T_0 - T) \tag{11.30}$$

where $\Delta S^{t \to m}$ is the transformational entropy change and has the value −3.2 J/mol · K below 100 °C. The transformation temperature T_0 of a free particle is reduced to M_s for an embedded particle. Cooling to M_s initiates the transformation without application of stress. Cooling to T intermediate between T_0 and M_s does not initiate transformation; a stress sufficient to produce an interaction term ΔU_I to bring ΔF_0 to zero or less is necessary to initiate transformation according to

$$\Delta F_0 = \Delta S^{t \to m}(M_s - T) - \Delta U_I \tag{11.31}$$

The interaction energy is

$$\Delta U_I = \sigma \varepsilon^T \tag{11.32}$$

where σ is the applied stress and ϵ^T is the dilatant strain. Requiring ΔF_0 to be zero and solving for the stress gives the critical stress for transformation as

$$\sigma_c = \Delta S^{t \to m}(M_s - T)/\varepsilon^T \tag{11.33}$$

Becher and Swain use the result of Marshall, Drory, and Evans, equation (11.13), to obtain

$$\Delta G^T = \frac{2V_T B}{\Delta S^{t \to m}(M_s - T)}(K_I \varepsilon^T)^2 \tag{11.34}$$

From equation (10.21) the critical stress intensity factor is

$$K_C^2 = (K_0 + \Delta K^T)^2 = \frac{E}{(1 - \nu^2)}(G_0 + \Delta G^T) \tag{11.35}$$

where K_0 and G_0 refer to crack extension in the matrix in the absence of transformation and E is Young's modulus. Solving gives

$$\Delta K = \left(\frac{E \, \Delta G/(1 - \nu^2)}{1 + 2\, K_0/(\Delta K)} \right)^{1/2} \tag{11.36}$$

Substituting an average value of 1.25 for $1 + 2K_0/(\Delta K)$ gives

$$\Delta K^T \approx 0.8 \left[\left(\frac{E}{1 - \nu^2} \right) \Delta G^T \right] \tag{11.37}$$

Substituting for ΔG^T from (11.13) and for w from (11.7) gives

$$\Delta K^T = 1.1 \left[\frac{E \, V_T \, B}{(1 - \nu^2) \Delta S^{T \to m} (M_s - T)} \right]^{1/2} (K_I \varepsilon^T) \tag{11.38}$$

with

$$B = \frac{3^{1/2} (1 + \nu)^2}{12\pi} \tag{11.39}$$

This gives a value of ΔK^T that is dependent on the applied stress intensity factor. Taking the case $K_I = K_0$ for crack advance leads to

$$\frac{K_0}{\Delta K^T} = 0.9 \left[\frac{\Delta S^{t \to m} (M_s - T)}{1.6[B/(1 - \nu^2)^2]\, E \, V_T (\varepsilon^T)^2} \right]^{1/2} - 1 \tag{11.40}$$

The value of ΔK^T thus depends on both the temperature and the martensitic start temperature. The half-width of the transformation zone is

$$w^{-1/2} = \left[\frac{\sigma_c (1 - \nu^2)}{K_0 \, B^{1/2}} \right] - \left[\frac{(0.8 \sigma_c \, E \, V_T \, \varepsilon^T)^{1/2}}{K_0} \right] \tag{11.41}$$

Substituting for σ_c gives the dependence of w on the temperature and the martensitic start temperature as

$$w^{-1/2} = \frac{(-\nu^2)[\Delta S^{t \to m} (M_s - T)^{1/2}}{B^{3/2} K_0 \varepsilon^T} \left[[\Delta S^{t \to m} (M_s - T)]^{1/2} - 1.1 \varepsilon^T \left(\frac{B E V_T}{(1 - \nu^2)^2} \right)^{1/2} \right] \tag{11.42}$$

These results show that parameters affecting the martensitic start temperature should have a strong influence on both the toughness achieved at a given temperature and the temperature range over which the transformation makes a contribution to the toughness. The factors influencing the martensitic start temperature include the type and amount of solutes (as these affect the phase diagram) and the size of the tetragonal grains. For example, adding ceria to tetragonal zirconia decreases the M_s temperature. The effect of grain size on transformation temperature is shown in Figure 11.4 from Becher and Swain (1992) for 12 mol % ceria–tetragonal zirconia ceramics. This figure shows a systematic increase in the martensitic start temperature as the grain size is increased from 1.3 μm to 8 μm. Several investigators have discussed the nucleation of monoclinic grains (Heuer, Claussen, Kriven, and Ruhle, 1982; Chen and Chiao, 1983) and local tensile stress has been shown by Hannink (1983) and Hannink and Swain (1983) to promote the transformation.

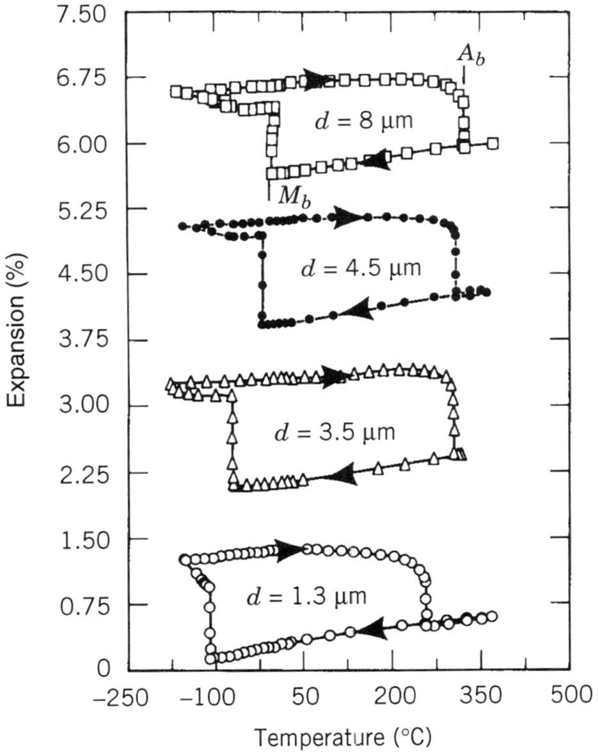

Figure 11.4 Thermal expansion hysteresis in 12 mol % ceria tetragonal zirconia ceramics. (From P. F. Becher and M. V. Swain, "Grain-Size Dependent Transformation Behavior in Polycrystalline Tetragonal Zirconia," *J. Am. Cerm. Soc.*, Vol. 75, 493–502, 1992. Reprinted by permission of the American Ceramic Society.)

Becher and Swain introduce the effect of local tensile stress from thermal expansion by assuming that the internal stress at a distance r from the corner of a grain of diameter d will be given by

$$\sigma_{\mathrm{TEA}} \frac{d}{r} \tag{11.43}$$

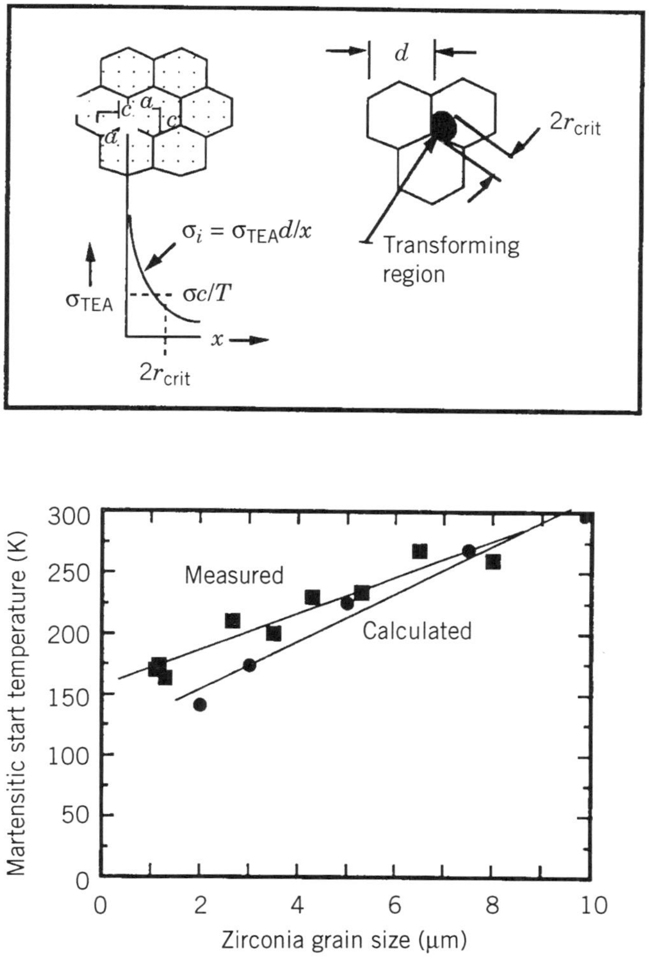

Figure 11.5 *Top:* Local stress concentrations at grain corners due to thermal expansion anisotropy. *Bottom*: The effect of these stress on the martensite start temperature. (From P. F. Becher and M. V. Swain, "Grain-Size Dependent Transformation Behavior in Polycrystalline Tetragonal Zirconia," *J. Am. Cerm. Soc.*, Vol. 75, 493–502, 1992. Reprinted by permission of the American Ceramic Society.)

Qualitatively similar stress behavior was obtained in a detailed analysis by Rao, Tsakalakos, and Cannon (1992). Equation (11.24) for the critical applied stress for transformation then becomes

$$\sigma = \sigma_c^* - \sigma_{\text{TEA}} \frac{d}{r} = \frac{\Delta S^{t \to m}(M_s - T)}{\varepsilon^T} \tag{11.44}$$

where

$$\sigma_c^* = \frac{\Delta S^{t \to m}(M_s^0 - t)}{\varepsilon^T} \tag{11.45}$$

is the critical transformation stress when no internal stress is present and $M_s{}^0$ is the martensitic start temperature in the absence of stress. The martensitic start temperature is then

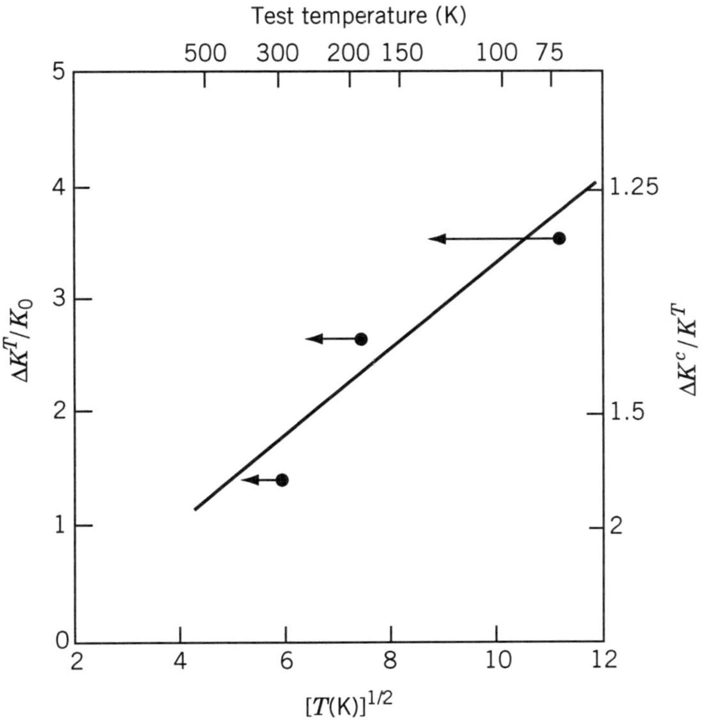

Figure 11.6 Dependence of transformation toughening on temperature. (From P. F. Becher and M. V. Swain, "Grain-Size Dependent Transformation Behavior in Polycrystalline Tetragonal Zirconia," *J. Am. Cerm. Soc.*, Vol. 75, 493–502, 1992. Reprinted by permission of the American Ceramic Society.)

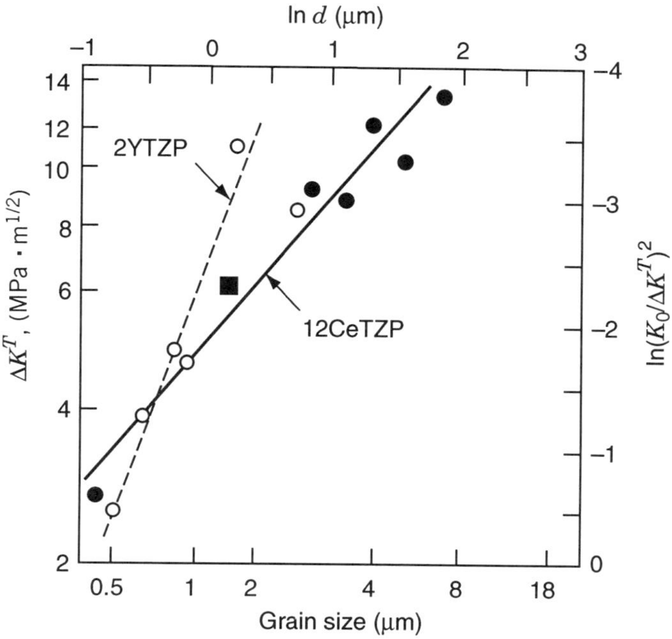

Figure 11.7 Dependence of transformation toughening on grain size. (From P. F. Becher and M. V. Swain, "Grain-Size Dependent Transformation Behavior in Polycrystalline Tetragonal Zirconia," *J. Am. Cerm. Soc.*, Vol. 75, 493–502, 1992. Reprinted by permission of the American Ceramic Society.)

$$M_s = M_s^0 - \sigma_{\mathrm{TEA}} \frac{\varepsilon^T}{r} \Delta S^{t \to m} d \tag{11.46}$$

Using $\sigma_{\mathrm{TEA}} = E\Delta\alpha\,\Delta T$ with $\Delta T = T_{sr} - M_s$, where T_{sr} is the temperature below which thermal stresses cannot relax, and defining

$$a = E\,\varepsilon^T \Delta \frac{\alpha}{r_{\mathrm{crit}}} \Delta S^{t \to m} \tag{11.47}$$

yields

$$M_s = \frac{M_s^0 - aT_{sr}d}{1 - ad} \tag{11.48}$$

For 12 mol % CeO_2 in T-ZrO_2 the result is a nearly linear dependence of M_s on d as shown in Figure 11.5.

The effect of grain size on ΔK^T can be treated similarly by substituting for M_s in terms of d. The result is

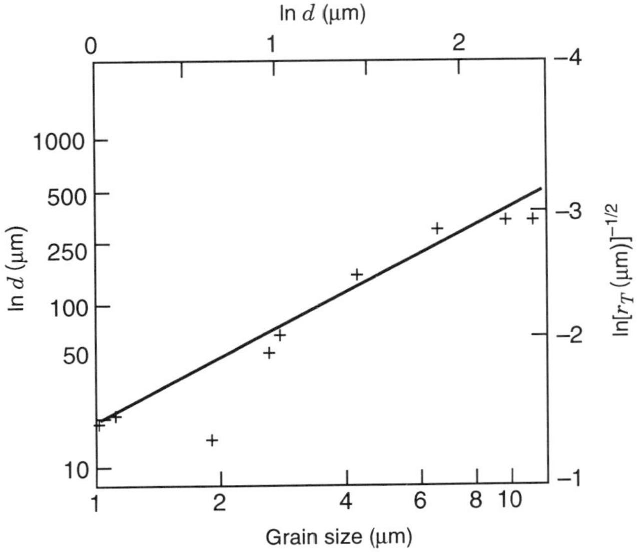

Figure 11.8 Dependence of transformation zone width on grain size. (From P. F. Becher and M. V. Swain, "Grain-Size Dependent Transformation Behavior in Polycrystalline Tetragonal Zirconia," *J. Am. Cerm. Soc.*, Vol. 75, 493–502, 1992. Reprinted by permission of the American Ceramic Society.)

$$\frac{K_0}{\Delta K^T} = 0.9\left(\frac{\Delta S^{t\to m}[(M_s^0 + d\sigma_{TEA}\varepsilon^T/(r\,\Delta S^{t\to m})) - T](1-\nu^2)}{B\,E\,V_T(\varepsilon^T)^2}\right)^{1/2} - 1 \qquad (11.49)$$

This equation predicts that as the temperature decreases below M_s the toughening due to transformation should increase. This equation also predicts an increase in transformation toughening as the grain size increases. These predicted effects are confirmed by the data of Becher and Swain shown in Figures 11.6 and 11.7.

Still another effect predicted by the treatment of Becher and Swain is that the width of the transformed zone should increase as the temperature is decreased. This prediction is confirmed by their data as shown in Figure 11.8.

CHAPTER 12

MECHANICAL PROPERTIES OF CONTINUOUS-FIBER-REINFORCED CERAMIC-MATRIX COMPOSITES

12.1 INTRODUCTION

The mechanical behavior of composites is an essential aspect of the study of the mechanical behavior of ceramics for several reasons. First, polycrystalline ceramics usually involve a grain boundary phase and are thus a two-phase composite. The properties of the ceramic, both fracture and creep, can strongly be influenced by the composite nature of the material. Second, efforts to toughen ceramics have led to the class of ceramic-matrix composites in which a reinforcing phase, usually fibrous but sometimes plate-shaped, is added. Third, ceramic fibers are combined with other materials to make metal-matrix composites or polymer-matrix composites. The basic behavior of fiber-reinforced composites is accordingly treated in this chapter. The term composites will be used to refer to fiber-reinforced materials here.

The potential benefits of reinforcing glass with strong fibers have been recognized for a long time (Sambell, Bowen, and Phillips, 1972) and extensive development has been stimulated by the work of the United Technologies group (Prewo and Brennan, 1980; Prewo, 1989). The development of fibers made predominantly of silicon carbide (Yajima, Hayashi, Omori, and Okamura, 1976) has greatly stimulated work on fiber-reinforced ceramic-matrix composites. A good historical review is given by Lewis (1996) and an overall general treatment by Chawla (1993). Lewis' paper is especially valuable for its discussion of the need to consider all failure modes and to achieve more cost-effective ceramic-matrix composites.

The conceptual mechanisms of toughening of ceramic-matrix composites have been discussed by R. W. Rice (1981, 1985). Brennan and Prewo (1982) demonstrated silicon carbide-reinforced glass-ceramic-matrix composites with high strength and toughness. Extensive experimental and theoretical developments followed. Basic references to theory include Kelly and MacMillan (1986), Aveston, Cooper, and Kelly (1971), Marshall, Cox, and Evans (1985), Budiansky, Hutchinson, and Evans (1986), Evans and Marshall (1989), Thouless, Sbaizero, Sigl, and Evans (1989), Cao, Bischoff, Sbaizero, Ruhle, Evans, Marshall, and Brennan (1990), and Campbell, Ruhle, Dalgleish, and Evans (1990). Useful reviews of toughening, including toughening in composites, can be found in Evans (1990), Shah and Ouyang (1991), and Steinbrech (1992).

12.2 ELASTIC BEHAVIOR OF COMPOSITES

Consider first a composite made up of a matrix containing continuous fibers aligned in the direction of applied stress as shown in Figure 12.1. We shall deal later with the question of the stress necessary to propagate the crack; at present we assume the crack is not present and examine the elastic behavior under tensile stress parallel to the fibers. A load P_c is applied. Of this load P_f is carried by the reinforcing fibers and P_m by the matrix, giving

$$P_c = P_f + P_m \tag{12.1}$$

Using σ and A to represent stress and area and using subscripts c, f, and m to indicate the composite, fibers, and matrix, respectively, this load equation becomes

$$\sigma_c A_c = \sigma_f A_f + \sigma_m A_m \tag{12.2}$$

or

$$\sigma_c = \sigma_f \frac{A_f}{A_c} + \sigma_m \frac{A_m}{A_c} \tag{12.3}$$

Letting V denote volume fraction and Vol denote volume and noting that the area fractions are equal to the volume fractions for either a composite made up of aligned continuous fibers or a composite made up of randomly dispersed short fibers, that is,

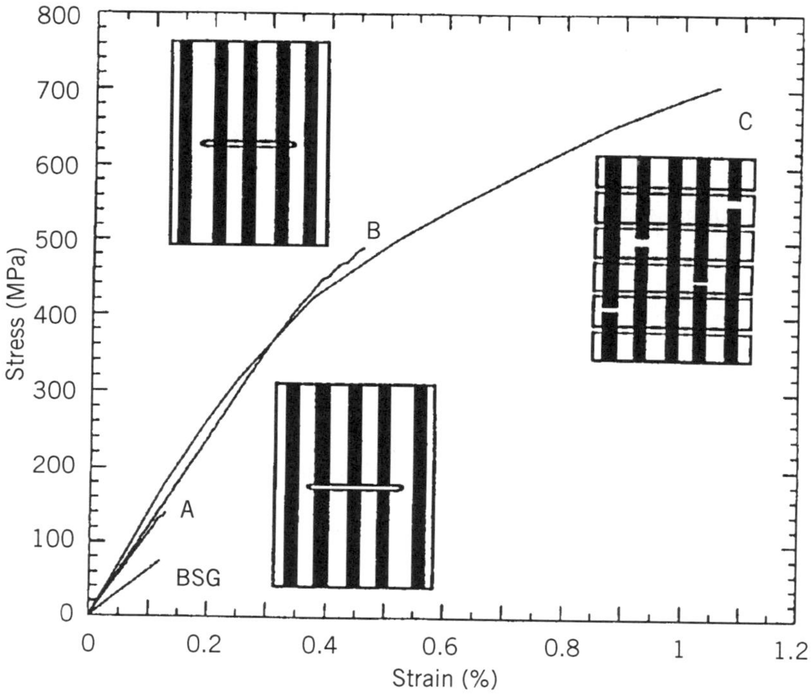

Figure 12.1 Stress–strain curves for composites. The curve labeled BSG is for unreinforced borosilicate glass. A is for a composite with a strong bond to the fibers. B and C are for composites with weaker bonds. (From P. G. Karandi-kar, T.-W. Chou, and A. Parvizi-Majidi, "Mechanical Properties," in *Handbook on Continuous Fiber Reinforced Ceramic Matrix Composites*, R. L. Lehman, technical ed., S. K. El-Rahaiby, managing ed., and J. B. Wachtman, American Ceramic Society ed., Purdue University/American Ceramic Society, 1996. Reprinted by permission of the American Ceramic Society.)

$$\frac{A_f}{A_c} = \frac{Vol_f}{Vol_c} = V_f \tag{12.4}$$

and

$$\frac{A_m}{A_c} = \frac{Vol_m}{Vol_c} = V_m \tag{12.5}$$

gives

$$\sigma_c = \sigma_f V_f + \sigma_m V_m \tag{12.6}$$

It is interesting to see how the average stress in the composite compares to the stress in the matrix. This can be done by noting that $V_f + V_m = 1$ and by equating the strains in the composite, fibers, and matrix, $\epsilon_c = \epsilon_f = \sigma_f/E_f = \sigma_m/E_m = \epsilon_m$, to obtain

$$\sigma_c = \sigma_f V_f + \sigma_m (1-V_f) = \sigma_m[1 + V_f(\frac{E_f}{E_m} - 1)] \tag{12.7}$$

The Young's modulus of the composite is obtained by writing equation (12.6) as

$$E_c\varepsilon_c = E_f \varepsilon_f V_f + E_m \varepsilon_m V_m \tag{12.8}$$

and using the equality of strains to get

$$E_c = E_f V_f + E_m V_m \tag{12.9}$$

Thus the elastic modulus of the composite is given simply by the rule of mixtures.

The ratio of the total load carried by the composite to the load carried by the fibers can be expressed as

$$\frac{P_c}{P_f} = \frac{\sigma_c A_c}{\sigma_f A_f} = \frac{E_c}{E_f}\frac{1}{V_f} \tag{12.10}$$

Using the expression for the elastic modulus of the composite gives

$$\frac{P_c}{P_f} = 1 + \frac{E_m}{E_f}\left(\frac{1}{V_f} - 1\right) \tag{12.11}$$

The equations of this section allow one to make predictions about the behavior of uniaxial composites, but they should be used with the understanding that they are approximate because no account has been taken of the variation of stress and strain locally around the ends of fibers. Also, the above treatment assumes an uncracked matrix. We take up the question of matrix cracking in the next section.

The above discussion of elastic deflection of composites with aligned, continuous reinforcements is adequate for our present purpose. However, we note that a more complete discussion of elastic behavior of such composites requires five independent elastic constants characteristic of transversely isotropic solids. These constants can be chosen in various ways. A useful choice (Christensen and Waal, 1972) is E_{11}, Young's modulus in the direction of fiber alignment; ν_{12}, Poisson's ratio for stress in the aligned direction; K_{23}, the plane strain bulk modulus in the plane containing the transverse axes; G_{12}, the shear modulus for a plane including the aligned direction; and G_{23}, the shear modulus for distortion in the transverse direction. The relation of the elastic moduli of the composite to the elastic properties of the fibers and matrix has been treated by Hashin and Rosen (1964).

12.3 FRACTURE BEHAVIOR OF COMPOSITES WITH CONTINUOUS, ALIGNED FIBERS

Much of the interest in ceramic-matrix composites arises from the improved fracture behavior that can be achieved with fiber reinforcement. The fracture behavior is illustrated in Figure 12.1 from Karandikar, Chou, and Parvizi-Majidi (1994). Experiments show very different behavior depending on whether the brittle matrix is reinforced or not and on whether the reinforcing fibers are coated with a nonreactive material. The stress–strain curve labeled BSG is for unreinforced borosilicate glass. The behavior is elastic up to sudden and complete failure at a strain of about 0.1%. The stress–strain curve labeled A is for borosilicate glass reinforced with uncoated Nicalon SiC parallel and continuous fibers. Again the behavior is elastic until complete brittle failure occurs at about the same 0.1% strain but at appreciably higher stress. The higher fracture stress is reasonable from equation (12.7), which predicts that the stress in the composite just before failure is greater than the stress in the matrix as long as $E_f > E_m$ and $V_f > 0$. Brittle failure of the composite occurs when the stress in the matrix reaches the fracture stress for the matrix provided that the sharp crack propagates through the fiber as shown in the lower drawing in Figure 12.1. If, however, the crack passes around the fiber rather than through it, a bridged crack results as shown in the upper left-hand drawing in Figure 12.1. This is the case for the stress–strain curve labeled B, which is for the same composite except that the fibers have been coated with a nonreactive coating. Much higher stress and strain values at failure are evident. Finally, the curve labeled C is for the case of multiple cracks and some fiber failure accompanied by energy dissipation as the fibers are drawn out of the matrix. Nearly ten times as much strain and six times as much stress at failure occur than for the unreinforced glass.

For cases B and C, where the fibers continue to carry the load after partial or complete matrix failure, the fibers must have sufficient strength. The requirement is that

$$V_f \sigma_{fu} \geq V_f \sigma_{fi} + V_m \sigma_{mu} \tag{12.12}$$

or

$$V_f \geq \frac{1}{1 + \dfrac{\sigma_{fu}}{\sigma_{mu}} - \dfrac{\sigma_{fi}}{\sigma_{mu}}} \tag{12.13}$$

where the second subscript u indicates ultimate strength and σ_{fi} is the stress in the fiber just before failure of the matrix. Assuming equality of strain in the matrix and fiber just before matrix failure gives

$$\sigma_{fi} = \frac{E_f}{E_m}\sigma_{mu} \tag{12.14}$$

If the Poisson ratios of the fiber and matrix are equal, the minimum volume fraction for the fibers to carry the load for a fully cracked matrix is then

$$V_f \geq \frac{1}{1 + \dfrac{\sigma_{fu}}{\sigma_{mu}} - \dfrac{E_f}{E_m}} \tag{12.15}$$

As an example, lithium aluminosilicate glass with E_m = 88 GPa and σ_{mu} = 0.17 GPa reinforced with SiC with E_f = 180 GPa and σ_{fu} = 2.5 GPa requires $V_f \geq 7.5\%$.

Matrix cracking without fiber failure in the plane of the crack can occur only if the concentrated stress at the head of the crack does not propagate the crack through the fiber. The alternative process is crack propagation along the interface between the fiber and the matrix at right angles to the primary crack plane. That is, partial debonding of the fiber from the matrix is required for the fiber to remain unbroken as the matrix crack passes around it. He and Hutchinson (1989) have shown that this requires $\gamma_i/\gamma_f \leq 0.25$, where γ_i is the fracture surface energy for the fiber–matrix interface and γ_f is the fracture surface energy for crack propagation through the fiber.

The above treatment assumed a single value for the ultimate strength of the fibers, although in reality a distribution of values is observed. Evidently the process of fracture in a composite involves many more factors than in a monolithic material and it is to be expected that a complete theory of fracture of composites will be complex. A minimal list of factors that should enter into such a theory has been given by Zok, Evans, and Mackin (1996) and is repeated in Table 12.1.

12.4 COMPLETE MATRIX CRACKING OF COMPOSITES WITH CONTINUOUS, ALIGNED FIBERS

It is easiest to consider a composite made up of a matrix reinforced with parallel, continuous fibers of uniform radius. The usual case is that of fibers which are stronger than the matrix. Under an increasing tensile stress parallel to the fibers, the matrix will eventually crack. Aveston, Cooper, and Kelly (1971) treated the matrix cracking stress as a property of the composite and derived an expression for it. Their treatment was subsequently extended to a

TABLE 12.1 Constituent Properties of Ceramic-Matrix Composites

Property	Typical Values
Matrix fracture energy, γ_m	5–50 J/m^2
Matrix Young's modulus, E_m	50–200 mPa
Matrix Poisson's ratio, ν	0.25–0.30
Fiber Young's modulus, E_f	180–415 mPa
Fiber Poisson's ratio, ν_f	0.25–30
Fiber strength, σ_{fu}	1.2–3 GPa
Fiber Weibull parameter, m	5–20
Fiber–matrix debond energy, γ_i	0–5 J/m^2
Fiber–matrix sliding stress, τ	1–200 mPa
Fiber–matrix misfit strain, $\delta\epsilon$	0–2 × 10^{-3}

Source: F. W. Zok, A. G. Evans, and T. J. Mackin, "The Mechanical Properties of Fiber-Reinforced Ceramic-Matrix Composites," in *Handbook on Continuous Fiber Reinforced Ceramic Matrix Composites,* R. L. Lehman, technical ed., S. K. El-Rahaiby, managing ed., and J. B. Wachtman, American Ceramic Society ed., Purdue University/American Ceramic Society, 1996. Reprinted by permission of the American Ceramic Society.

full fracture mechanics treatment to deal with the stress necessary to propagate a crack of variable length (Marshall, Cox, and Evans, 1985; McCartney, 1987). The earlier Aveston–Cooper–Kelly result for matrix cracking stress was shown to be the limit for long cracks of the stress necessary to propagate a crack.

The fiber stress just before matrix cracking, σ_{fi}, can be related to the matrix stress just before cracking, σ_{mu}, from the equality of strains to give

$$\sigma_{fi} = \sigma_{mu} \frac{E_f}{E_m} = \varepsilon_{mu} E_f \tag{12.16}$$

After a crack passes through the matrix but leaves the fibers unbroken, the fiber stress in the plane of the crack becomes

$$\sigma_{fc} = \sigma_{fi} + \sigma_{mu} \frac{V_m}{V_f} = \sigma_{fi} \left(1 + \frac{E_m V_m}{E_f V_f}\right) = \sigma_{fi} (1 + \eta) \tag{12.17}$$

where the latter equalities come from (12.16) and defining

$$\eta = \frac{E_m V_m}{E_f V_f} \tag{12.18}$$

The fiber–matrix interface shown in Figure 12.2 is assumed to have a shear strength τ. Near the matrix crack the fiber will slip along a length x such that the total interface force on the fiber along this distance causes a fiber stress equal to the increase of the fiber stress calculated above. Thus

$$\frac{2\pi r \tau x}{\pi r^2} = (\sigma_{fc} - \sigma_{fi}) = \eta \sigma_{mu} \frac{E_f}{E_m} = \eta\, \varepsilon_{mu}\, E_f \tag{12.19}$$

where ϵ_{mu} is the strain in the matrix just prior to cracking. Thus the fiber slips along a distance

$$x = \frac{E_f \varepsilon_{mu} \eta r}{2\tau} = \left(\frac{V_f}{V_m}\right)\frac{\sigma_{mu} r}{2\tau} = \left(\frac{E_m V_m}{V_f}\right)\frac{\varepsilon_{mu}}{2\tau} r \tag{12.20}$$

The matrix stress is thus zero at the crack and rises linearly with distance from the matrix crack to reach its original value σ_m at a distance x, as shown in Figure 12.2. Correspondingly, the fiber stress has the value σ_{fc} at the crack and drops linearly with distance from the matrix crack to reach its original value σ_{fi} at a distance x.

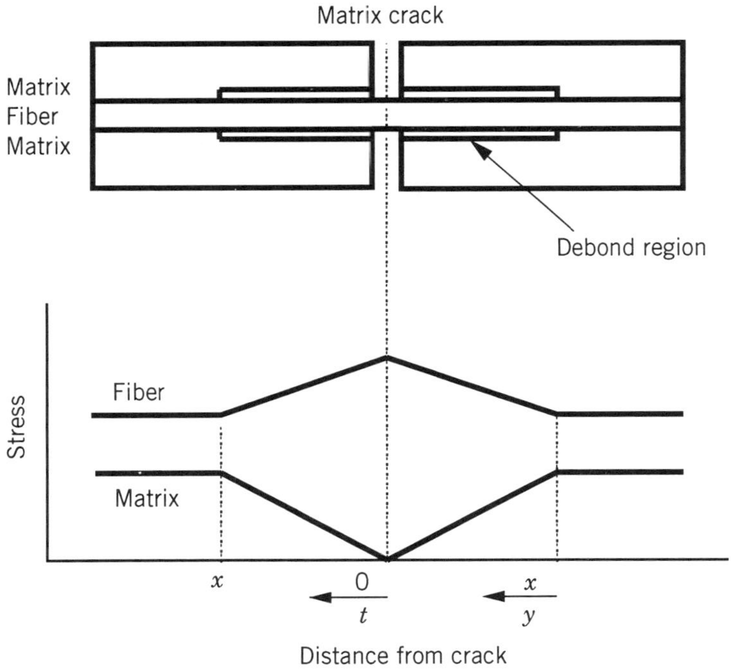

Figure 12.2 Debonding and stress distribution in composite.

One can now consider the behavior of the composite under constant load, evaluate the work done and energy conversions that take place, and take an energy balance as in the Griffith theory to obtain the matrix cracking strain and stress. Five changes in energy must be considered: (1) the work, ΔW, done by the external force as the composite expands as the crack runs; (2) the decrease, ΔU_m, in elastically stored energy in the matrix due to stress reduction within the distance x; (3) the increase, ΔU_f, in elastically stored energy in the fiber due to stress increase within the distance x; (4) the dissipation, ΔU_s, of energy associated with sliding against the friction force; and (5) the energy, $2\gamma V_m$, required for creation of the crack surface. We consider each of these energies as energy per unit area perpendicular to the fibers.

The work done by the external force can be obtained from considering the motion of the composite against the force per unit area of composite, $A_f\sigma_f/A_c = V_f\sigma_f$, as the strain in the fibers changes from the initial to final value. At the crack the change in strain will be

$$\Delta\varepsilon = \frac{\sigma_{fc} - \sigma_{fi}}{E_f} = \frac{\eta\, \sigma_{fi}}{E_f} = \eta\, \varepsilon_{mu} \tag{12.21}$$

The average change in strain over the distance x is half this value. The motion is equal to the average change in strain times x or

$$\text{motion} = \frac{x\eta\, \varepsilon_m}{2} \tag{12.22}$$

The work done per unit area is the force per unit area times distance times two to take into account both sides of the crack or, using (12.16), (12.17), and (12.20),

$$\Delta W = 2\left(\frac{x\eta\, \varepsilon_{mu}}{2}\right) V_f \sigma_{fc} = x\eta\, \varepsilon_{mu}\, V_f (1+\eta)\varepsilon_{mu}\, E_f = E_f V_f \varepsilon_{mu}^2\, x\, \eta(1+\eta)$$

$$= \frac{E_f E_m V_m \varepsilon_{mu}^3}{2\tau}\, r\, \eta\, (1+\eta) \tag{12.23}$$

Here the work done is taken as positive since it provides additional energy that could contribute to crack opening.

The matrix undergoes a relaxation leading to a loss in elastically stored energy. This energy change is taken as positive in terms of being energy available to go into crack creation. The matrix stress at the crack surface is zero and rises linearly to a value σ_m at a distance x so that the stress at a distance t from the crack will be

$$\sigma_m(t) = \sigma_{mu} \frac{t}{x} \quad (\text{for } 0 \leq t \leq x) \tag{12.24}$$

The elastically stored energy at any point t is

$$\frac{\sigma(t)\,\varepsilon(t)}{2} = \frac{[\sigma(t)]^2}{2E_m} \tag{12.25}$$

The original energy stored per unit area in the two regions of thickness x under uniform stress σ_m is

$$\frac{xV_m}{E_m}(\sigma_m)^2 \tag{12.26}$$

The energy after relaxation stored in the two regions of thickness x is

$$\frac{xV_m}{E_m}(\sigma^2)_{\text{ave}} \tag{12.27}$$

where the average or the square of the stress is taken over the length x and is

$$(\sigma^2)_{\text{ave}} = \frac{1}{x}\int_0^x \left(\sigma_{mu}\,\frac{t}{x}\right)^2 dt = \frac{(\sigma_{mu})^2}{3} \tag{12.28}$$

The resulting value of the decrease in elastic energy stored in the matrix is

$$\Delta\, U_m = \frac{2}{3} E_m\, V_m\, (\varepsilon_{mu})^2\, x = \frac{E_f\, E_m V_m}{3\tau}\, \varepsilon_{mu}^3\, r\, \eta \tag{12.29}$$

where the decrease is taken as positive because it is energy available to create a new crack surface.

After matrix cracking the fiber experiences a stress from (12.17) given by

$$\sigma_{ff} = \sigma_{fi} + \frac{y(\sigma_{fc} - \sigma_{fi})}{x} = \sigma_{fi} + \eta\, \sigma_{fi}\, \frac{y}{x} \quad (\text{for } 0 \leq y \leq x) \tag{12.30}$$

Where $y = t - x$ runs from 0 to x, as shown in Figure 12.2. The change in elastically stored energy in the fibers in the two regions of thickness x adjacent to the crack that is available to propagate the crack is

$$\Delta\, U_f = \frac{2V_f x}{2E_f}\, [(\sigma_{fi}^2) - (\sigma_{ff}^2)_{\text{ave}}] \tag{12.31}$$

The average of the stress squared is taken in the same manner as was done above for the matrix

$$(\sigma_{ff}^2)_{ave} = \frac{1}{x}\int_0^x \left(\sigma_{fi} + \eta\sigma_{fi}\frac{y}{x}\right)^2 dy \tag{12.32}$$

giving

$$(\sigma_{ff}^2)_{ave} = (\sigma_{fi})^2 \left[1 + \eta + \frac{\eta^2}{3}\right] \tag{12.33}$$

The result for the change in elastic energy in the fibers is

$$\Delta U_f = -E_f V_f (\varepsilon_{mu})^2 x \eta \left(1 + \frac{\eta}{3}\right) = -\frac{E_f E_m V_m}{2\tau} \varepsilon_{mu}^3 r \eta \left(1 + \frac{\eta}{3}\right) \tag{12.34}$$

where the minus sign is appropriate because the increase in energy stored in the fibers reduces the energy available for a new crack surface.

As the matrix relaxes within the distance x from the crack after cracking and the fiber picks up the load, the matrix surface layer slides back while the fiber slides forward against the frictional force τ so that energy is dissipated. The relative displacement of the fiber and matrix at a point y can be calculated by noting that $\Delta\epsilon_f - \Delta\epsilon_m = du/dy$, where u is displacement. The change in strain resulting from the relaxation of stress in the matrix is, using (12.16) and (12.30),

$$\Delta\varepsilon_f - \Delta\varepsilon_m = \frac{\Delta\sigma_f}{E_f} - \frac{\Delta\sigma_m}{E_m} = \frac{\sigma_{ff} - \sigma_{fi}}{E_f} - \frac{\Delta\sigma_m}{E_m} = \frac{\eta\sigma_{fi}}{E_f}\frac{y}{x} - \left(\frac{-\sigma m}{E_m}\frac{y}{x}\right)$$

$$= \varepsilon_{mu}(1 + \eta)\frac{y}{x} \tag{12.35}$$

The displacement as a function of y then is

$$u(y) = \int_0^y \Delta\varepsilon\, dy = \varepsilon_{mu}(1+\eta)\int_0^y \frac{y\,dy}{x} = \frac{\varepsilon_{mu}(1+\eta)}{2x} y^2 \tag{12.36}$$

A segment dy experiences a force $2\pi r\tau\, dy$ and dissipates energy $2\pi r\tau\, u(y)\, dy$ The energy dissipated per area πr^2 at the interface is then

$$\Delta U_s = \frac{2V_f}{2\pi r^2}\int_0^x 2\pi r \tau \left(\frac{\varepsilon_{mu}(1+\eta)}{2x}\right) y^2 dy \tag{12.37}$$

The result is, using (12.20),

$$\Delta U_s = \frac{2 V_f \tau \varepsilon_{mu} x^2 (1+\eta)}{3r} = \frac{E_f E_m V_m}{6\tau} \varepsilon_{mu}^3 r \eta (1+\eta) \tag{12.38}$$

The Griffith energy balance condition then is

$$2\gamma V_m + \Delta U_s \leq \Delta W + \Delta U_m + \Delta U_f \tag{12.39}$$

which reduces to

$$2\gamma V_m + \Delta U_s \leq \frac{2}{3} E_c \varepsilon_m^2 x \eta \tag{12.40}$$

This in turn becomes

$$\varepsilon_{mu} \leq \left[12\tau \gamma_m \left(\frac{V_f^2}{rV_m}\right) \left(\frac{E_f}{E_c E_m^2}\right) \right]^{1/3} \tag{12.41}$$

This is equivalent to a stress on the composite, σ_{cmc}, required for matrix cracking of

$$\sigma_{cmc} = E_c \varepsilon_{mu} = \left[\frac{12\tau \gamma V_f^2 E_f E_c^2}{r V_m E_m^2} \right]^{1/3} \tag{12.42}$$

12.5 PROPAGATION OF SHORT, FULLY BRIDGED CRACKS

The above treatment of Aveston, Cooper, and Kelly (1971) leads to a strength for matrix cracking that is not explicitly dependent on the size of flaws. Their treatment does not start with an initial flaw size nor does it consider the condition to propagate it as a function of size. Instead, their treatment assumes a single value of matrix strength and corresponding maximum matrix strain. Marshall, Cox, and Evans (1985) and McCartney (1987) considered the propagation of a crack of given initial size in the matrix. We summarize the treatment of Marshall, Cox, and Evans.

A long, fully bridged crack is shown in Figure 12.3. The length is c and the crack opening displacement far from the tip is $2u_0$. This limiting separation is approached at a distance c_0 from the tip. For cracks with $c > c_0$ the matrix cracking stress should be that calculated by Aveston, Cooper, and Kelly. For $c < c_0$ the stress for crack propagation should depend on the crack length as for monolithic materials. The crack in the composite can be considered to be formed as a crack in a monolithic material (i.e., cutting through the fibers as well as the matrix) followed by tractions $T(x)$ for each fiber to pull fiber ends back to their original positions, where x is the distance from the open end of the crack. One can approximate the effect of these tractions on the crack surface by replacing them with a continuous stress $p(x) = T(x)V_f$. The crack surfaces then experience a total stress of $\sigma_\infty - p(x)$, where σ_∞ is the distant applied stress. The total stress intensity factor for a straight-through crack in a homogeneous material with the average properties of the composite is then given by equation (5.63) as

$$K^L = 2\left(\frac{c}{\pi}\right)^{1/2} \int_0^1 \frac{[\sigma_\infty - p(X)]dX}{(1 - x^2)^{1/2}} \tag{12.43}$$

where $X = x/c$. For a penny-shaped crack the stress intensity factor is

$$K^L = 2\left(\frac{c}{\pi}\right)^{1/2} \int_0^1 \frac{[\sigma_\infty - p(X)]XdX}{(1 - x^2)^{1/2}} \tag{12.44}$$

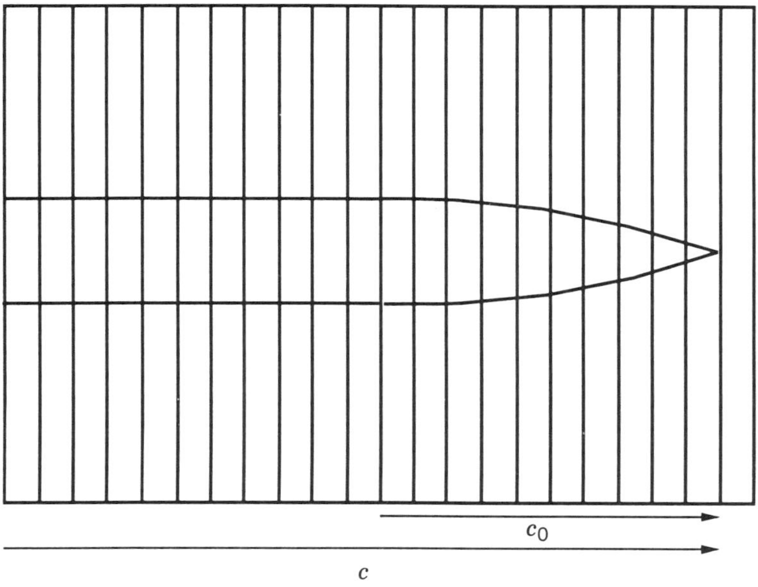

Figure 12.3 Fully bridged matrix crack in composite.

The stress intensity factor in the matrix, K^M, scales with the Young's modulus so that

$$K^L = K^M \frac{E_c}{E_m} \tag{12.45}$$

The condition for crack growth in the matrix is that K^M attain a critical value ($K_C{}^M$) so that the condition for matrix crack growth can be expressed as

$$K^L = K_C^L = K_C^M \frac{E_c}{E_m} \tag{12.46}$$

An analysis similar to that of the Aveston–Cooper–Kelly treatment gives the value of the stress p as a function of the crack half-opening u as

$$p = 2[u\,\tau\, V_f^2 E_f(1+\eta)/r]^{1/2} \tag{12.47}$$

The crack half-opening is given by Sneddon and Lowengrub (1969) for a straight crack as

$$u(X) = \frac{4(1-\nu^2)c}{\pi E_c} \int_X^1 \frac{s}{(s^2 - X^2)^{1/2}} \int_0^1 \frac{[\sigma_\infty - p(t)]dt}{(s^2 - t^2)^{1/2}}\, ds \tag{12.48}$$

and for a penny-shaped crack as

$$u(X) = \frac{4(1-\nu^2)c}{\pi E_c} \int_X^1 \frac{1}{(s^2 - X^2)^{1/2}} \int_0^1 \frac{[\sigma_\infty - p(t)]t\,dt}{(s^2 - t^2)^{1/2}}\, ds \tag{12.49}$$

Solving for the stress to propagate short cracks thus requires solving the equations for p and u followed by solving for σ_∞ from the K^L equation with K^L set equal to $K_C{}^L$. Marshall, Cox, and Evans (1985) performed a numerical solution and also obtained an approximate analytical solution. The latter gives a good approximation and also provides physical insight. The approximate analytic solution is obtained by assuming that the crack profile for $x < c - c_0$ is that of a simple, unbridged crack with the same value of K^L. That is, the crack opening displacement is assumed to be

$$u(x) = \frac{2(1-\nu^2)K^L c^{1/2}(1 - x^2/c^2)^{1/2}}{E_c \pi^{1/2}} \tag{12.50}$$

The stress is then

$$p(x) = [\alpha K^L c^{1/2}(1-x^2/c^2)^{1/2}]^{1/2} \tag{12.51}$$

where

$$\alpha = \frac{8(1-\nu^2)\,\tau\, V_f E_f (1+\eta)}{E_c r\, \pi^{1/2}} \tag{12.52}$$

This solution for $u(x)$ is exact for x near c but is not correct for small x because the crack opening approaches a constant value for a bridged crack. This value is obtained by putting $p = \sigma_\infty$ in (12.47) to obtain

$$u_0 = \frac{\sigma_\infty^2 r}{4\tau V_f^2 E_f (1+\eta)} \tag{12.53}$$

The transition crack length c_0 is obtained by taking $u = u_0$ and $x = 0$ in the approximate equation for u, (12.50), and is

$$c_0 = \frac{\sigma_\infty^4}{\alpha^2 (K^L)^2} \tag{12.54}$$

For fully bridged cracks longer than c_0 the approximate crack opening is constant for $x < c - c_0$ and is given by the above approximate near crack tip solution for $x > c - c_0$. Using this approximate crack opening displacement in (12.43) and (12.44), the stress intensity factor becomes

$$K^L = \Omega\, \sigma_\infty\, c^{1/2} - I \left(\frac{4\alpha}{\pi}\right)^{1/2} (K^L)^{1/2}\, c^{3/4} \tag{12.55}$$

where Ω and I are dimensionless constants. For a straight crack $\Omega = \pi^{1/2}$ and $I = 1.20$, while for a penny-shaped crack $\Omega = 2/\pi^{1/2}$ and $I = 2/3$. Substituting $K^L = K_C^L$ gives the stress necessary to propagate a crack with $c \leq c_0$ as

$$\sigma_\infty = \frac{K_C^L}{\Omega c^{1/2}} + \left(\frac{4\,\alpha\, K_C^L I^2}{\pi\, \Omega^2}\right)^{1/2} c^{1/4} \tag{12.56}$$

This can be put in normalized form as

$$\frac{\sigma}{\sigma_{\text{norm}}} = \left(\frac{1}{3}\right)\left(\frac{c}{c_{\text{norm}}}\right)^{-1/2} + \left(\frac{2}{3}\right)\left(\frac{c}{c_{\text{norm}}}\right)^{1/4} \tag{12.57}$$

where

$$c_{\text{norm}} = \left(\frac{\pi K_C^L}{I^2 \alpha} \right)^{2/3} \tag{12.58}$$

and

$$\sigma_{\text{norm}} = \left(\frac{6\, I^{2/3}}{\pi^{1/2} \Omega} \right) \left(\frac{(1 - \nu^2)(K_C^M)^2\, \tau\, E_f\, v_f^2\, V_m\, (1 + \eta)^2}{E_m r} \right)^{1/3} \tag{12.59}$$

For $c > c_0$ the stress to propagate a matrix crack is independent of crack length. Marshall, Cox, and Evans show that for straight cracks $\sigma_0 = 1.02\sigma_m$ and $c_0 = 1.88\ c_m$, while for penny-shaped cracks $\sigma_0 = \sigma_m$ and $c_0 = c_m$.

The stress to propagate a fully bridged matrix crack in a composite with continuous, aligned fibers according to this approximate solution is shown in Figure 12.4 as the curve labeled MCE, fully bridged. The stress to propagate a crack in the unreinforced matrix is shown in comparison. Both plots are in terms of σ_∞/σ_m as a function of c/c_m. The stress for the unreinforced matrix decreases without limit as the crack length increases according to the usual fracture mechanics result for material with a resistance to fracture that is independent of crack length. The stress required for matrix crack propagation in the composite initially decreases with crack length but levels off at the Aveston–Cooper–Kelly (ACK) stress for long cracks. This provides a degree of insensitivity to flaws for the composite. That is, if stresses on the composite

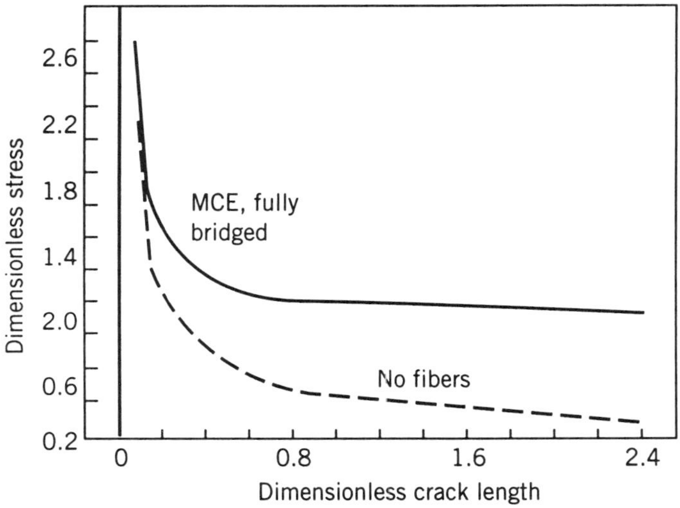

Figure 12.4 Stress to propagate a fully bridged matrix.

are kept below the ACK stress the composite will not fail from any size of matrix crack.

The above results can also be expressed in terms of a toughening due to bridging (Marshall and Evans, 1986). That is, the stress intensity factor at the crack tip can be written as

$$K^L = K_\infty - K_p \tag{12.60}$$

where K_p is the contribution due to the fibers and the sign has been chosen to make it a positive quantity. At matrix fracture $K^L = K_C$ and for a penny-shaped crack

$$K_\infty = \Omega\, \sigma_\infty\, c^{1/2} \tag{12.61}$$

and

$$K_p = \left(\frac{16\alpha}{9\pi}\right)^{1/2} K_C^{1/2}\, c^{3/4} \tag{12.62}$$

The applied stress to propagate a matrix crack for $c < c_0$ can thus also be written approximately as

$$\sigma_c = \frac{K_C}{\Omega\, c^{1/2}} + \left(\frac{16\alpha K_C}{9\pi\Omega^2}\right)^{1/2} c^{1/4} \tag{12.63}$$

12.6 PROPAGATION OF PARTIALLY BRIDGED CRACKS

The treatment of the previous section assumes fully bridged cracks. In practice, the fibers may not be strong enough to remain unbroken over the full length of the crack and so may remain unbroken only some distance back from the crack front. Variation in fiber strength introduces a further complication, which we will consider later. In the present section the treatment of Marshall and Evans (1986) is followed in which a uniform strength is assumed leading to a region of crack bridging extending back from the crack tip a distance d until a critical length d^* is reached beyond which there will be a completely unbridged portion further from the crack tip, as shown in Figure 12.5. For this case at fracture

$$K_p = \left(\frac{16\alpha}{9\pi}\right)^{1/2} K_C^{1/2}\, [d^*(2 - d^*/c)]^{3/4} \tag{12.64}$$

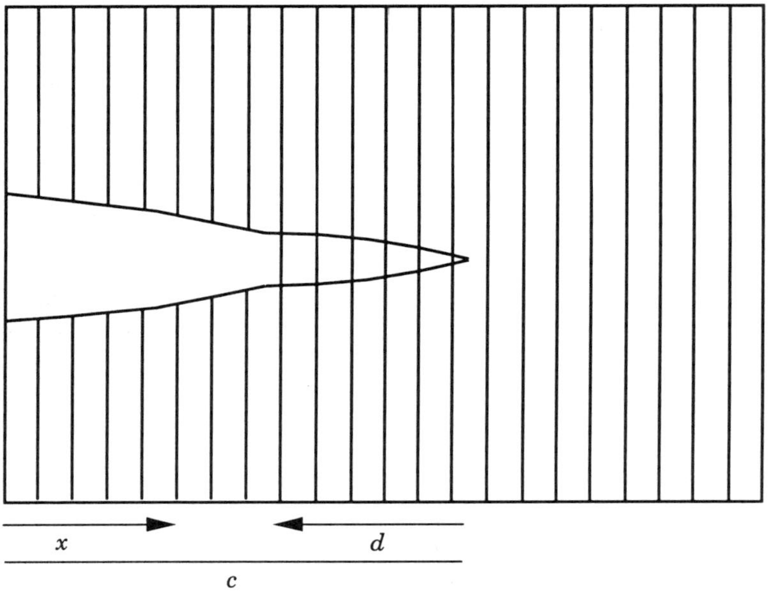

Figure 12.5 Partially bridged crack.

The value of d is obtained by requiring $p(x) = SV_f$ at $x = c - d^*$, which gives

$$d^*(2 - d^*/c) = \frac{(S\,V_f)^4}{(\alpha K_C)^2} \qquad (12.65)$$

This gives a toughening contribution from the partial bridging of

$$K_P = \left(\frac{4}{3\,\pi^{1/2}}\right)\frac{(SV_f)^3}{\alpha K_C} \qquad (12.66)$$

and a dimensionless stress for matrix crack propagation of a partially bridged crack of

$$\frac{\sigma}{\sigma_m} = \frac{1 + 2(SV_f\,/\,\sigma_M)^3}{3(c\,/\,c_m)} \qquad (12.67)$$

Note that in this case "partial bridging" means that the length of the bridged portion of the crack has been determined by the applied stress. A crack that was initially completely unbridged (such as a sawn notch) would propagate some distance before the bridged length built up to the extent that the fibers at the rear would fail. The normalized stress for such a partially bridged crack is shown in Figure 12.6. A short matrix crack that was initially fully bridged

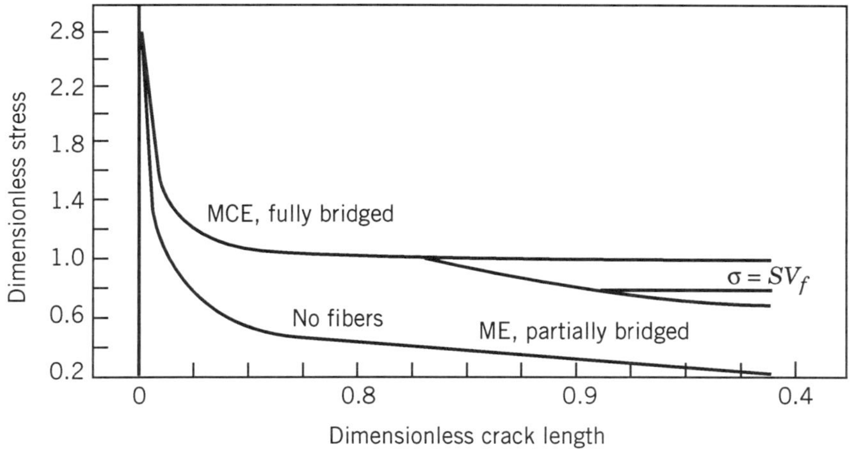

Figure 12.6 Stress to propagate a partially bridged matrix crack.

would extend under the stress given by the MCE plot. As progressively longer, fully bridged cracks are considered, the MCE plot continues to apply as long as $\sigma/\sigma_m < SV_f/\sigma_m$ and would continue to be fully bridged so that the horizontal portion of the MCE plot would be reached. If, however, $SV_f/\sigma_m < 1$ a point on the MCE plot will be reached where $c/c_m = (SV_f/\sigma_m)^4$, where the fibers most distant from the crack tip can no longer sustain the load and a partially bridged crack develops. Then the plot labeled ME would apply. Over the range of application of the MCE plot, the stress required for crack propagation drops with the reciprocal square root of crack length in the ME approximation, but the stress for crack propagation in the composite would be higher than for the same crack length in the unreinforced matrix, as shown in Figure 12.6. For large cracks that approach a constant crack opening displacement the approximate treatment no longer applies and the entire composite will fail at $\sigma = SV_f$, so that the ME curve no longer applies below this stress, as indicated by the horizontal line in Figure 12.6.

Another case of interest is that of an initially unbridged crack of length c_0 that is stressed and develops a bridged portion of length d so that $c = c_0 + d$. The stress for further propagation is given by Marshall and Evans (1986) as

$$\frac{\sigma}{\sigma_m} = \frac{1}{3}\left(\frac{c_m}{c}\right)\left\{1 + 2(d/c_m)^{3/4}\,[2 - (d/c_m)/(c/c_m)]^{3/4}\right\} \tag{12.68}$$

This result can be expressed equivalently as a T-curve behavior with a crack resistance K_R that varies with crack length according to

$$\frac{K_R}{K_C} = 1 + 2(d/c_m)^{3/4}\,[2 - (d/c_m)/(c/c_m)]^{3/4} \tag{12.69}$$

The result is that crack growth will be stable for $c_0/c_m \geq 0.15$ and $d < d^*$. As the stress is increased stable crack growth occurs until $d = d^*$ when fiber failure occurs and catastrophic failure happens. Further detailed discussion of fracture behavior is given by Marshall and Evans (1986).

12.7 ADDITIONAL TREATMENTS OF CRACK BRIDGING EFFECTS

An extensive literature has subsequently developed on various aspects of bridging and its effects on fracture of continuous, aligned, brittle-matrix composites. The subject is too large and complex for full treatment here. We mention some of the important papers and the effects treated and then give a brief summary of two important effects not treated so far here: the effects of residual stress and the effect of statistical distribution of strength of fibers including the breaking of some fibers at a distance from the crack plane and subsequent toughening due to fiber pullout.

Both Budiansky (1993) and Butler and Fuller (1993) have provided excellent general reviews of the theory of toughness in ceramic-matrix composites. A review of progress in continuous-fiber ceramic composites is available (Lehman, El-Rahaiby, and Wachtman, 1996) in which two chapters are particularly pertinent (Karandikar, Chou, and Parvizi-Majidi, 1996; Zok, Evans, and Mackin, 1996). Further treatments of matrix fracture and related effects have been given by Budiansky, Hutchinson, and Evans (1986) and McCartney (1987). Marshall and Evans (1988) have considered the effect of residual stress on the toughness of composites and Sigl and Evans (1989) have considered the effect of residual stress and frictional sliding. Budiansky and Amazigo (1989) treated toughening by frictionally constrained fibers. Debonding properties of residually stressed composites were treated by Charalambides and Evans (1989). Cox (1991) and Cox and Lo (1992) have considered the effect of external factors such as load ratio, notch, and scale effects on the propagation of cracks in composites. The effect of a statistical distribution of fiber strengths on behavior of composites has been treated by Cox, Marshall, and Thouless (1989), Knowles and Yang (1991), and Cao, Bischoff, Sbaizero, Ruhle, Evans, Marshall, and Brennan (1990). The interaction of fiber and transformation toughening has been treated by Cui (1992) and Cui and Budiansky (1993).

The results from various theories including effects of residual stress, fiber sliding properties, and statistical distribution of strength have been summarized concisely by Cao, Bischoff, Sbaizero, Ruhle, Evans, Marshall, and Brennan (1990). These authors state that long cracks will propagate at the stress given by the Aveston–Cooper–Kelly formula modified by a term for the initial stress

$$\sigma_{cmc} = \left[\frac{12\tau\gamma_m V_f^2 E_f E_c^2}{r V_m E_m^2}\right]^{1/3} - q\frac{E_c}{E_m} \tag{12.70}$$

Here allowance has been made for an axial residual stress in the matrix, q, resulting from fiber–matrix misfit, with $q > 0$ indicating tensile stress. Steady-state matrix cracking should occur when

$$\sigma_{cmc} < \frac{S_b}{V_f} \tag{12.71}$$

where S_b is the fiber bundle strength. The condition for the matrix cracking stress equation to hold is that the crack length, c, obey

$$c > c_c = \left[\frac{\pi}{4\,(2/3)^{4/3}}\right]\left[\frac{\gamma_m^{1/2}\, E_m^{3/2}\, V_m^2\, r\,(1+\eta^{-1})}{\tau V_f^2\, E_f\,(1-\nu^2)}\right]^{2/3} \tag{12.72}$$

When c is smaller than c_c, a higher stress is needed to propagate the crack. As the crack length increases, the stress required for propagation drops until the length reaches c_c.

If the stress is increased above σ_{mc}, a set of parallel cracks develops. The crack spacing reaches a saturation value d, when the stress within the uncracked blocks is smaller than the matrix blocks can support. This saturation spacing is related to the sliding stress between the fiber and matrix, τ, by

$$\tau = \lambda\left[\frac{V_m^2\, \gamma_m\, E_f\, E_m\, r^2}{V_f\, E_c\, d^3}\right]^{1/2} \tag{12.73}$$

where $\lambda = 1.337(3/4)^{1/2} = 1.34$. The matrix cracking stress can be expressed in terms of the saturation crack spacing through

$$\sigma_{cmc} = (6\lambda)^{1/3}\left[\frac{V_v\, \gamma_m\, E_f\, E_c}{E_m\, d}\right]^{1/2} - q\frac{E_c}{E_m} \tag{12.74}$$

The ultimate strength is the stress required to cause fiber bundle failure and is given by (Evans and Marshall, 1989; Cao, Bischoff, Sbaizero, Ruhle, Evans, Marshall, and Brennan, 1990)

$$\sigma_u = V_f\, \hat{S} \exp\left\{\frac{[1-(1-\tau d/r\hat{S})^{m+1}]}{(m+1)[1-(1-\tau d/r\hat{S})^{m}]}\right\} \tag{12.75}$$

with

$$\left(\frac{r\hat{S}}{\tau d}\right)^{m+1} = \frac{(A_0/2\pi rL)(rS_0/\tau d)^m}{[1-(1-\tau d/r\hat{S})^m]} \tag{12.76}$$

where S_0 and m are the Weibull parameters, A_0 is another scale parameter (usually set equal to 1 m^2), and L is the specimen gage length. The effect of the sliding stress τ on σ_u appears directly and also through its effect on the crack spacing d, while the effect of residual stress is present indirectly through its effect on τ.

12.8 ADDITIONAL STATISTICAL TREATMENTS

Cho, Holmes, and Barber (1992) have carried out a theoretical analysis that shows that the matrix strength statistics have a notable effect on the distribution of matrix cracks in continuous fiber-reinforced composites. A significant size effect is predicted.

Yang and Knowles (1992) have reviewed the generally accepted Aveston–Cooper–Kelly theory that cracking of the matrix of a composite will continue until the matrix is separated into blocks each of a length between x and $2x$, where

$$x = \frac{V_m \sigma_{mu} R}{2V_f \tau} = \frac{\sigma_{cmc} R}{2V_f \tau}\left(\frac{1}{1+\dfrac{E_f V_f}{E_m V_m}}\right) \tag{12.77}$$

They consider that the matrix has a Weibull distribution of strength as a function of the length L in the direction of the reinforcing fibers

$$P_s(\sigma,L) = \exp\left[-\frac{L}{L_0}\left(\frac{\sigma}{\sigma_0}\right)^m\right] \tag{12.78}$$

and find a distribution of crack spacings beginning at x and extending to values as high as $8x$ for $m = 4$ and as high as $4x$ for $m = 20$.

Thouless and Evans (1988) allowed for fiber strength variation by using the Weibull distribution. The result is that fibers generally fail some distance from the crack plane and some do not break at all. They obtain an increase in toughness as measured by the required strain energy release rate given by

$$\Delta G = \Delta G_b + \Delta G_p \tag{12.79}$$

where the subscripts b and p refer, respectively, to the unbroken bridging fibers and the broken fibers that contribute to toughness through frictional resistance to pullout. Their theory gives

$$\Delta G_b \propto \left[\frac{r^{m-5}}{\tau^{m-2}}\right]^{1/(m+1)} \tag{12.80}$$

and

$$\Delta G_p \propto \left[\frac{r^{m-3}}{\tau^{m-2}}\right]^{1/(m+1)} \tag{12.81}$$

For the range of m values generally found ($m > 5$), it is thus desirable to have large radius and small friction for high toughness.

12.9 SUMMARY OF FIBER TOUGHENING MECHANISMS

Campbell, Ruhle, Dalgleish, and Evans (1990) classified the toughening mechanisms associated with fibers in ceramic matrices into four categories and gave an equation with a term for each. Their expression is presented in the context of a discussion of whisker toughening but is more general and provides insight into fiber or whisker toughening. The contribution to the strain energy release rate, G_c, imparted by fibers or whiskers is

$$\frac{\Delta G_c}{V_f d_{db}} \approx \frac{S^2}{E_f} - E e_T^2 + \frac{4\gamma_i}{rV^m} + \frac{\tau}{d_{db}} \sum_i \frac{h_i^2}{r} \tag{12.82}$$

Here S is the strength of the fiber, r is the fiber radius, V_m is the volume fraction of the matrix, E_f is Young's modulus of the fiber, γ_i is the interface fracture energy, and h_i is the pullout length of the ith fiber.

The first term, S^2/E_f, is the strain energy stored in the fiber over the debonded length on both sides of the matrix crack before the fiber fails. This energy contributes positively to the toughening because it is dissipated as acoustic waves.

The second term is the residual strain energy within the debond length of the composite caused by the misfit strain e_T. This strain energy is lost when the fiber fails and so is taken by Campbell et al. as contributing to the energy needed for crack propagation and thus reduces the toughness.

The third term is the energy needed to create the debond fracture surface and the fourth term is the pullout contribution.

12.10 OTHER FAILURE MECHANISMS IN CONTINUOUS, ALIGNED-FIBER COMPOSITES

Marshall and Evans (1985) have investigated the failure mechanisms in unidirectional ceramic-fiber, ceramic-matrix composites. For tensile loading parallel to the fibers, failure occurs in several stages as the above theory predicts: multiple matrix cracking followed by fiber fracture and pullout. In flexural loading a complex failure process occurs. The matrix cracks begin at the tension face and propagate only about halfway through the bar and the crack openings on the tensile surface do not become large enough to cause complete fiber pullout before compressive failure occurs on the compressive surface. Evidently care is necessary in using flexure testing to assess mechanical behavior of composites or in applying theories developed for uniform tension to members in bending.

Aligned-fiber composites are usually made in successive layers (lamina). Interlaminar strengths are an inherent weakness. For this reason three-dimensional arrangements of fibers made by textile machine techniques and then impregnated with the matrix material have been tried.

A serious limitation on the high-temperature use of ceramic-matrix composites made with nonoxide reinforcing fibers is the oxidation of the fibers that typically occurs at temperatures above about 1000 °C.

12.11 TENSILE STRESS–STRAIN CURVE OF CONTINUOUS, ALIGNED-FIBER COMPOSITES

A model describing the tensile stress–strain curve of ceramic–fiber composites has been developed by Cao and Thouless (1990). They used a composite of SiC fibers in a lithium aluminum silicate glass-ceramic matrix with the properties given in Table 12.2 The general type of stress–strain curve expected for a brittle-matrix composite reinforced with aligned, continuous fibers is shown is Figure 12.7 and has the features of matrix cracking followed by rising stress to the ultimate strength, in turn followed not by sudden failure but by a period of decreasing stress with further strain as the weaker fibers break before ultimate failure of the composite. Their theoretical model gives the results shown in Figure 12.8, which also displays the experimental results. The model describes the behavior quite well up to the stage of fiber pullout. Their model did not take into account the load supported by the broken fibers during pullout. Despite this limitation, the comparison suggests that the behavior of this class of composites is now well understood and well described by quantitative theory.

A subsequent treatment of tensile constitutive behavior of ceramic-matrix composites that relates this behavior to constituent properties has been given by Evans, Domergue, and Vagaggini (1994). These authors point out that linear elastic fracture mechanics as developed for metals cannot be applied directly because the failure mechanism is not that of propagation of a dominant mode I crack. An improved approach is that of large-scale bridging mechanics in which large-scale bridging of matrix cracks by fibers is treated. It needs, however, to be augmented by continuum damage mechanics (Hayhurst, Leckie, and Evans, 1991). These authors classify fracture behavior from a notch in a

TABLE 12.2 Material Properties for SiC–Glass-Ceramic Composite

Property	Value
Fiber Young's modulus, E_f	200 MPa
Fiber radius, r	8 μm
Fiber volume fraction, V_f	0.5
Matrix Young's modulus, E_m	85 GPa
Statistical property, S_0	2.988 MPa
Statistical property, m	2.1
Interface sliding stress, τ	2 mPa
Matrix crack spacing, l	400 μm
Matrix cracking stress, σ_m	290 MPa
Matching compliance, C	0.093 m/MN
Cross-sectional area of composite, A_c	2.18 mm by 2.48 mm
Gage length, L	15 mm

ceramic-matrix composite into three categories. Class I is the model considered earlier in the present chapter in which fibers break some distance from the plane of a single dominant crack and then undergo pullout as the crack advances. Class II involves matrix cracking with crack branching without fiber failure. Class III involves shear damage by multiple matrix cracking at an angle to the notch plane. The results are too complex to present here. A model for

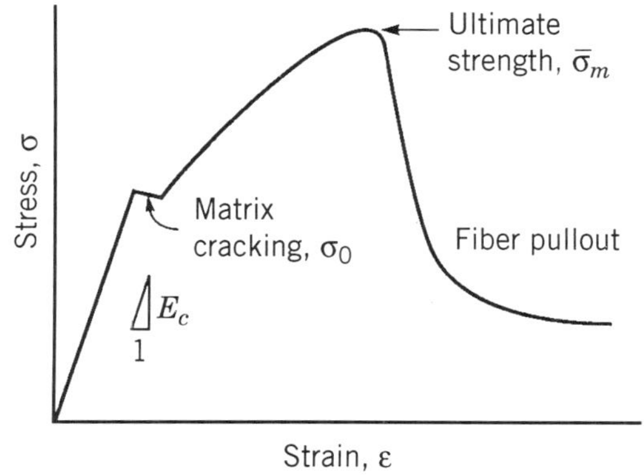

Figure 12.7 Schematic tensile stress–strain curve for a "tough" fiber composite. (From H. N. Cao and M. D. Thouless, "Tensile Tests of Ceramic-Matrix Composites: Theory and Experiment," *J. Am. Ceram. Soc.*, Vol. 73, 2091–2094, 1990. Reprinted by permission of the American Ceramic Society.)

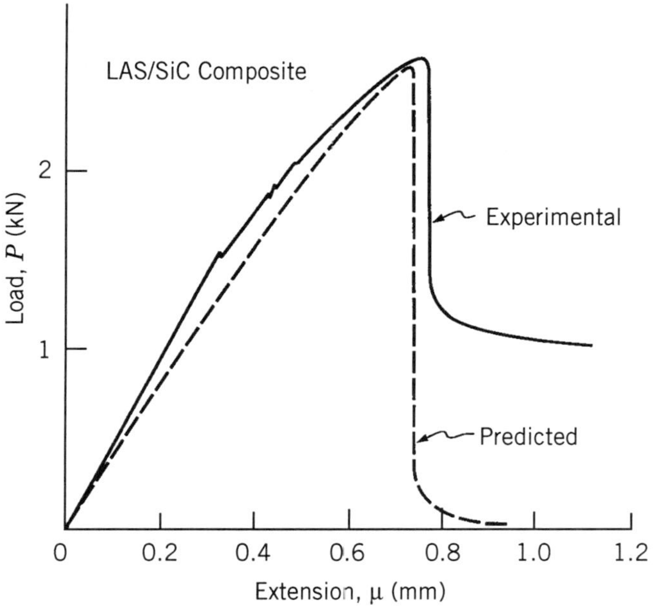

Figure 12.8 Comparison of predicted and experimental load–displacement curves. (From H. N. Cao and M. D. Thouless, "Tensile Tests of Ceramic-Matrix Composites: Theory and Experiment," *J. Am. Ceram. Soc.*, Vol. 73, 2091–2094, 1990. Reprinted by permission of the American Ceramic Society.)

delamination cracking in laminated ceramic-matrix composites has been developed by Charalambides (1991).

12.12 LAMINATED COMPOSITES

The fracture behavior discussed in this chapter focused on propagation of cracks perpendicular to the direction of parallel, continuous reinforcing fibers. It is evident that cracks propagating parallel to the fibers would not experience the toughening and strengthening imparted to cracks perpendicular to the fibers. That is, the benefits of reinforcement of parallel fibers apply primarily to cracks in one set of parallel planes. Attempts to extend these benefits to fracture on other inclined planes involve making a three-dimensional woven structure of reinforcing fibers or making a laminated structure from layers of continuous, parallel fiber composites with the fiber directions oriented differently in successive layers. The latter family of laminated composites is discussed briefly in the present chapter. As would be expected, models for mechanical behavior in terms of the properties and arrangement of the constituents become quite complex mathematically. No attempt is made here to present such models, partly because they appear usually to have only partial success in describing behavior as local fractures develop. Only brief mention of some selected work is given to indicate qualitatively the directions being pursued.

The elastic behavior of composites with continuous, parallel fibers was treated in Section 12.2 in a manner sufficient for the purpose of discussing basic models of fracture behavior. Such composites are actually characterized elastically by nine independent elastic constants referred to a set of axes, with one axis along the direction of the continuous fibers (Hearmon, 1961; Chawla, 1993). For tilted axes these constants transform into a full set of constants. The relation of elastic constants of a composite to the elastic constants of the components is an extensive subject and is not treated here. The elastic deflection of composites of general orientation is also complex and is beyond our scope.

Study of laminated ceramic–ceramic composites with regard to damage mechanism and the effect of internal damage on elastic behavior, toughness, and strength is in a relatively early stage. Some individual results are briefly described here to give an idea of the approaches being investigated.

Sbaizero and Evans (1986) made symmetric (0°/90°) composites composed of SiC fibers in a lithium aluminosilicate glass matrix and tested them in both tension and flexure. Several types of local failure taking place before overall failure were identified: matrix cracking, delamination cracks, and fiber bundle failure. Beyerle, Spearing and Evans (1992) made further studies of the same composite system. Basic matrix cracking models gave a good description of the incidence of cracking. However, attempts to model the effects of internal cracking on elastic properties were less successful.

Russo, Harmer, Chan, and Miller (1992) have made a three-layer laminated composite with the intention of improving both strength and toughness. All three layers were made of alumina with 20 vol. % aluminum titanate, termed AST20. Both external surface layers were processed to have fine and homogeneous microstructures having high strength but relatively low toughness and little R-curve behavior. The interior layer was fabricated to have inhomogeneous microstructure with lower strength but having pronounced R-curve behavior. They succeeded in producing composites having high toughness and flaw tolerance without sacrificing strength in the small-flaw regime. The design is intended for situations in which tensile stress is a maximum at the surface.

A somewhat similar concept to that of Russo et al. (1992) of a hard surface combined with a tough interior was used by Chen and Mecholsky (1993), who made laminated composites with five layers of alumina and four layers of nickel. They found that the strength and toughness of the composites were greatly improved in comparison to that of monolithic alumina. The indentation strength values of the laminates was approximately constant for indentation loads between 10 and 100 N. The results suggest that such a composite would have good damage tolerance.

CHAPTER 13

MECHANICAL PROPERTIES OF WHISKER-, LIGAMENT-, AND PLATELET-REINFORCED CERAMIC-MATRIX COMPOSITES

13.1 INTRODUCTION

We have discussed aligned-fiber, ceramic-matrix composites in Chapter 12 and the technique of improving properties in two directions through lamination. An approach to imparting good mechanical properties in all directions is to use relatively short fibers distributed in random or near-random orientations. In this way good mechanical properties have been achieved in aluminum oxide reinforced with silicon carbide whiskers for which isotropic toughness values as high as 9 MPa · $m^{1/2}$ have been reached (Becher, Hsueh, Angelini, and Tiegs, 1988). Toughening in this class of materials was briefly summarized in Chapter 11. We present more detail in the present chapter. The processes causing toughening in whisker-reinforced ceramics are thought to be fundamentally the same as those in ceramic-matrix composites reinforced with aligned, continuous fibers, but modified in detail by the short length and random orientation of the whiskers and by differences in the whisker–matrix interfaces as compared to typical fiber–matrix interfaces. In random whisker-reinforced ceramics, toughening is thought to be caused by a combination of crack deflection, whisker bridging, and whisker pullout.

Whisker pullout apparently is less important because of the short whisker lengths (usually < 50 μm). A model developed by Becher and his associates fits many experimental observations and is described in the next section.

Good lists of basic references are given in Becher, Hsueh, Angelini, and Tiegs (1988, 1989), Becher (1991), and Evans (1990).

13.2 A MODEL FOR WHISKER TOUGHENING

In this discussion we follow the treatment of Becher, Hsueh, Angelini, and Tiegs (1988), denoted BHAT, and adopt their notation. Parameters describing the formation of a crack with a bridged zone of length D_B are shown in Figures 13.1 and 13.2 from BHAT. The fibers have radius r, are spaced a center-to-center distance λ apart, and debond a distance l on either side of the crack. The crack opening displacement is $2u$. The forces imposed by the bridging fibers are on average equivalent to a stress $\sigma(x)$, where x is the distance along the crack from the tip. At the end of the bridging zone $x = D_B$ and $u = u_{max}$. The toughening due to bridging is

$$\Delta K = 2\left(\frac{c}{\pi}\right)^{1/2} \int_0^{D_B} \frac{\sigma(x)}{[c^2 - (c-x)^2]^{1/2}}\, dx \tag{13.1}$$

for a through crack of total length $2c$ in an infinite medium. When $c >> D_B$ this becomes

$$\Delta K = \left(\frac{2}{\pi}\right)^{1/2} \int_0^{D_B} \frac{\sigma(x)}{x^{1/2}}\, dx \tag{13.2}$$

If the closure stress is taken as constant, $\sigma(x) = \sigma^c$, analogous to the Dugdale–Barrenblat zone treatments, the result is

$$\Delta K = 2\sigma^c \left(\frac{2D_B}{\pi}\right)^{1/2} \tag{13.3}$$

This constant closure stress approximation is used by BHAT, who point out that assuming that the closure stress varies linearly from zero at the crack tip to σ^c at the end of the bridging zone would cause $2\sigma^c$ to be replaced by $2\sigma^c/3$.

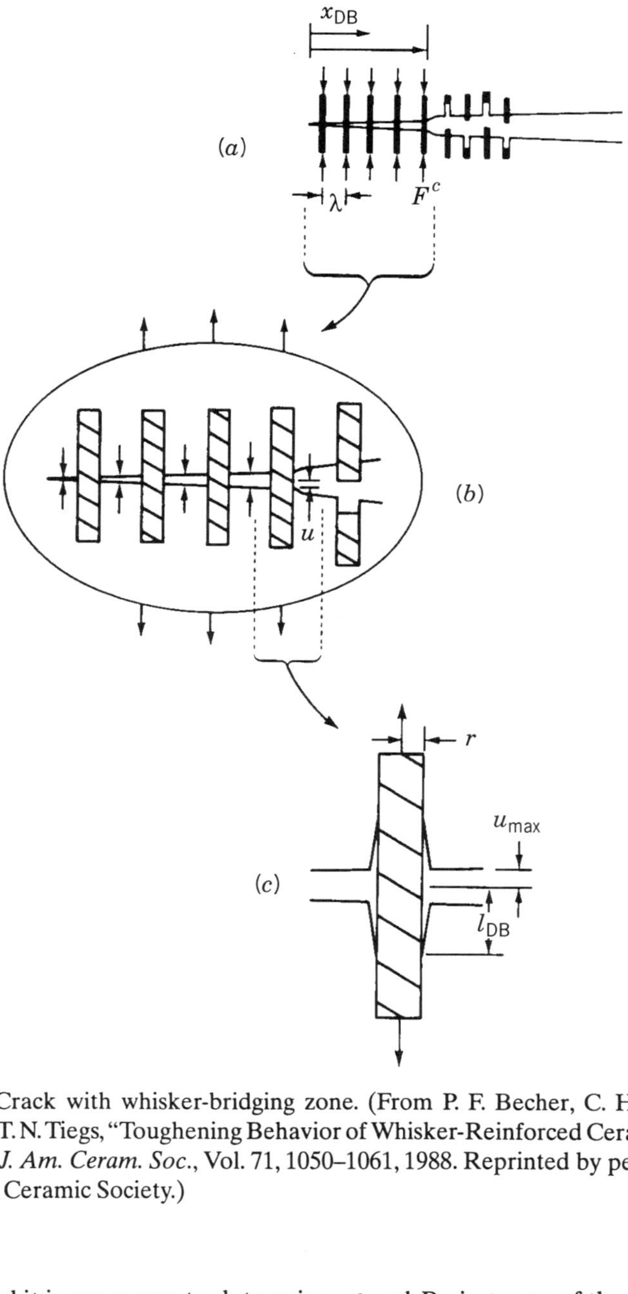

Figure 13.1 Crack with whisker-bridging zone. (From P. F. Becher, C. H. Hsueh, P. Angelini, and T. N. Tiegs, "Toughening Behavior of Whisker-Reinforced Ceramic Matrix Composites," *J. Am. Ceram. Soc.*, Vol. 71, 1050–1061, 1988. Reprinted by permission of the American Ceramic Society.)

To proceed it is necessary to determine σ^c and D_B in terms of the properties of the composite. The closure stress is given by the whisker strength σ_f^w times the volume fraction of whiskers v_f

$$\sigma^c = \sigma_f^w v_f \tag{13.4}$$

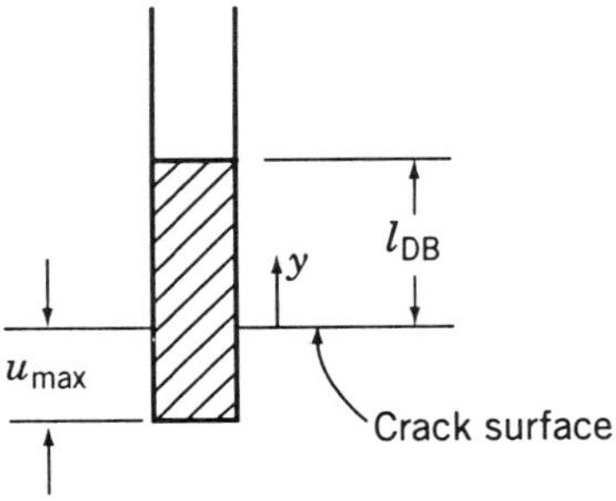

Figure 13.2 Debonding at end of bridging zone. (From P. F. Becher, C. H. Hsueh, P. Angelini, and T. N. Tiegs, "Toughening Behavior of Whisker-Reinforced Ceramic Matrix Composites," *J. Am. Ceram. Soc.*, Vol. 71, 1050–1061, 1988. Reprinted by permission of the American Ceramic Society.)

The crack opening displacement from linear elastic fracture mechanics for plane stress conditions is

$$u = \left(\frac{8}{\pi}\right)^{1/2} \frac{K(1-\nu^2)x^{1/2}}{E^c} \tag{13.5}$$

where E^c is Young's modulus for the composite, ν is Poisson's ratio, and $K = K^0 + dK$ is the appropriate stress intensity factor. Using equation (13.3) gives the crack opening displacement at the end of the bridging section:

$$u = \frac{8(1-\nu^2)\sigma^c D_B}{\pi E^c} + \frac{4(1-\nu^2)K^0 D_B^{1/2}}{E^c(2\pi)^{1/2}} \tag{13.6}$$

The maximum displacement that an unbonded length l^* of whisker supports is

$$u_{max}^w = \sigma_f^w\, l^*/E^w \tag{13.7}$$

The debond length from Budiansky, Hutchinson, and Evans (1986) is given by

$$l_{DB} = r\left(\frac{G^m}{6G^i}\right) \tag{13.8}$$

where G^m and G^i are the required mechanical energy release rates for the matrix and the interface, respectively. Approximating $l^* = l_{DB}$ and setting $u_{max}{}^w$ equal to the crack opening displacement at the end of the debonding zone gives

$$D_b^{1/2} = \frac{[(K^0)^2 + 2E^c v_f (\sigma_f^w)^2\, r\, G^m/(3G^i(1-\nu^2)\,E^w)]^{1/2} - K^0}{4(2)^{1/2} v_f \sigma_f^w/\pi^{1/2}} \tag{13.9}$$

Substituting (13.9) and (13.4) into (13.3) gives

$$\Delta K^{wr} = \frac{[(K^0)^2 + 2(\sigma_f^w)^2 \, v_f \, r \, E^c \, G^m/(3(1 - \nu^2)E^w \, G^i)]^{1/2} - K^0}{2} \tag{13.10}$$

It is noteworthy that the critical stress intensity for the unreinforced matrix, K^0, appears in the toughening expression. For weak toughening [i.e., when $(K^0)^2$ is large compared to the other term in the square root], the toughening can be approximated as

$$\Delta K^{wr} \approx \frac{(\sigma_f^w)^2 v_f \, r \, E^c \, G^m}{6(1 - \nu^2) \, E^w \, G^i \, K^0} \tag{13.11}$$

For strong toughening [i.e., when $(K^0)^2$ is small compared to the other term in the square root], one can approximate this result as

$$\Delta K^{wr} \approx \sigma_f^w \left(\frac{v_f \, r}{6(1 - \nu^2)} \frac{E^c}{E^w} \frac{G^m}{G^i} \right)^{1/2} \tag{13.12}$$

which clearly shows the influence of the various parameters on the toughening.

An analysis of toughening from the mechanical energy release point of view was also presented by BHAT. It produces the same dependence on composite parameters as (13.10) but gives a different numerical constant, presumably because of the approximate nature of the treatments. Their energy analysis is given here to emphasize the equivalence of the treatments when the Irwin equation is properly used, as discussed in Section 10.4. This treatment proceeds by calculating the sum of the changes in the elastic energy stored in the unbonded portion of the whisker, ΔU_e, and the energy required for frictional sliding of the unbonded portion of the whisker, ΔU_s. Assuming constant interfacial shear stress to slide the whisker, the force to slide the whisker at a distance y from the crack plane is $2\pi r(l_{DB} - y)\tau$ and the force exerted by the whisker is $\pi r^2 \sigma(y)$, giving

$$\sigma(y) = \frac{2(l_{DB} -)\tau}{r} \tag{13.13}$$

The elastic energy stored in the whisker is

$$dU_e = \frac{\pi r^2 \sigma^2(y) \, dy}{2E^w} = \frac{2\pi(l_{DB} - y)^2 \, \tau^2 \, dy}{E^w} \tag{13.14}$$

The work done by an element dy in sliding against τ is

$$dU_s = 2\pi r\,\tau\, u\, dy \tag{13.15}$$

where the displacement of the fiber relative to the matrix is

$$u = \int_y^{l_{DB}} \varepsilon(y)\, dy = \int_y^{l_{DB}} \frac{\sigma(y)}{E^w}\, dy \tag{13.16}$$

giving

$$u = \frac{(l_{DB} - y)^2\, \tau}{E^w\, r} \tag{13.17}$$

Substituting (13.17) into (13.15) gives

$$dU_s = \frac{2\pi (l_{DB} - y)^2\, \tau^2\, dy}{E^w} = dU_e \tag{13.18}$$

The energy required per unit area of crack surface to stretch and slide the whiskers is

$$\Delta G = 2N\int_0^{l_{DB}} (dU_e + dU_s) = 4N\int_0^{l_{DB}} dU_e = \frac{4\, N\, \pi\, \tau^2\, l_{DB}^3}{3E^w} \tag{13.19}$$

where N is the number of whiskers per unit area. The stress in the whisker is greatest at $y = 0$ and the toughness is greatest when this stress is equal to the fiber strength σ_f^w so that

$$\sigma_f^w = \frac{2 l_{DB}\, \tau}{r} \tag{13.20}$$

and

$$\Delta G = \frac{(\sigma_f^w)^2\, v_f\, r\, G^m}{18\, E^w\, G^i} \tag{13.21}$$

The equivalent toughening ΔK is obtained by using (10.25) and is

$$\Delta K^{wr} = \{(K^0)^2 + (\sigma_f^w)^2\, v_f\, r\, E^c\, G^m/(18(1 - \nu^2)E^w\, G^i]\}^{1/2} - K^0 \tag{13.22}$$

the numerical constant is different but the same dependence on the composite parameters is obtained as for the direct stress intensity calculation, equation (13.10). As before, this expression can be approximated for weak toughening as

$$\Delta K^{wr} \approx \frac{(\sigma_f^w)^2 v_f r\, E^c\, G^m}{18(1-\nu^2)\, E^w\, G^i\, K^0} \tag{13.23}$$

and for strong toughening as

$$\Delta K^{wr} \approx \sigma_f^w \left(\frac{(v_f)6r}{18(1-\nu^2)} \frac{E^c}{E^w} \frac{G^m}{G^i} \right)^{1/2} \tag{13.24}$$

Comparing equations (13.23) and (13.24) with (13.11) and (13.12), one sees that the factor 6 obtained in the direct stress intensity calculation has been replaced by the factor 18 in the result obtained from the energy calculation. BHAT discuss some of the approximations made in the treatments that may be responsible for the difference. However, the important point is that the

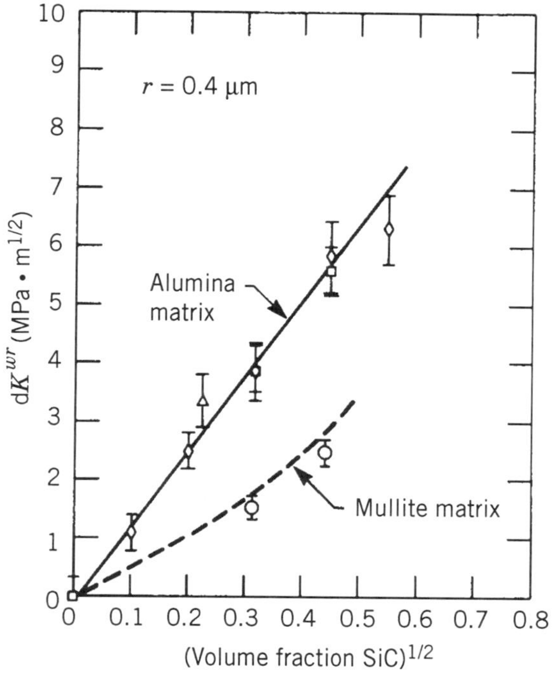

Figure 13.3 Toughening as a function of the square root of the volume fraction of whickers. (From P. F. Becher, C. H. Hsueh, P. Angelini, and T. N. Tiegs, "Toughening Behavior of Whisker-Reinforced Ceramic Matrix Composites," *J. Am. Ceram. Soc.*, Vol. 71, 1051–10601, 1988. Reprinted by permission of the American Ceramic Society.)

same functional dependence is obtained. The approximate nature of the calculation is not likely to give an accurate numerical coefficient in either case.

The main predictions of this model have been confirmed experimentally by BHAT, as we illustrate with Figures 13.3 through 13.7 from their work. Figure 13.3 shows alumina and mullite matrices reinforced with 0.4-μm radius SiC whiskers with low surface oxygen content. The model's prediction of linearity in $v_f^{1/2}$ is confirmed for alumina. Curvature in the plot for mullite is attributed to the effect on E^c for the mullite case of $E^w >> E^m$. The model also predicts a linear dependence on $(E^c/E^m)^{1/2}$ and this is borne out by the results shown in Figure 13.4, which are for 20 vol. % of 0.4-μm whiskers with low surface oxygen content in matrices of glass, mullite, and alumina that have progressively increasing values of E^m. The model predicts that the toughness should increase with $r^{1/2}$ for a constant volume fraction of whiskers and this is shown to be obeyed for 20 vol. % SiC whiskers in alumina in Figure 13.5. The toughness data for two sets of whiskers having, respectively, low and high surface oxygen content fall on parallel lines. The data can be rationalized by assuming a lower τ value for the high surface oxygen content as is reasonable. Another interesting experimental result is shown in Figure 13.6, which shows the effect of varying the grain size in an alumina matrix on the overall tough-

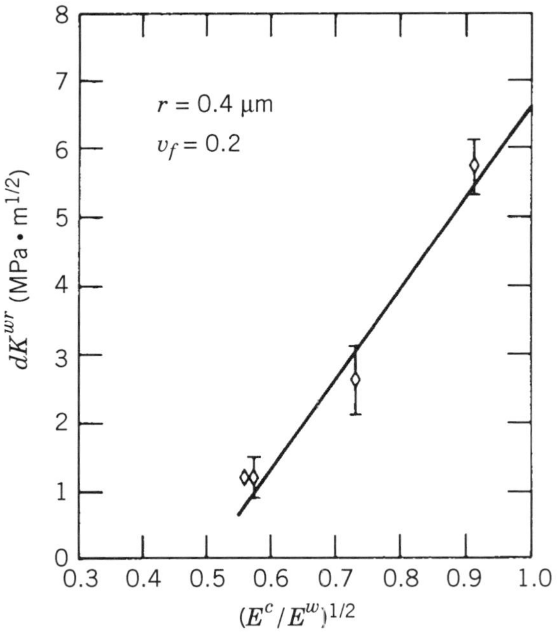

Figure 13.4 Toughening as a function of the square root of the Young's modulus ratio. (From P. F. Becher, C. H. Hsueh, P. Angelini, and T. N. Tiegs, "Toughening Behavior of Whisker-Reinforced Ceramic Matrix Composites," *J. Am. Ceram. Soc.*, Vol. 71, 1051–1060, 1988. Reprinted by permission of the American Ceramic Society.)

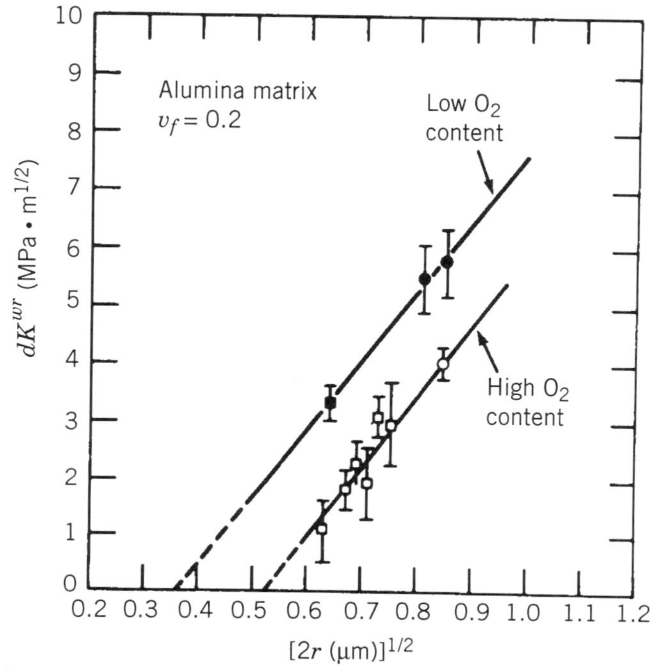

Figure 13.5 Toughening as a function of the square root of whisker radius. (From P. F. Becher, C. H. Hsueh, P. Angelini, and T. N. Tiegs, "Toughening Behavior of Whisker-Reinforced Ceramic Matrix Composites," *J. Am. Ceram. Soc.*, Vol. 71, 1050–1061, 1988. Reprinted by permission of the American Ceramic Society.)

ness of composites with varying volume fractions of 0.4-μm SiC whiskers. A series of parallel straight lines were obtained as a function of $v_f^{1/2}$. The toughness of the unreinforced alumina increases with increasing grain size and provides the base upon which the toughness due to fiber reinforcement is added.

The authors were able to determine the pullout length l_{PO} experimentally and took this to be equal to the debond length. They obtained values of $\beta = l_{DB}/r = l_{PO}/r$ for low-surface-oxygen-content 0.4-μm SiC whiskers of 5 in alumina matrices, 1–2 in mullite matrices, and 0.5–1 in glass matrices. Using their model they estimated the fracture strength of the whiskers in the composites to be in the range 7–11 GPa. This is consistent with strength values of 8–9 MPa measured directly on free, longer (>>5 mm) whiskers with a larger diameter (≈ 4μm). They accordingly took a value of 10 MPa for whiskers in their calculations. Figure 13.7 was calculated from equation (13.10) using the β values indicated; the calculated curves fit the data well. The authors used their theory to predict the combinations of r and v_f that would be required for β values of 5 or 10 to give toughness values of 10 or 20 MPa. The results are shown in Figure 13.8.

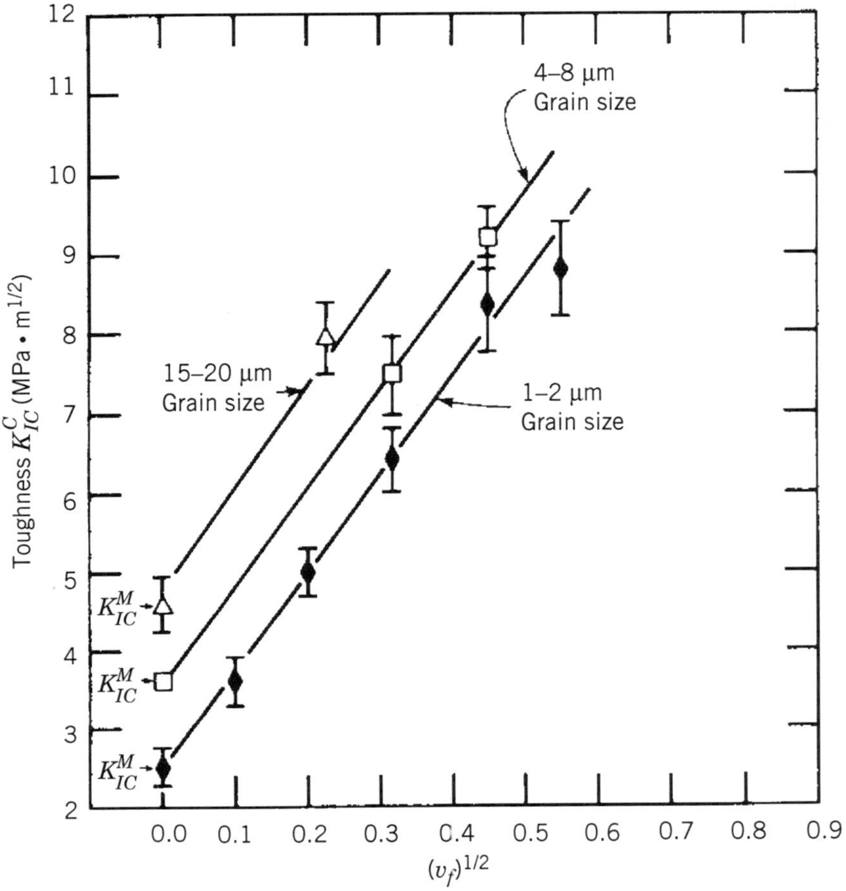

Figure 13.6 Toughness as a function of the square root of the volume fraction of whiskers. (From P. F. Becher, C. H. Hsueh, P. Angelini, and T. N. Tiegs, "Toughening Behavior of Whisker-Reinforced Ceramic Matrix Composites," *J. Am. Ceram. Soc.*, Vol. 71, 1050–1061, 1988. Reprinted by permission of the American Ceramic Society.)

Despite its success, the above model has some limitations. The model takes the number of whiskers per unit area involved in bridging equal to v_f, which is a reasonable assumption for parallel fibers but not for randomly oriented whiskers. An empirical constant A_{fb} equal to the area fraction of the whiskers effectively bridging the crack can be introduced into the theory. Another limitation of the above model is that the equations take no account of the possibility of whiskers breaking at some distance from the crack plane and pulling out against the sliding stress and thus dissipating additional energy. Fibers having a distribution of strengths would be expected to fail at various points near the crack plane and so undergo pullout. This is in fact observed. In Chapter 12 we mentioned a treatment by Thouless and Evans (1988) for

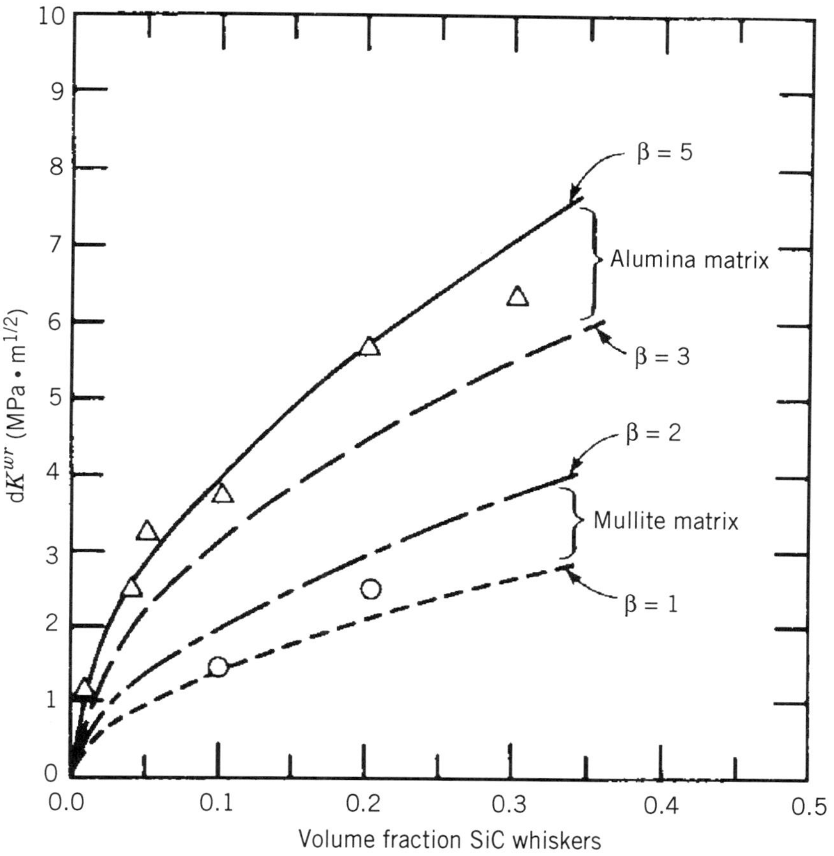

Figure 13.7 Toughness values calculated with selected β values. (From P. F. Becher, C. H. Hsueh, P. Angelini, and T. N. Tiegs, "Toughening Behavior of Whisker-Reinforced Ceramic Matrix Composites," *J. Am. Ceram. Soc.*, Vol. 71, 1050–1061, 1988. Reprinted by permission of the American Ceramic Society.)

composites with aligned, continuous fibers that considers the statistics of this process. Becher (1991) treats the effect of breaking and pullout by assuming that an area fraction A_{PO} break at an average distance l_{PO} from the crack and undergo pullout. Becher's treatment for this case gives the increase in the critical mechanical energy release rate ΔG^{PO} (discussed by him in terms of J) caused by pullout as

$$\Delta G_{PO} = A_{PO}\,\tau\left(\frac{l_{PO}}{r}\right) l_{PO} \tag{13.25}$$

and the corresponding increase in the critical stress intensity factor as

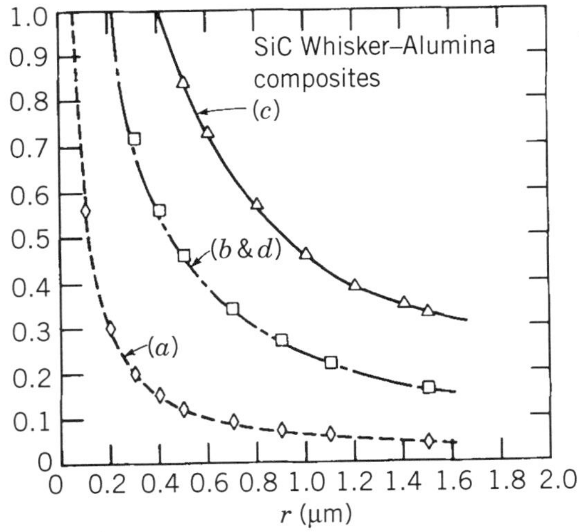

Curves	Parameters
d	dK^{wr} 20 MPa • $m^{1/2}$ σ_f^w = 20 GPa, β = 5
c	dK^{wr} 20 MPa • $m^{1/2}$ σ_f^w = 20 GPa, β = 10
b	dK^{wr} 10 MPa • $m^{1/2}$ σ_f^w = 10 GPa, β = 5
a	dK^{wr} 5 MPa • $m^{1/2}$ σ_f^w = 10 GPa, β = 5

Figure 13.8 Combinations of fiber radius and volume fraction needed for detected toughness values. (From P. F. Becher, C. H. Hsueh, P. Angelini, and T. N. Tiegs, "Toughening Behavior of Whisker-Reinforced Ceramic Matrix Composites," *J. Am. Ceram. Soc.*, Vol. 71, 1050–1061, 1988. Reprinted by permission of the American Ceramic Society.)

$$\Delta K_{PO} = \frac{(E_c A_{PO} \tau r)^{1/2} l_{PO}}{r} \tag{13.26}$$

From these equations the ratio of contribution to the critical energy release rate from pullout to that from frictional bridging is

$$\frac{\Delta G_{PO}}{\Delta G_{FB}} = 3 \frac{l_{PO}}{r} \frac{\tau E_f}{(\sigma_f^w)^2} \frac{A_{PO}}{A_{FB}} \frac{l_{PO}}{l_{DB}} \tag{13.27}$$

It is evident that the relative effects of bridging without breakage and breakage with pullout depend strongly on l_{PO}/r, l_{PO}/l_{db}, and A_{PO}/A_{FB}. Becher has estimated that for SiC whisker-reinforced alumina $\Delta G_{PO} \approx 1.5\Delta G_{FB}$, while for SiC whiskers in mullite $\Delta G_{PO} \approx 0.1\Delta G_{FB}$ and for SiC in soda-lime glass ΔG_{PO} is negligible.

13.3 COMBINED TOUGHENING MECHANISMS IN WHISKER-REINFORCED COMPOSITES

The simultaneous action of microcracking, load transfer, bridging, and crack deflection mechanisms of toughening in whisker-reinforced ceramic-matrix composites has been considered by Giannakopoulos and Breder (1991). They

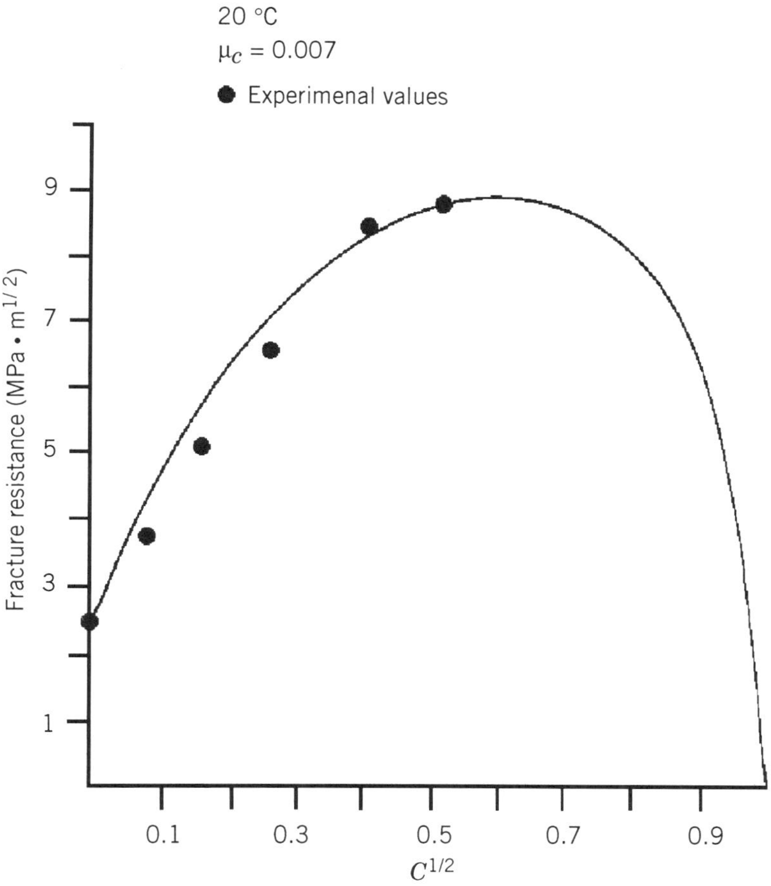

Figure 13.9 Toughness of SiC–alumina composites as a function of the volume fraction of whiskers. (From A. E. Giannakopoulos and R. Breder, "Synergism of Toughening Mechanisms in Whisker-Reinforced Ceramic-Matrix Composites," *J. Am. Ceram. Soc.*, Vol. 74, 194–202, 1991. Reprinted by permission of the American Ceramic Society.)

assume that the load transfer between the matrix and the whiskers is due to Coulomb friction, model the whiskers as bridging springs, carry out a J-integral calculation, and use the Irwin relation to obtain the stress intensity factor of the toughened composite. Without reviewing their model in detail, we present their final result. They use K_t for the crack tip stress intensity factor (i.e., the untoughened value) and K_∞ for the applied stress intensity value (i.e., the measured toughness) and obtain

$$\frac{1-\nu^2_{2-D}}{E_{2-D}}K_\infty^2 = (1-v_f)\,\frac{1-\nu_m^2}{E_m}\left(1+\frac{15\pi+56}{30\pi}\left(\frac{E_m}{E_s}-1\right)\right)DK_t^2 + V \qquad (13.28)$$

Using this equation they calculate the fracture resistance for an alumina reinforced with 25 vol. % whiskers. The results are shown in Figure 13.9 from their paper.

The relative effects of contributing mechanisms in combined toughening can change with temperature according to White and Guazzone (1991), who studied crack growth resistance in SiC-whisker-reinforced alumina at 20, 1200, and 1400 °C. They found that a frontal zone mechanism, probably microcracking, dominates at room temperature. At 1200 °C a following wake zone mechanism of crack face whisker-bridging is dominant.

13.4 LIGAMENT-REINFORCED CERAMIC-MATRIX COMPOSITES

Interest in ceramics toughened by metal inclusions has been stimulated by the high toughness values achieved, although their use is limited by the temperature and oxidation limits of the reinforcing metal ligaments. Flinn, Lo, Zok, and Evans (1993) presented a literature survey and developed a model that gave good fits to experimental data on alumina containing aluminum ligaments. These specimens had 28 vol. % aluminum in the form of fibers about 19 μm in diameter. Initial toughness values were about 3 MPa · $m^{1/2}$ but increase to about 35 MPa · $m^{1/2}$ after crack extensions of 1.5–2.5 mm.

13.5 PLATELET-REINFORCED CERAMIC-MATRIX COMPOSITES

Ceramic-matrix composites toughened with platelets are generally not as tough as those toughened with whiskers. Chou and Green (1993) studied hot pressed alumina and found that the average fracture toughness could be increased from 4.3 MPa · $m^{1/2}$ for unreinforced alumina to 7.1 MPa · $m^{1/2}$ for alumina reinforced with 30 vol. % SiC platelets. They concluded that crack deflection is the dominant toughening mechanism.

CHAPTER 14

CYCLIC FATIGUE OF CERAMICS

14.1 INTRODUCTION

Cyclic fatigue is the loss of strength of material under cyclic loading. It can lead to failure at stresses well below the short-time failure stress. It has long been known to be a very important cause of failure in metals and so has been studied extensively in metals. More recently, cyclic fatigue has been recognized to occur in ceramics including ceramics with microstructures designed for enhanced toughness under noncyclic stress. Cyclic fatigue work on metals is very briefly reviewed in this chapter before turning to fatigue in ceramics. In the metallurgical literature the term fatigue is used rather than cyclic fatigue (Suresh, 1991).

The term cyclic fatigue is used here to distinguish strength deterioration essentially dependent on the cyclic nature of loading from the type of subcritical crack propagation treated in Chapter 8. The latter takes place under constant stress. This constant-stress effect is sometimes termed "static fatigue." If the stress is cyclically varied from some value to a lower value, subcritical crack propagation of this static fatigue type can still take place at a rate that varies during the cycle but is still dependent only on the tensile stress acting instantaneously. A treatment of the effect of static fatigue type slow crack propagation under alternating stress was presented by Evans and Fuller (1974) and under arbitrary stress–time history by Jakus and Ritter (1981). Cyclic fatigue type strength loss is a different phenomenon than static fatigue type slow crack propagation and can be a very strong effect. The mechanisms of cyclic fatigue in ceramics are still under investigation. For ceramics toughened by crack bridging it is thought that the periodic reduction of stress causes partial crack closure and damages the crack bridging elements, leading to a higher critical stress intensity at the crack tip under the next maximum stress during the stress cycle and so to a loss of strength.

14.2 CYCLIC FATIGUE OF METALS

The most common type of failure of structural metallic alloys is fatigue failure so that a large body of work on metallic fatigue exists (Hertzberg, 1989; Courtney, 1990). Some of the basic features of metallic fatigue are briefly summarized here because phenomenologically similar behavior is observed in ceramics, so that the existing framework used to describe metallic fatigue is also used to describe fatigue in ceramics. It should be borne in mind that the underlying processes are not necessarily the same in ceramics, however.

Fatigue tests are usually conducted under periodically varying stress or strain. A simple and common situation is sinusoidally varying stress, as shown in Figure 14.1. The minimum stress may be less than zero (i.e., compression) as shown or may simply be a smaller value of tensile stress. Both the maximum stress, σ_{max}, and the difference in stress, $\Delta\sigma = \sigma_{max} - \sigma_{min}$, are important parameters determining the fatigue behavior. A typical measurement is to determine the number of cycles required for failure as a function of one of the stress parameters such as the maximum stress. The results when plotted are

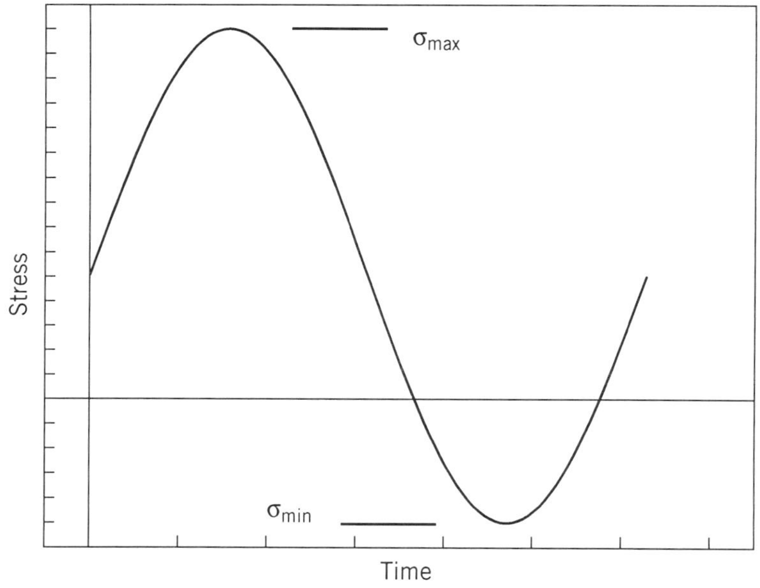

Figure 14.1 Cyclic stress for fatigue testing.

termed S-N curves. These curves typically show a large increase in number of cycles required for failure as the stress is decreased. For stresses below a critical value, no failure may occur over any practically observable number of cycles, suggesting the existence of a fatigue limit. Another important observation is that the presence of a notch generally drastically reduces the number of cycles required for failure.

Fatigue in metals is generally considered to be a three-stage process: Figure 14-2 shows (I) fatigue crack initiation, (II) fatigue crack propagation, and (III) final failure. Microscopic examination of a fatigue fracture surface typically shows a region of parallel markings (termed clam shell markings) corresponding to the slow and intermittent crack propagation characteristic of the second stage of fatigue. The subsequent portion of the fracture surface is typically rough and fibrous and is associated with the final fracture stage.

The initial stage of fatigue crack nucleation in metals is not fully understood in terms of detailed mechanisms but is believed to be associated with localized plastic deformation. It generally begins at an external surface or at some internal surface such as the interface with an inclusion. The direction of the development of the initial fatigue crack is determined by the direction of slip rather than the direction of maximum tensile stress. Fatigue crack initiation requires that at least a part of the stress cycle consist of tensile stress. Fatigue failure in metals apparently does not take place under stress cycles that are entirely compressive.

Stage II of fatigue crack propagation is characterized by crack growth perpendicular to the direction of maximum tensile stress. The growth rate is

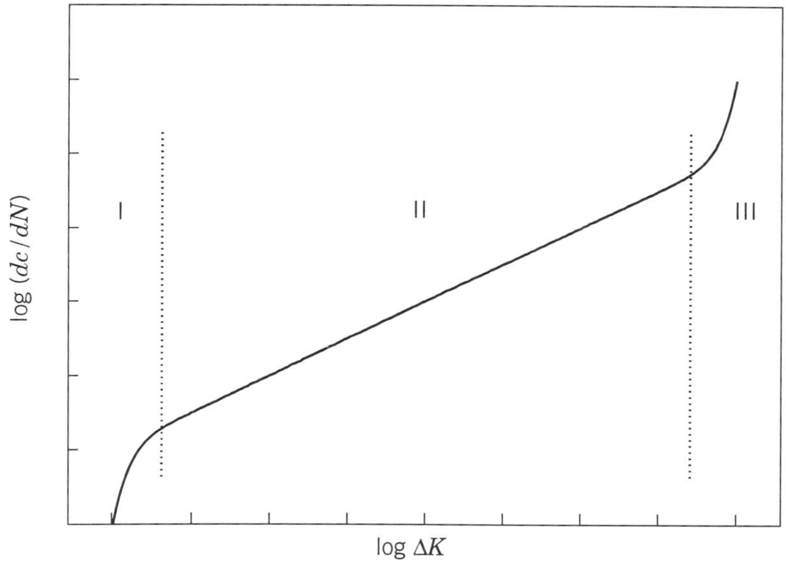

Figure 14.2 Schematic of crack growth rate during cyclic fatigue.

controlled primarily by the difference between the maximum and minimum tensile stress rather than by the average stress.

Stage III of fatigue failure is an overload failure that occurs when the remaining cross section can no longer sustain the maximum tensile stress.

Because stage II is a crack propagation process, it is plausible that the stress intensity factor should be the appropriate controlling variable rather than the stress. Careful studies have shown that this is so. The rate of crack growth in stage II is usually found to obey the Paris law

$$\frac{dc}{dN} = A\,(\Delta K)^{m_f} \tag{14.1}$$

where c is the crack length, N is the number of cycles of stress, $\Delta K = Y(\sigma_{max} - \sigma_{min})c^{1/2}$, and A and m_f are empirical parameters that generally depend on the material, the environment, the frequency, the temperature, and the ratio $\sigma_{min}/\sigma_{max}$. Here m_f is not the Weibull m. The fatigue parameter m_f typically has values in the range 2 to 4 for metals but can be much greater for ceramics.

A schematic plot of log dc/dN as a function of log ΔK is shown in Figure 14.2. The Paris law (straight-line portion of the figure) is a good approximation of the observed behavior over most of the range of ΔK, but for small values the crack growth rate typically drops off sharply and suggests the existence of a threshold value, ΔK_{th}, below which fatigue crack growth does not occur. This threshold value depends, however, on $\sigma_{min}/\sigma_{max}$.

The Paris law can be used to predict the number of cycles required for failure for a constant σ_{max} and constant $\sigma_{min}/\sigma_{max}$ by using the relation

$$\Delta K = Y(\Delta\sigma)c^{1/2} \tag{14.2}$$

and following the crack length until the stress intensity value under maximum stress is sufficient to cause fast fracture. In general, Y varies with crack length so that numerical integration of the crack length is required. When Y is independent of crack length (as in the case of a circular flaw far from the surface of the specimen), the integration can be carried out in closed form and the result (Hertzberg, 1989) for the number of cycles to failure, N_f, is

$$N_f = \frac{2}{(m_f-2)AY^{m_f}(\Delta\sigma)^{m_f}}\left[\frac{1}{a_i^{(m_f-2)/2}} - \frac{1}{a_f^{(m_f-2)/2}}\right] \quad \text{for } m \neq 2 \tag{14.3}$$

where a_i is the initial fatigue crack length, a_f is the crack length at failure (i.e., the length at which the stress intensity factor under maximum stress is sufficient to cause fast failure), and $\Delta\sigma = \sigma_{max} - \sigma_{min}$.

14.3 CYCLIC FATIGUE IN CERAMICS

Until recently ceramics were thought not to undergo fatigue at or near room temperature (apart from the effects of subcritical crack propagation of the type discussed in Chapter 8). This belief was apparently based on the fact that fatigue in metals is associated with plastic deformation and the assumption that sufficient plastic deformation to cause fatigue does not occur in most ceramics at room temperature. Fatigue at room temperature has now been found in many ceramics, although the mechanisms appear to involve degradation of the toughening elements of microstructure rather than plastic deformation. A comprehensive review of fatigue in ceramics was presented by Ritchie and Dauskardt (1991) and extensive references to work on fatigue on ceramics were also given by Dauskardt, James, Porter, and Ritchie (1992) and by Hu and Mai (1992).

Fatigue in many ceramics (including zirconia, graphite, alumina, silicon nitride, silica glass-ceramics, LAS–SiC_f, Al_2O_3–SiC, and laminated graphite–pyrolytic carbon composites) has been demonstrated by comparing static fatigue curves with S-N curves on the same material. The time to failure under constant stress was determined as a function of stress in the static fatigue experiment. From these results the predicted time to failure from the static fatigue process alone was obtained by integrating. The predicted time is longer for a particular maximum stress under the assumption of alternating stress than under a constant stress equal to the maximum alternating stress. The time to failure under alternating stress was measured and plotted as a function of the maximum stress in the S-N experiments. Techniques used were zero-tension and tension–tension cantilever bending and three- and four-point bending. Tension–compression testing was done in rotary bending, uniaxial push–pull or tension–torsion testing. The failure times in the alternating stress experiments were found to be shorter than in the static fatigue experiments, indicating the occurrence of a true cyclic fatigue effect.

The indicated fatigue limit determined from S-N curves was generally much less than the strength in simple loading to failure. For MS grade (mid-toughness) zirconia, the fatigue limit as determined with 10^8 cycles and $\sigma_{min}/\sigma_{max} = -1$ was about 50% of the single-cycle strength. An alumina was found to fail at 10^8 cycles at a strength of 25–40% of the single-cycle strength.

Fatigue crack propagation studies on ceramics have been done in two modes: long crack and short crack studies. The long crack studies (typically with crack length about 3 mm) were done with through cracks in precracked specimens, such as single-edge-notched specimens in three- and four-point bending or tapered double-cantilever beam specimens, or compact tension specimens. Crack lengths were monitored optically or by measuring the resistance of a conducting coating. Short crack experiments (typically with cracks less than 250 μm in length) were done by bending beams using surface cracks introduced from a notch or by indentation. The long crack propagation studies gave results that could be described by the Paris law, but the exponent m varied between 12 and 40, which is much larger than the values of 2–4 typical of metals. The short crack studies gave much greater crack growth rates than the long crack studies at equivalent applied stress intensity values. Also, the crack growth takes place in the short crack studies at stress intensities below the threshold value required in long crack studies (Ritchie and Dauskardt, 1991).

Typical results of long crack cyclic fatigue studies on zirconia are shown in Figures 14.3, 14.4, and 14.5 from Dauskardt, Marshall, and Ritchie (1990). Toughness curves (T-curves) for several grades of zirconia are shown in Figure 14.3. The corresponding crack growth rate curves showing Paris law behavior are shown in Figure 14.4. Transformation toughening greatly increases the ΔK value required to produce a given rate of crack growth per cycle. Proof that the effect is truly cyclic fatigue is shown in Figure 14.5, which contrasts the crack velocity in cyclic tests with that measured under constant stress. The growth rates in cyclic fatigue are up to seven orders of magnitude greater than those under the corresponding constant stress.

14.4 MECHANISMS OF CYCLIC FATIGUE IN CERAMICS

Mechanisms of cyclic fatigue in ceramics have not yet been definitely established. The possible mechanisms have been classified into two categories

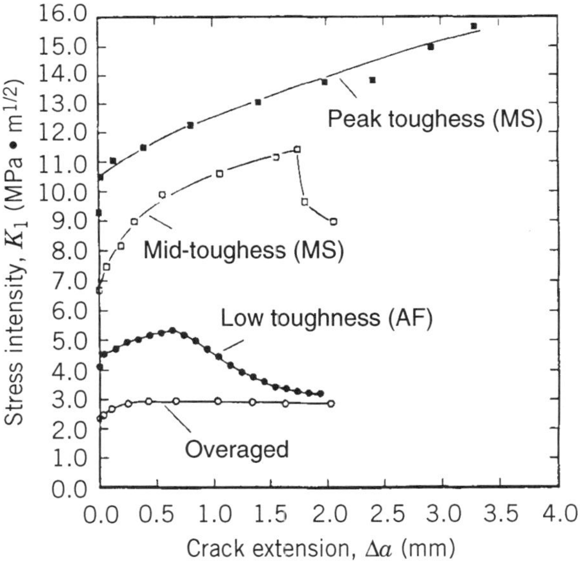

Figure 14.3 Fracture toughness behavior of Mg–PSZ subeutectoid aged to a range of toughness values. (From R. H. Dauskardt, D. B. Marshall, and R. O. Ritchie, "Cyclic Fatigue-Crack Propagation in Magnesia-Partially-Stabilized Zirconia Ceramics," *J. Am. Ceram. Soc.*, Vol. 73, 893–903, 1990. Reprinted by permission of the American Ceramic Society.)

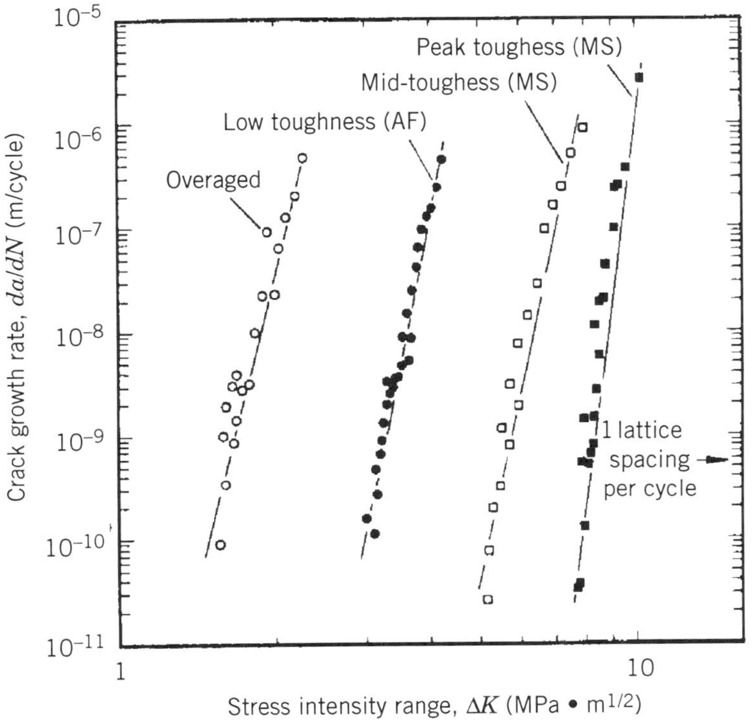

Figure 14.4 Cyclic fatigue crack growth behavior for Mg–PSZ subeutectoid aged to a range of toughness values. (From R. H. Dauskardt, D. B. Marshall, and R. O. Ritchie, "Cyclic Fatigue-Crack Propagation in Magnesia-Partially-Stabilized Zirconia Ceramics," *J. Am. Ceram. Soc.*, Vol. 73, 893–903, 1990. Reprinted by permission of the American Ceramic Society.)

(Ritchie and Dauskardt, 1991): intrinsic and extrinsic. The intrinsic mechanism would involve the creation of a fatigue-damaged microstructure ahead of the crack tip that would produce a crack-advance mechanism operating uniquely in cyclic fatigue. Possible mechanisms are alternating crack tip blunting and resharpening or, for whisker-reinforced composites, whisker breakage.

Extrinsic mechanisms may not involve a change in the crack propagation mechanism under cyclic fatigue conditions. Instead, the unloading may change the value of the crack tip stress intensity factor under the next maximum stress by reducing the crack tip shielding. Thus cyclic fatigue might reduce the crack tip shielding effect of transformation toughening or crack tip bridging.

More than one mechanism of fatigue may be active simultaneously. For example, Jacobs and Chen (1994) have found evidence in silicon nitride for both environmentally sensitive fatigue combined with an environmentally insensitive component apparently associated with damage to the crack wake shielding zone.

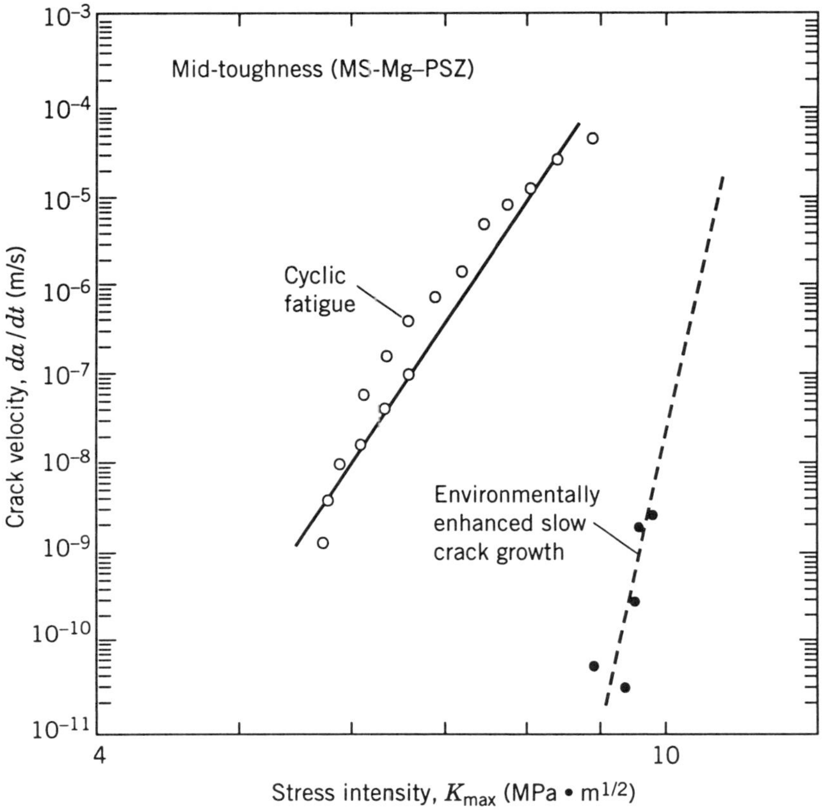

Figure 14.5 Subcritical crack growth behavior in mid-toughness Mg–PSZ, showing comparison of cyclic and static conditions. (From R. H. Dauskardt, D. B. Marshall, and R. O. Ritchie, "Cyclic Fatigue-Crack Propagation in Magnesia-Partially-Stabilized Zirconia Ceramics," *J. Am. Ceram. Soc.*, Vol. 73, 893–903, 1990. Reprinted by permission of the American Ceramic Society.)

14.5 CYCLIC FATIGUE BY DEGRADATION OF CRACK BRIDGES

Definite evidence for cyclic fatigue by degradation of bridging grains in alumina has been presented by Lathabai, Rödel, and Lawn (1991). Tension–tension cyclic loading tests were conducted on a coarse grained (23–35 μm average grain size) alumina that shows T-curve behavior by grain-interlock bridging. Fatigue effects were observed with both small cracks introduced with an indenter and with large cracks using compact tension specimens. Bridge degradation was identified using a special device for in situ observation of crack propagation in the electron microscope.

Hu and Mai (1992) have developed a compliance method of determining the bridging stress and applied this method to a study of alumina ceramics under monotonic and cyclic loading. They found that the bridging stress after

4.3×10^5 cycles of stress with $\sigma_{min}/\sigma_{max} = -1$ was much smaller than that obtained with monotonic loading. They conclude that the frictional grain bridges are damaged by the compressive part of the cycle. They thus conclude that the use of an R-curve (T-curve) determined in monotonic loading is not appropriate for cyclic fatigue conditions.

14.6 SHORT CRACK FATIGUE OF CERAMICS

Studies of cyclic fatigue in ceramics using long crack techniques are clearly important in understanding basic fatigue phenomena in ceramics and data obtained in such experiments may apply to design considerations in some circumstances. However, the same problem arises as with other uses of long crack data. Namely, ceramics having only the naturally occurring short cracks are generally used for load-bearing applications so that fatigue data with short cracks is needed. Two approaches are being explored: cyclic loading studies under Hertzian loading conditions and under pointed indenters.

Reece and Guiu (1991) studied indentation fatigue of a high-purity alumina bioceramic in various alcohols, toluene, and simulated physiological environments with high water content. They applied a load P to a pointed indenter, removed the load, and then repeatedly applied a load $P' \leq P$ and determined the number of repeated loadings necessary to cause chipping. They constructed plots of the number of cycles to failure as a function of P' analogous to the usual S-N fatigue curves. The resulting curves showed that the presence of water considerably decreased the number of cycles to failure. However, the length of crack produced by the preload was independent of environment. The results suggest that relative humidity and chemical activity of water alone cannot fully account for the observed behavior. Gui, Li, and Reece (1992) studied cyclic fatigue in alumina and concluded that cyclic loads cause degradation of the strength of bridging ligaments.

Lawn and his collaborators have done extensive work on short crack fatigue of ceramics using Hertzian loading (Guibertau, Padture, Cai, and Lawn, 1993; Guiberteau, Padture, and Lawn, 1994; Lawn, Padture, Cai, and Guiberteau, 1994; Cai, Kalceff, and Lawn, 1994; Lawn, Padture, Guiberteau, and Cai, 1994). A spherical ball of radius r, Young's modulus E', and Poisson's ratio ν' is pressed with a force P against a flat specimen with Young's modulus E and Poisson's ratio ν. The deflection of the ball and substrate causes a circular contact of radius a. The mean compressive stress immediately under the ball is

$$\sigma = P / \pi a^2 \tag{14.4}$$

If the deflection is purely elastic, the Hertz theory of contact stresses can be used to obtain

$$a = \left(\frac{4Pkr}{3E}\right)^{\frac{1}{3}} \tag{14.5}$$

where

$$k = \frac{9}{16}\{(1 - v^2) + (1 - (v')^2(\frac{E}{E'})]\} \tag{14.6}$$

Under these conditions the compressive stress in the center just under the ball is 3/2 of σ from (14.4). Substituting P from (14.5) into (14.4) gives

$$\sigma = \frac{3E}{4\pi k}\left(\frac{a}{r}\right) \tag{14.7}$$

Lawn and his collaborators treat σ as indentation stress and a/r as indentation strain. The Hertz theory predicts a linear elastic curve of indentation strain as a function of indentation stress as shown by (14.7). Experimental results on several polycrystalline aluminas are given in Figure 14.6, which shows an initially linear portion as predicted. At higher stresses greater strain occurs corresponding to multiple subsurface cracking. The stress field below the surface is complex but is the sum of an isostatic compression and a shear stress in a roughly spherical volume below the contact circle. The isostatic pressure

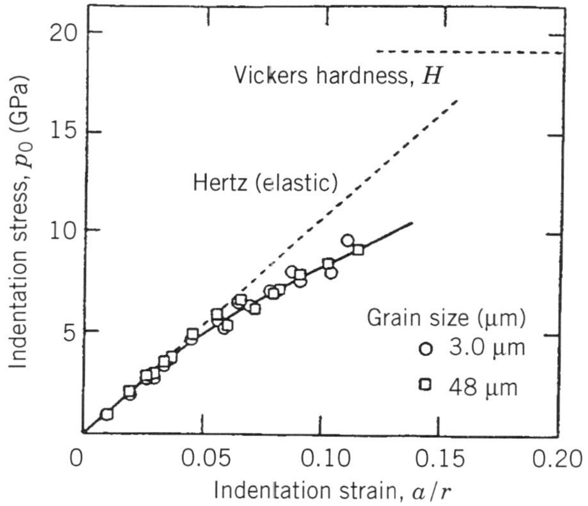

Figure 14.6 Indentation stress–strain curve on alumina. (From F. Guiberteau, N. P. Padture, and B. R. Lawn, "Effect of Grain Size on Hertzian Contact Damage in Alumina," *J. Am. Ceram. Soc.*, Vol. 77, 1825–1831, 1994. Reprinted with permission of the American Ceramic Society.)

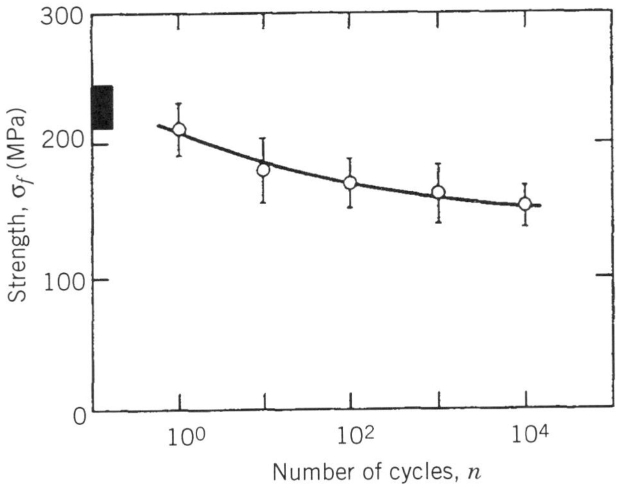

Figure 14.7 Biaxial strength of 23-μm grain-size polycrystalline alumina as a function of the number of cycles of loading with a sphere of radius 3.18 mm and load of 1000 N. (From F. Guiberteau, N. P. Padture, H. Cai, and B. R. Lawn, "Indentation Fatigue: A Simple Cyclic Hertzian Test for Measuring Damage Accumulation in Polycrystalline Ceramics," *Philos. Mag.*, Vol. A68, 1003–1016, 1993. reprinted by permission of the American Ceramic Society.)

allows the shear stress to reach a value sufficient to cause inelastic shear deformation to take place in alumina at room temperature (see Chapter 23) predominantly by twinning and possibly also by slip under high shear stress. The resulting grain boundary stress concentrations, acting in concert with thermal anisotropy stresses, cause grain boundary microcracking. Some amount of this type of damage occurs upon a single loading and unloading. Lawn and his collaborators have used repeated cycles of loading to study localized cyclic fatigue from Hertzian loading. Figure 14.7 shows that a single loading on the ball causes a decrease in the biaxial strength (average of four measurements) and that repeated cycles of loading and unloading on the ball cause further decrease in the subsequently measured biaxial strength. Damage from Hertzian loading is greatest for large grain size materials. Such damage is relatively insensitive to preexisting flaws because the shear deformation process generates flaws.

Lawn and co-workers conclude that the strategy for microstructural design to reduce damage from contact with blunt objects is to use small grain size and reduce the grain size distribution to eliminate the occasional large grains. This is in contrast to the strategy for increasing long crack toughness in materials that exhibit toughening by grain-interlock toughening. For the latter situation large grain size is favored for high toughness.

14.7 IMPLICATIONS OF CYCLIC FATIGUE FOR DESIGN WITH CERAMICS

Earlier chapters, including Chapter 8 on the static fatigue type of slow crack propagation, have presented the basis for a flaw-tolerant method of design with ceramics under load. In this approach the existence of cracks with an initial maximum length is assumed. Minimal safe life under a chosen stress can then be predicted based on the fracture mechanics approach to the static fatigue type of slow crack propagation. Recent results on cyclic fatigue of ceramics indicate that this procedure may drastically overestimate the safe life under stress when a significant component of time-varying stress is present. For this situation one might turn to lifetime predictions based on the Paris law. However, two serious difficulties arise. First, the Paris law m_f exponents for ceramics are much larger than for metals and apparently depend on microstructure in a way that is not well understood, making it difficult to choose a value to use. Second, the behavior of short cracks in cyclic fatigue in ceramics is different from that of long cracks and introduces additional uncertainties into any prediction scheme. Prediction of safe life of ceramics under cyclic fatigue conditions appears to be a problem on the frontier of research at the time of writing (1995).

CHAPTER 15

THERMAL STRESS AND THERMAL SHOCK IN CERAMICS

15.1 INTRODUCTION

Temperature gradients in ceramics cause stresses to develop because of the constraining effect of adjacent elements of the body. The theory of heat flow and temperature distribution resulting from given external conditions applied to a sample is well developed (Carslaw and Jaeger, 1959). The theory of thermal stress resulting from heat flow is also well developed (Boley and Weiner, 1960). If the thermal stress is large enough it can cause crack propagation, which may be either stable or unstable depending on the temperature distribution. If the crack propagation is stable, the result is a body with longer cracks and the strength is progressively lowered as the severity of the shock is increased. If the crack propagation is initially unstable but stops as the strain is relieved, the body experiences a discontinuous change in strength, as discussed later. Some cases have been worked out for specific cracks and temperature distributions, but the general problem is complex. Experimental results of single and multiple thermal shock experiments for a variety of ceramics have been presented by Lewis and Rice (1981). In this chapter we present some theoretical treatments based on the extensive work of Hasselman (1969, 1985). Before giving these results we give an example illustrating the magnitude that thermal stress can reach.

15.2 MAGNITUDE OF THERMAL STRESS

A simple example is that of a bar or cylinder constrained at the ends and heated in the center. This will approximate a furnace tube clamped at both ends in a water-cooled metal housing and heated in the middle. The resulting stress value indicates why some flexibility allowing movement of the tube ends is required in practice. If the heated length is L and the temperature is T while the temperature of the rigidly clamped ends is T_0, the change in length is $\Delta L = \alpha(T - T_0)L$, where α is the average thermal expansion coefficient for the temperature range involved. This is a strain of $\epsilon = \alpha(T - T_0)$ and produces a stress of $E\alpha(T - T_0)$. For $\alpha = 7 \times 10^{-6}$, $E = 400$ GPa, and $T - T_0 = 1000$ K, the resulting thermal stress is 2800 MPa, which is well above the strength of most ceramics. The example is crude but indicates that thermal stresses can be large and destructive. Allowing for thermal stresses in design with ceramics is thus crucial in devices that experience thermal gradients.

Such gradients can be constant in time as in cases of steady-state heat flow, or can change rapidly with time, as when a furnace or heat engine is shut off. Failure resulting from the latter situation is termed failure from thermal shock.

15.3 FIGURES OF MERIT FOR VARIOUS THERMAL STRESS CONDITIONS

Hasselman (1985) has considered various situations in which thermal stresses arise and can cause harmful effects if too large. Such harmful effects include not only crack failure (local or general) but also elastic instability. That is, there are various failure criteria such as fracture or elastic instability. For each combination of geometry and failure criterion, a particular combination of properties of the material appears in the expression for the maximum allowable temperature difference for the conditions assumed. This combination is termed a figure of merit for the material. Examples given by Hasselman are summarized below.

1. Flat plate externally constrained to prevent in-plane thermal expansion and of sufficient thickness to prevent thermoelastic instability (buckling) and subject to a uniform temperature change ΔT. This is the same problem discussed above. The maximum allowable temperature difference is

$$\Delta T_{max} = \frac{\sigma_f(1 - \nu)}{\alpha E} \tag{15.1}$$

where σ_f is the strength in tension σ_{ft} for $\Delta T < 0$ and is the strength in compression σ_{fc} for $\Delta T > 0$. Thus for the maximum allowable temperature difference one needs maximum strength, minimum thermal expansion, and minimum elastic modulus.

2. Thin rod constrained in axial direction as above but subject to thermoelastic instability. The maximum allowable temperature difference without thermoelastic instability is

$$\Delta T_{max} = \frac{C \pi^2 I}{\alpha A L^2} \tag{15.2}$$

where C is a constant that depends on the rotational constraints at the end of the rod, L is the length, I is the cross-sectional moment of inertia, and A is the cross-sectional area. As far as material properties go, maximizing ΔT simply means minimizing the thermal expansion.

3. Infinitely long hollow concentric cylinder undergoing radially outward heat flow.

$$\Delta T_{max} = \frac{\sigma_f(1-\nu)}{\alpha E}\left(2 \ln\left(\frac{b}{a}\right)\right) \times \left(1 - \frac{2a^2}{b^2 - a^2} \ln\left(\frac{b}{a}\right)\right)^{-1} \tag{15.3}$$

where a and b are the inner and outer radii of the tube. The figure of merit of the material for maximum temperature difference is $\sigma_f(1 - \nu)/\alpha E$. If the tube is used as a heat exchanger the heat flow is

$$Q = \frac{4\pi(1-\nu)K}{\alpha E}\left(1 - \frac{2a^2}{b^2 - a^2} \ln\left(\frac{b}{a}\right)\right)^{-1} \tag{15.4}$$

where K is the thermal conductivity. The figure of merit of the material for maximum heat flow is $\sigma_f(1 - \nu)K/\alpha E$.

4. Solid circular cylinder of radius b initially at uniform temperature subject to an instantaneous decrease in temperature. The heat transfer is assumed to be linearly convective with a heat transfer coefficient of h.

$$\Delta T_{max} = \frac{1.45\,\sigma_{ft}(1-\nu)}{\alpha E}\left(1 + \frac{3.41K}{bh}\right) \tag{15.5}$$

For a large convective heat transfer rate compared to conduction (i.e., $3.41K/bh \ll 1$)

$$\Delta T \approx \frac{1.45\sigma_{ft}(1-\nu)}{\alpha E} \tag{15.6}$$

so that high strength together with small thermal expansion and small Young's modulus is desirable for this condition. For larger conduction compared to convection (i.e., $3.41K/bh >> 1$)

$$\Delta T \approx \frac{4.94\sigma_{ft}(1-\nu)K}{\alpha Ebh} \tag{15.7}$$

so that high strength and thermal conduction combined with low values of thermal expansion and Young's modulus are desirable for this condition.

5. Solid circular cylinder initially at a uniform temperature subjected to an instantaneous decrease in the temperature of the surrounding medium with infinite boundary conductance.

$$\Delta T_{max} = \frac{\sigma_{ft}(1-\nu)}{\alpha E}\left\{\left(\frac{K_1\rho_1 c_1}{K_2\rho_2 c_2}\right)^{1/2} + 1\right\} \tag{15.8}$$

where the subscripts 1 and 2, respectively, designate the cylinder and the surrounding medium, K is thermal conductivity, ρ is density, and c is specific heat. For large thermal conductivity [i.e., $K_1\rho_1c_1/K_2\rho_2c_2)^{1/2} >> 1$], the material figure of merit is $\sigma_{ft}(1 - \nu)(K\rho c)^{1/2}/\alpha E$. For small thermal conductivity [i.e., $(K_1\rho_1c_1/K_2\rho_2c_2)^{1/2} << 1$], the material figure of merit is $\sigma_{ft}(1 - \nu)/\alpha E$.

6. Solid opaque circular cylinder with emissivity ϵ at a low initial temperature suddenly heated by black-body radiation with no convective cooling. The maximum radiant heat flux to avoid tensile fracture at the center is

$$q_{max} = \frac{\sigma_{ft}(1-\nu)K}{\alpha E\varepsilon\mu a^2} \tag{15.9}$$

where μ is the absorption coefficient. The figure of merit for the material under these conditions is $\sigma_{ft}(1 - \nu)K/\alpha E\epsilon\mu$.

Hasselman discusses these and other examples and points out that the same figures of merit for the material apply to other shapes with the same thermal boundary conditions. In all cases a low thermal expansion is desirable. In most cases a low value of Young's modulus is desirable.

15.4 CRACK PROPAGATION UNDER THERMAL STRESS

The above examples describe conditions to avoid failure. In some circumstances, such as thermal shock, failure may not be complete. That is, cracks may propagate some distance and then stop as they outrun the thermal stress field.

In a classic paper, Hasselman (1969) presented an analysis that is quite useful in understanding the effect of thermal shock in lowering the strength of ceramics when it does not lead to complete failure. He assumed randomly distributed, circular cracks of radius c with a crack density of N per unit volume. He assumed a solid body constrained in all directions and uniformly cooled through a temperature difference ΔT. This produces a triaxial tensile stress given by

$$\sigma = \frac{\alpha E \Delta T}{1 - \nu} \tag{15.10}$$

or equivalently a triaxial tensile strain given by

$$\varepsilon = \alpha \Delta T \tag{15.11}$$

The elastically stored energy per unit volume is

$$Ue = \frac{\sigma^2}{2B} = \frac{\varepsilon^2 E^2}{2B(1 - 2\nu)^2} = \frac{3}{2}\frac{\varepsilon^2 E}{1 - 2\nu} \tag{15.12}$$

where B is the bulk modulus and the latter equality follows from $B = E/3(1 - 2\nu)$. Using the Walsh (1965) expression for the bulk modulus of a body with cracks gives

$$E = E_0\left(1 + \frac{16(1 - \nu^2)Nc^3}{9(1 - 2\nu)}\right)^{-1} \tag{15.13}$$

or

$$U_e = \frac{3(\alpha \Delta T)^2 E_0}{2(1 - 2\nu)}\left(1 + \frac{16(1 - \nu^2)Nc^3}{((1 - 2\nu)}\right)^{-1} \tag{15.14}$$

The surface energy is

$$U_s = 2\pi N c^2 \gamma \tag{15.15}$$

Using the Griffith condition for this system [i. e., $d(U_e + U_s)/dc = 0$] gives

$$\Delta T_C = \left(\frac{\pi\gamma(1 - 2\nu)^2}{2E_0\alpha^2(1 - \nu^2)}\right)^{1/2}\left(1 + \frac{16(1 - \nu^2)Nc^3}{9(1 - 2\nu)}\right) c^{-1/2} \tag{15.16}$$

This expression can be simplified for the two cases of short and long cracks. For short cracks the second term in parentheses is small compared to 1 so that

$$\Delta T_C = \left(\frac{\pi \gamma (1-2\nu)^2}{2E_0 \alpha^2 (1-\nu^2)} \right)^{1/2} c^{-1/2} \tag{15.17}$$

while for long cracks the expression becomes

$$\Delta T_C = \left(\frac{128\pi\gamma(1-\nu)N^2c^5}{81\alpha^2 E_0} \right)^{1/2} \tag{15.18}$$

The expression (15.16) for critical temperature difference can be converted into an expression for critical thermal strain by dividing the equation by the first parentheses. The result is plotted in Figure 15.1 from Hasselman (1969) to give the critical thermal strain to initiate crack propagation as a function of crack half-length.

An expression for the crack length after propagation can be derived by assuming that cracks extend until all the energy released is converted into

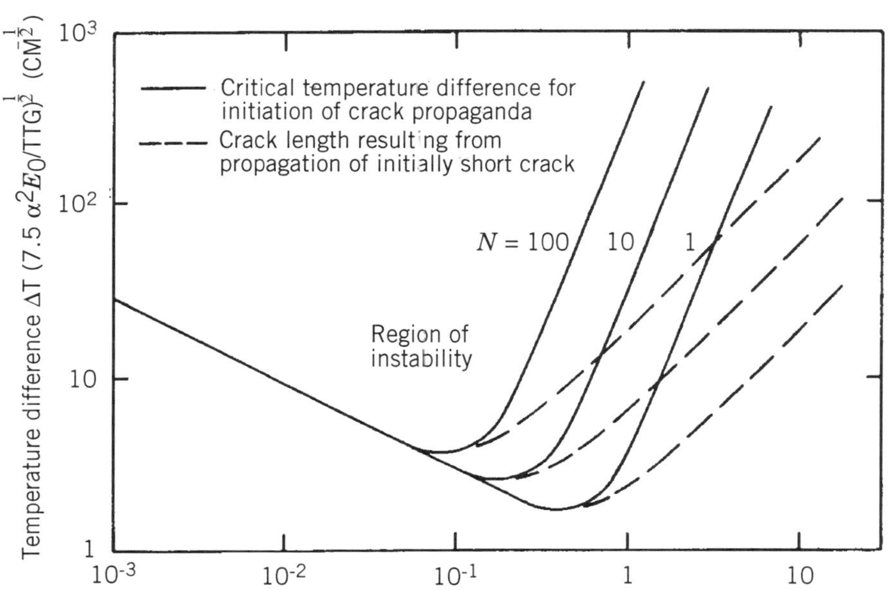

Figure 15.1 Minimum theraml strain ($\alpha\,\Delta T$ normalized to give units of square root of length) required to initiate crack propagation as a function of crack length and crack density N for Poisson's ratio = 0.25. (From D. P. H. Hasselman, "Unified Theory of Thermal Shock Fracture Initiation and Crack Propagation in Brittle Ceramics," *J. Am. Ceram. Soc.*, Vol. 52, 600–604, 1969. Reprinted with permission of the American Ceramic Society.)

crack surface energy. This is done by setting the difference of (15.14) written for initial and final crack lengths equal to the difference of (15.15) written for initial and final crack lengths:

$$\frac{3(\alpha\,\Delta\,T_C)^2 E_0}{2(1-2\nu)}\left\{\left[1+\frac{16(1-\nu)Nc_0^3}{9(1-2\nu)}\right]^{-1}-\left[1+\frac{16(1-\nu)Nc_f^3}{9(1-2\nu)}\right]^{-1}\right\}$$
$$= 2\pi N\gamma(c_f^2 - c_0^2) \tag{15.19}$$

For initial crack lengths that are small compared to the final crack length, the terms in c_0 in equation (15.19) can be dropped and equation (15.16) can be used to eliminate ΔT_C to give the final crack length as

$$c_f = \left(\frac{3(1-2\nu)}{8(1-\nu^2)c_0 N}\right)^{1/2} \tag{15.20}$$

The final crack lengths for initially short cracks after the critical temperature difference is exceeded are shown by the dashed lines in Figure 15.1. An important point is that after propagation from an imposed thermal strain that equals or slightly exceeds the critical value, the crack will be stable under subsequent imposition of somewhat larger thermal strains. This prediction is shown in Figure 15.2 from Hasselman (1969). As the size of the initial temperature difference is increased there is no crack propagation and no reduction in specimen strength until a critical value ΔT_C is exceeded, upon which the cracks extend kinetically to a new length, giving the specimen a lower strength. Exposing the specimen to a subsequent, slightly greater temperature difference causes no further crack propagation or further decrease in strength. If the weakened specimen is thermally shocked through a set of progressively larger temperature differences, eventually a critical value $\Delta T_C'$ is reached at which point further crack propagation and strength decrease do take place. Experimental data from Davidge and Tappin (1967) replotted by Hasselman (1969) is shown in Figure 15.3. The data fit Hasselman's general theoretical predictions quite well.

Specimens with initial crack lengths longer than the length corresponding to the minimum in the plot of Figure 15.1 should propagate stably. The predicted behavior of crack length and strength under increasing temperature difference is shown in Figure 15.4 from Larson, Coppola, Hasselman, and Bradt (1974). The left portion of this figure repeats the behavior shown in Figure 15.3 expected from the unstable propagation of initially short cracks. The right portion of Figure 15.4 shows the behavior expected for specimens with cracks longer than the length at the minimum of Figure 15.1. Figure 15.5 shows the expected behavior for a high-alumina refractory.

Analysis of these data was based on a treatment of thermal shock of the type given above that was modified to deal with the propagation of surface

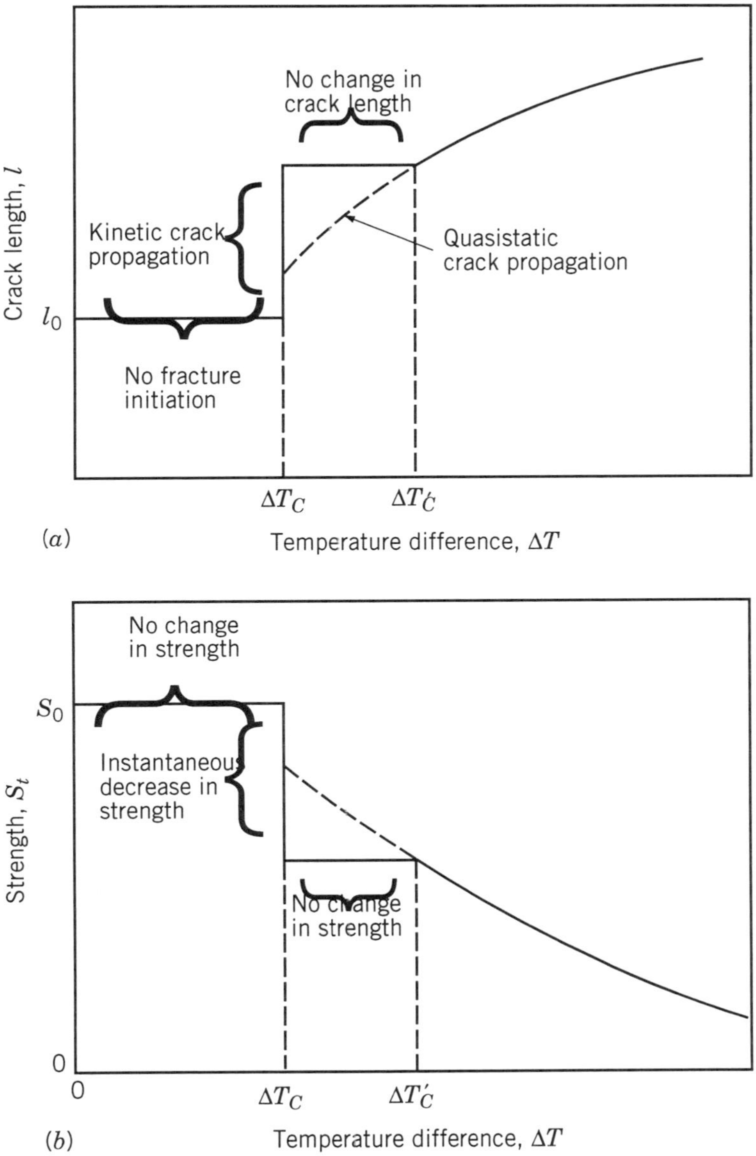

Figure 15.2 Crack length and specimen strength as a function of thermal history. (From D. P. H. Hasselman, "Unified Theory of Thermal Shock Fracture Initiation and Crack Propagation in Brittle Ceramics," *J. Am. Ceram. Soc.*, Vol. 52, 600–604, 1969. Reprinted with permission of the American Ceramic Society.)

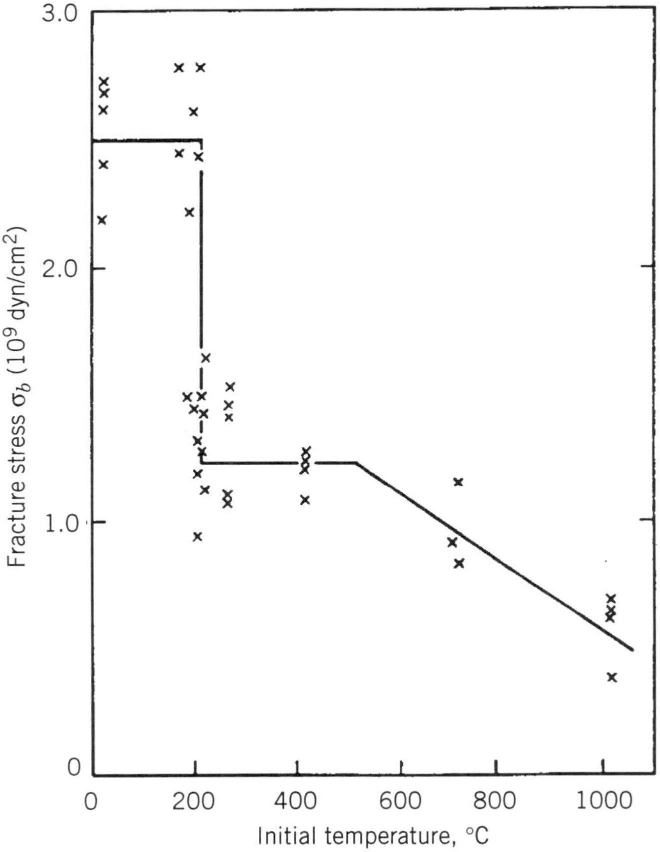

Figure 15.3 Strength at room temperature of 5-mm diameter alumina rods after quenching into water. (From D. P. H. Hasselman, "Unified Theory of Thermal Shock Fracture Initiation and Crack Propagation in Brittle Ceramics," *J. Am. Ceram. Soc.*, Vol. 52, 600–604, 1969. Reprinted with permission of the American Ceramic Society.)

cracks (Hasselman, 1971). This treatment uses an effective Young's modulus given by

$$E_{eff} = E(1 - 2\pi Nc^2)^{-1} \tag{15.21}$$

The treatment leads to a curve of the same shape as that in Figure 15.1 with the critical temperature difference given by

$$\Delta T_C = \left(\frac{2\gamma}{\pi\alpha^2 E_0 c_0}\right)^{1/2} (1 + 2\pi Nc_0^2) \tag{15.22}$$

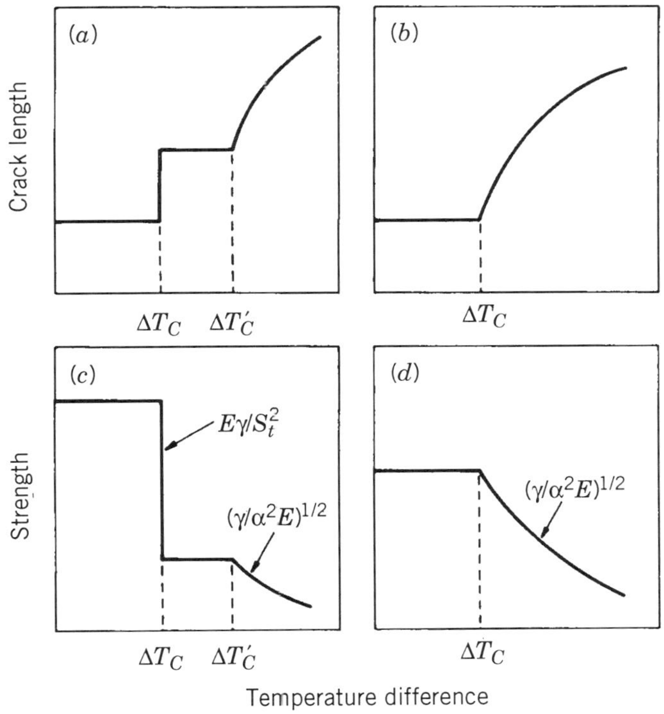

Figure 15.4 Crack propagation and strength for unstable (*a, c*) and stable (*b, d*) crack propagation conditions. (From D. R. Larson, J. A. Coppola, D. P. H. Hasselman, and R. C. Bradt, "Fracture Toughness and Spalling Behavior of High-Al_2O_3 Refractories," *J. Am. Ceram. Soc.*, Vol. 57, 417–421, 1974. Reprinted by permission of the American Ceramic Society.)

having a minimum at a crack length of

$$c_m = (6\pi N)^{-1/2} \tag{15.23}$$

and a final crack length for initially short cracks of

$$c_f = S_t^2/4N\gamma E = \frac{1}{4\pi N L_0} \tag{15.24}$$

For short cracks the relation between temperature difference and crack length is the same as in the previous model, but for long cracks the result is

$$\Delta T_C \approx \left(\frac{8\pi\gamma^2 c^3}{\alpha^2 E}\right)^{1/2} \tag{15.25}$$

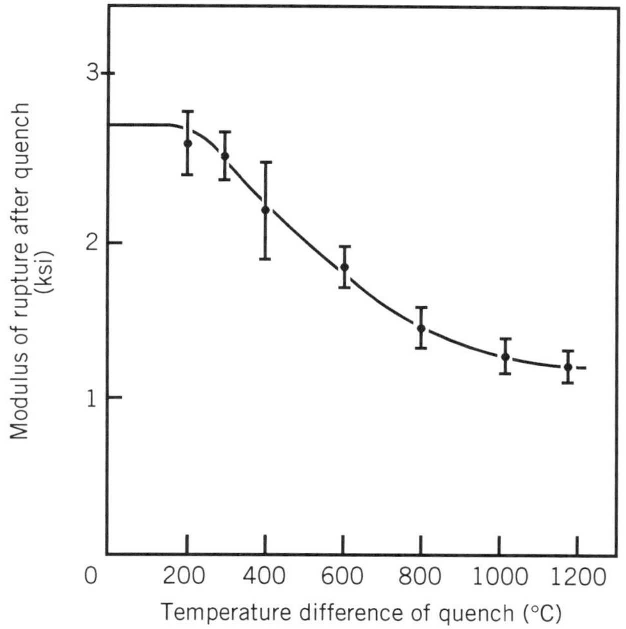

Figure 15.5 Strength of 90% alumina refractory quenched in water. (From D. R. Larson, J. A. Coppola, D. P. H. Hasselman, and R. C. Bradt, "Fracture Toughness and Spalling Behavior of High-Al_2O_3 Refractories," *J. Am. Ceram. Soc.*, Vol. 57, 417–421, 1974. Reprinted by permission of the American Ceramic Society.)

The strength retained after thermal shock was found to correlate with the fracture surface energy determined in a work of fracture measurement, as shown in Figure 15.6.

Lutz, Swain, and Claussen (1991) have shown that critical temperature for thermal shock is not very much different in "duplex ceramics" (dispersions of monoclinic zirconia particles in a matrix) but that the retained strength after exceeding the critical shock temperature is significantly improved compared with the matrix material.

15.5 MICROCRACKING CAUSED BY THERMAL STRESS

A polycrystalline material having anisotropic thermal expansion (or a two-phase material with different values of thermal expansion for the different phases) will develop internal stress when cooled from the sintering temperature. In some volume elements these stresses will be tensile and can cause localized cracking. It has been reported that this microcracking does not occur below a particular grain size characteristic of each material (Matsuo and Sasaki, 1966; Kuszyk and Bradt, 1973). A full theory requires treatment

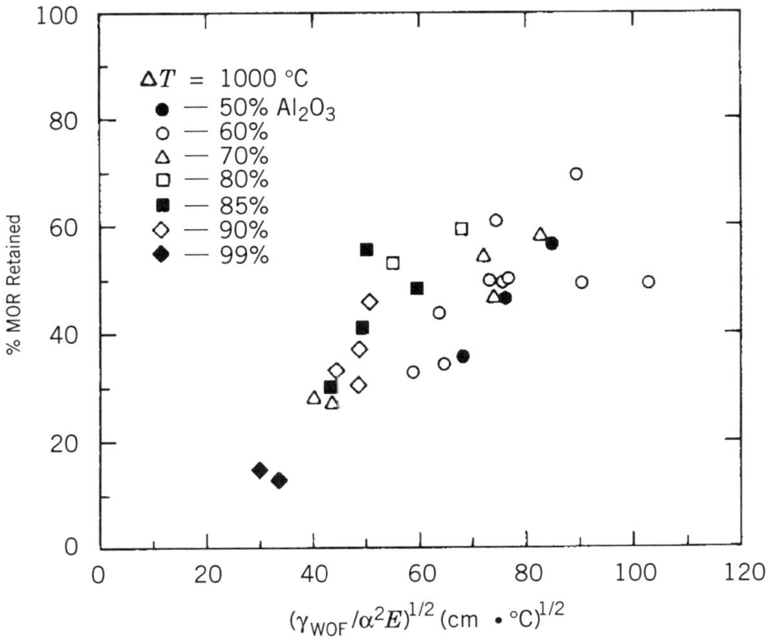

Figure 15.6 Retained strength of high-alumina refractories quenched into water as a function of a thermal-stress-resistance parameter. (From D. R. Larson, J. A. Coppola, D. P. H. Hasselman, and R. C. Bradt, "Fracture Toughness and Spalling Behavior of High-Al_2O_3 Refractories," *J. Am. Ceram. Soc.*, Vol. 57, 417–421, 1974. Reprinted by permission of the American Ceramic Society.)

of complex stress distributions and statistical effects. There is a long history of attempts to provide judiciously simplified theoretical models to deal with the essential physics (F. Clarke, 1964; Davidge and Tappin, 1968; Kuszyk and Bradt, 1973; Evans, 1978b; D. Clarke, 1980; Sridhar, Yang, Srolovitz, and Fuller, 1994).

The occurrence of a minimum size for microcracking caused by thermal stress requires explanation. One simple model considers an imbedded sphere of radius a having a thermal expansion difference $\Delta\alpha$ from the matrix at a temperature ΔT below the sintering temperature. The tangential tensile stress at the surface of the sphere is

$$\sigma = \frac{2\,G\,(\Delta\,\alpha)(\Delta\,T)(1+\nu)}{3(1-\nu)} \tag{15.26}$$

where G is the shear modulus and ν is Poisson's ratio of both matrix and sphere. It is noteworthy that the size of the sphere does not appear in the stress equation. Simply equating the thermal stress to a critical value does not give a grain size dependence.

Kuszyk and Bradt (1973) proposed a model that considers the conversion of the strain energy into surface energy. They assume an average strain energy SE per unit volume associated with a constrained grain of size l giving an initial total strain energy $SE\, l^3$. They assume that a crack of area l^2 and surface energy per unit area γ_f develops and completely relieves the strain energy. The change in energy upon cracking is

$$\Delta U = -SE\, l^3 + \gamma_f l^2 \tag{15.27}$$

They assume that SE in not a function of l. Considering progressively larger grain sizes, it is evident that a point is reached at which the energetically favorable situation is to have a crack develop and that the critical size is given by

$$l_{crit} \propto \gamma_f / SE \tag{15.28}$$

This model gives some rationale for the existence of a minimum grain size for microcracking but avoids discussion of the details of the process. Evans (1978) has noted that a proper theory should use as the critical condition for microcrack development "the energy transfer that occurs during the *crack increment* at which the strain energy *release* just exceeds the energy *dissipation*."

F. Clarke (1964) presented the first detailed theory, which was subsequently discussed by Davidge and Tappin (1968). He assumed a grain diameter of $2l$ and a boundary of length l with a pore of diameter c at each end (Figure 15.7). The thermal strain mismatch across the boundary was taken as $\epsilon = (\Delta\alpha)(\Delta T)$. He used an approximate stress distribution and calculated that the Griffith energy balance condition would be met at a critical value of half grain size l_c given by

$$l_c - c = \frac{24\, \gamma_{gb} (1 - \nu^2)}{E\varepsilon^2} \tag{15.29}$$

where γ_{gb} in the grain boundary fracture surface energy, E is Young's modulus, and ν is Poisson's ratio.

A more detailed treatment based on a four-grain model was given by D. Clarke (1980) following a method of calculation suggested by Evans (1978). This treatment more clearly shows the physical ideas involved and is briefly summarized here. The calculation is in two parts. First, the local stress distribution is calculated. Second, the applied stress intensity factor acting on a preexisting crack is calculated and set equal to the critical stress intensity factor.

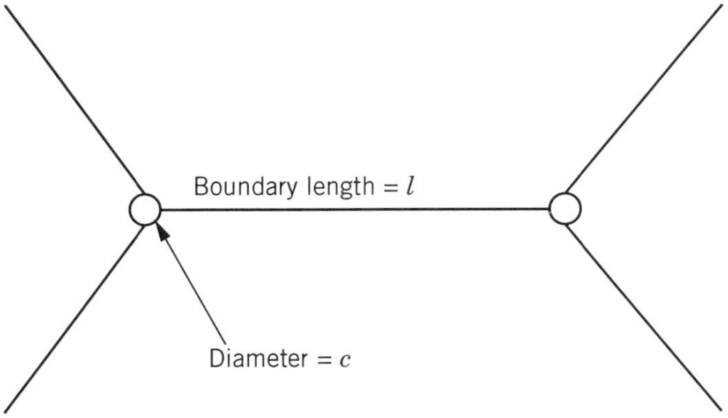

Figure 15.7 F. Clarke's model for fracture initiation at a grain boundary.

Clarke took a model of four square grains each of side length l (i.e., grain size of l rather than $2l$) and considered each to shear in a symmetrical way by an amount ϵ^s and to undergo a volume change ϵ^v. The relation of thermal expansion to volume change (or area change in two dimensions) and shear can be seen in a two-dimensional treatment as follows. Taking $\epsilon_c = \alpha_c \,\Delta T$ and $\epsilon_a = \alpha_a \,\Delta T$, the two-dimensional strain matrix is

$$\varepsilon = \begin{pmatrix} 0 & \varepsilon_a \\ \varepsilon_c & 0 \end{pmatrix} = \varepsilon_A \begin{pmatrix} 1 & 0 \\ 0 & 1 \end{pmatrix} + \varepsilon_S \begin{pmatrix} -1 & 0 \\ 0 & 1 \end{pmatrix} \tag{15.30}$$

where the area change is $\epsilon_A = (\alpha_a + \alpha_c)\,\Delta T/2$ and the shear change is $\epsilon_S = \Delta\alpha \,\Delta T/2$. He also assumed an initial flaw of size $2a$ and took $a = l/20$, which is equivalent to F. Clarke's assumption. He obtained a critical grain size for spontaneous microcracking of

$$l_c = \frac{32(1+\nu)^2 \pi^2 \gamma_{gb}}{E\left(\varepsilon^s K^s(a/l_c) + \varepsilon^v K^v(a/l_c)\dfrac{4(1+\nu)}{1-\nu}\right)^2} \tag{15.31}$$

D. Clarke calculated curves of applied stress intensity factors K^s and K^v from shear and volume change, respectively, as a function of a/l and from these curves took the critical values at $a/l = 0.05$ to be given by

$$K^s = \frac{\mu \varepsilon^s l^{1/2}}{2\,\pi(0.95)} \tag{15.32}$$

and

$$K^{\nu} = \frac{E\,\varepsilon^{\nu}\,l^{1/2}}{1(1-\nu)\pi(0.80)} \tag{15.33}$$

He applied the shear results to the case of thermal expansion anisotropy in alumina and found a critical grain size of 100 μm in good agreement with the value range of 75–100 μm found by Evans (1978). For aluminum titanate he assumed that γ_{gb} was in the range 1–5 J/m^2 and obtained a corresponding range of critical grain sizes of 0.9–4.5 μm in good agreement with the average value of 3 μm found experimentally by Kuszyk and Bradt (1973).

CHAPTER 16

FRACTOGRAPHY

16.1 INTRODUCTION

Fractography is the microscopic study of fracture surfaces and their interpretation in terms of fracture origin, failure stress, and time under stress before failure. Under favorable circumstances information on each of these aspects of fracture can be obtained. The results are often useful in design of materials, control of processing, and assessment of service conditions. Fractography is sometimes important in legal cases involving responsibility for failure. In practice, failure analysis is a mixture of art and science in which fractography is a critical component, but the experience and judgment of the analyst are important. Scientific principles based on fracture mechanics are involved. The present chapter presents a summary of such principles, but it is important to keep in mind the need for practical experience in the interpretation of fracture surfaces.

Recent reviews include Bradt and Tressler (1994), Mecholsky (1991), Michalske (1991), Varner (1991), Frechette (1990), Richerson (1986), and Mecholsky and Powell (1984). Important basic papers include Kirchner and Conway (1987a, 1987b), Kirchner and Kirchner (1979), Mecholsky, Gonzalez, and Freiman (1979), Mecholsky, Freiman, and Morey (1977), and Mecholsky, Rice, and Freiman (1974).

16.2 QUALITATIVE FEATURES OF FRACTURE SURFACES

The successive regions of a typical brittle fracture surface are shown in Figure 16.1 from Mecholsky, Rice, and Freiman (1974). These regions approximate semicircles for fracture originating from a half-penny-shaped surface crack in a specimen under uniform tension and are usually treated as such. The central region is the source of failure and is often approximately semicircular; it is represented in the figure by a half-ellipse with semiaxes a and b. This is surrounded by a smooth, flat region termed the mirror. Beyond the mirror region a ring of small radial ridges termed the mist region occurs. The radius of the boundary between the mirror and mist region is denoted by r_m. The mist region terminates in a region of bigger radial ridges. The radius of the boundary between the mist and hackle regions is denoted by r_h.

These regions are usually quite obvious in fracture surfaces of glass. They can often be discerned on the fracture surface of reasonably fine-grained polycrystalline ceramic fracture surfaces. These qualitative features can generally be used to find the fracture source at the center of the mirror region. This source can then be studied in higher magnification and various characterization techniques (including chemical analysis when the fracture source appears to be a region of different composition) can then be used to determine the nature of the source.

These features are thought to result from changes in the crack velocity as it grows under stress. Measurements of crack velocity in glass as a function of crack length indicate an approach to a steady-state value (about 1400 m/s for soda-lime glass after the crack reaches 2 cm in length). An equation that appears to give a reasonable fit to several sets of data is

$$v = v_m \left(1 - \frac{c_0}{c}\right) \tag{16.1}$$

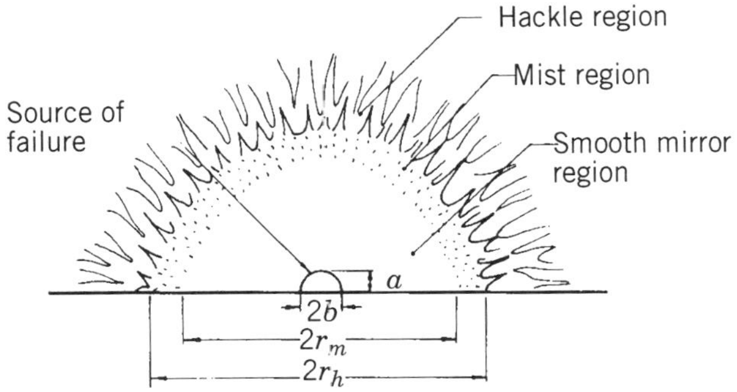

Figure 16.1 Schematic of glass fracture surface showing fracture mirror, mist region, and hackle region. (From J. J. Mecholsky, R. W. Rice, and S. W. Freiman, "Prediction of Fracture Energy and Flaw Size in Glasses from Measurements of Mirror Size," *J. Am. Ceram. Soc.*, Vol. 57, 440–443, 1974. Reprinted by permission of the American Ceramic Society.)

with $v_m = m(E/\rho)^{1/2}$, where m is about 0.4–0.5 (Doremus, 1982).

16.3 QUANTITATIVE FRACTOGRAPHY

For a given material fractography data generally obey the relation

$$\sigma\, r_j^{1/2} = M_j \tag{16.2}$$

where σ is the fracture stress, r_j is the mirror–mist (r_m), mist–hackle (r_h), or crack–branching (r_{cb}) boundary and M_j is the set of corresponding constants associated with these boundaries, respectively, for the given material (Mecholsky, personal communication, 1995). Note that r_j should be measured in a constant stress field, that is, for a flexure bar, along the tensile surface. A more general expression was developed by Kirchner and Kirchner (1979) and Kirchner and Conway (1987a, 1987b) and can be expressed succintly as

$$Y\sigma\, r_j^{1/2} = K_j \tag{16.3}$$

where r_j is now either the crack size, c [for $j = 0$; often taken as $c = (ab)^{1/2}$ for $a \neq b$], or the mirror–mist ($j = 1$), mist–hackle ($j = 2$), or crack–branching ($j = 3$) boundaries, K_j are the corresponding critical stress intensity constants, and Y_j are the geometric constants corresponding either to the crack or mirror boundaries. Note that $j = 0$ corresponds to equation (5.54).

Examples of plots of fracture stress as a function of the two mirror radii for various glasses are shown in Figure 16.2 from Mecholsky, Rice, and Freiman (1974). The initial flaw size equation (16.1) can be justified by using the Griffith equation in the form of (5.54)

$$\sigma = \left(\frac{2\gamma E}{Y^2 c}\right)^{1/2} \tag{16.4}$$

to give

$$\sigma c^{1/2} = \left(\frac{2E\gamma}{Y^2}\right)^{1/2} = \frac{K_{IC}}{Y} \tag{16.5}$$

Krohn and Hasselman (1971) used experiments on strength and critical stress intensity factor in soda-lime-silica glass to obtain $c = 0.10\, r_m$. Mecholsky,

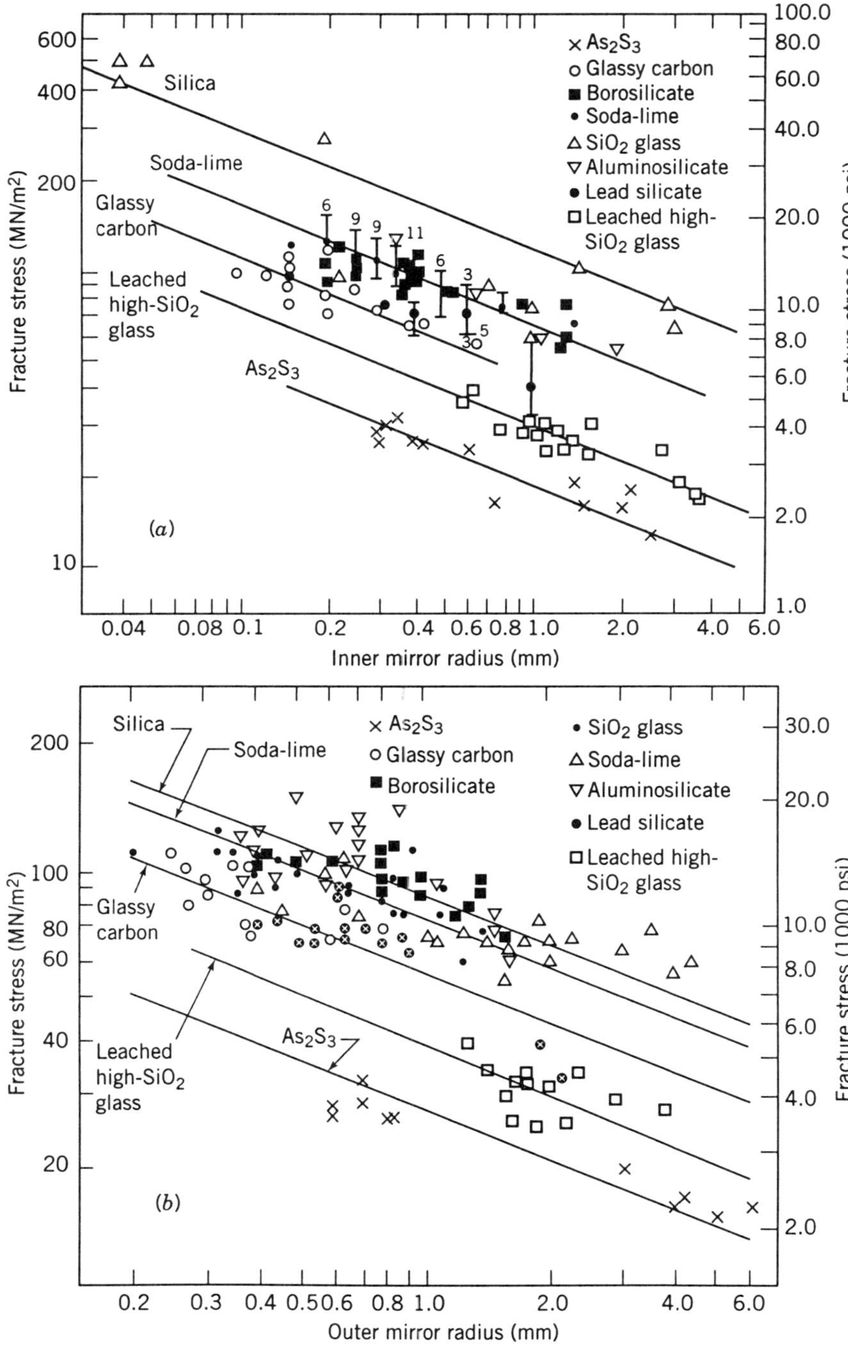

Figure 16.2 Strength as a function of (*a*) inner mirror (mist) radius and (*b*) outer (hackle) radius. (Mecholsky, Rice, and Freiman, 1974)

Rice, and Freiman (1974) studied a range of materials listed in Table 16.1. They measured the fracture surface energy using a constant moment, double cantilever beam technique. They also determined the strength in bending and measured c, r_m, and r_h. The ratios of mirror to flaw size generally are about 10; the two exceptions have over 20% porosity. The ratio of hackle radius to flaw size is generally about 13.

The fracture energies determined from the fractography measurements are compared with those determined in the double cantilever experiment in Table 16.2.

The above treatment does not consider the effect of slow crack growth on fractography but this was studied by Mecholsky, Gonzalez, and Freiman (1979). In dead-weight loading experiments at 50% relative humidity, the flaw size grew slowly from its initial value to a critical value and fracture then occurred. The ratio of fracture mirror radius to initial flaw size is a function of time but the ratio of mirror size to critical flaw size is a constant. The mirror constants were found to be the same in delayed and rapid fracture.

16.4 FRACTAL CONCEPTS IN FRACTOGRAPHY

The concept of fractals was introduced by Mandelbrot (1982) and has been applied to the description of various irregular outlines and surfaces including fracture surfaces (Mecholsky, and Freiman, 1991). A quantity D, called the fractal dimension, is used to characterize irregular objects. A perfectly flat square has $D = 2$, while a flat square with some superimposed bumps has a

TABLE 16.1 Flaw-to-Mirror Size Ratios

Glass	Mirror Radii (mm)		Flaw Size	Ratio	
	r_m	r_h	c	r_m/c	r_h/c
SiO_2	1.20	1.31	0.08	12.5	16.7
Aluminosilicate	0.71	0.86	0.06	12.5	14.3
Borosilicate	0.78	1.04	0.08	10.0	12.5
Soda-lime	0.59	0.85	0.05	12.5	16.7
Lead silicate	0.61	0.77	0.06	10.0	12.5
Leached high-SiO_2	0.75	1.40	0.04	14.3	25.0
Glassy C	0.18	0.32	0.04	4.6	7.7
As_2S_3	0.95	1.25	0.11	8.3	12.5
$0.3PbSe–0.7Ge_{1.5}As_{0.5}Se_3$	0.28	0.37	0.05	7.1	9.1
$Ge_{33}As_{12}Se_{35}$	0.75	1.10	0.07	10.7	15.7

TABLE 16.2 Calculated and Measured Fracture Energies

		Mirror Constant (MPa $\cdot$ m$^{1/2}$			Fracture Energy (J/m2)		
					From Radii Indicated		Measured
Glass	σ (MPa)	*E* (GPa)	A_m	A_h	r_m	r_h	
SiO_2	78.2	72.1	2.23	2.42	4.4	3.9	3.7
Aluminosilicate	97.4	89.1	2.14	2.40	3.3	3.6	3.7
Borosilicate	87.6	63.7	1.87	2.10	4.4	4.4	4.0
Soda-lime	83.7	73.4	1.92	2.21	3.2	3.2	3.5
Lead silicate	64.8	52.4	1.61	1.78	4.0	3.9	3.8
Leached high-SiO_2	28.2	16.9	0.91	1.19	2.2	1.9	2.5
Glassy C	75.1	24.8	1.17	1.67	9.7	10.8	8.5
As_2S_3	18.6	16.6	0.56	0.65	1.8	1.6	2.1
$0.3PbSe–0.7Ge_{1.5}As_{0.5}Se_3$	22.8	13.8	0.48	0.55	1.9	1.9	—
$Ge_{33}As_{12}Se_{35}$	19.8	22.1	0.55	0.65	1.0	1.0	—

fractal dimension $D = 2D^*$. A relatively flat object might have $D^* = 0.1$, while a very bumpy object almost filling a three-dimensional space might have $D^* = 0.9$. A technique for measuring D^* of a fracture surface was developed and used to show that

$$\frac{1}{D^*} \propto \frac{r_m}{c} \tag{16.6}$$

with a proportionality constant close to 1.

16.5 INTERLABORATORY COMPARISON IN FRACTOGRAPHY

A round robin on characterizing the fracture origin in a variety of ceramics was carried out under the VAMAS program (Swab and Quinn, 1995). The organizers concluded that the guidelines (Quinn, Swab, and Slavin, 1992) for characterizing fracture origins were adequate; the general attribute scheme of identity, location, and size worked well. There was a good consensus on location of the origin, but less so in determining the identity and size. Swab and

Quinn discuss the range of information available to the fractographer and consider that all of it must be used for success. One of the three methods in the ASTM standard to measure fracture toughness, the controlled surface flaw method discussed in Section 6.5, requires a fractographic measurement of the precrack size, which typically ranges from 50 to 100 μm deep.

CHAPTER 17

DISLOCATIONS AND PLASTIC DEFORMATION IN DUCTILE CRYSTALS

17.1 INTRODUCTION

Dislocations are the crystal defect primarily responsible for plastic deformation in crystals. They have a fundamental importance to studies of plasticity analogous to the fundamental importance of microcracks to brittle fracture. The study of dislocations forms a large part of the study of mechanical properties of metals. For ceramics dislocations are certainly important but do not occupy the same central place because of two facts. First, dislocations in many ceramics require for movement a stress higher than the fracture stress at moderate temperatures, typically below half the absolute melting point. Second, although dislocations do move at achievable stresses in some ceramics, in particular, the oxides with rock-salt structures such as MgO, the resulting plasticity is much more limited than in ductile metals and alloys. Dislocations in MgO interact with each other and with other microstructural features to cause failure at small plastic deformations in polycrystalline ceramics. On the whole then, in the study of mechanical properties of ceramics, dislocations take second place to fracture mechanics and the study of plastic effects associated with point defects, grain boundaries, and glassy phases.

Dislocations, however, do occur in ceramics and are important in hardness indents and in wear, in some cases, and do come into play at high temperatures in ceramics. Accordingly, they are treated here. The basic aspects of disloca-

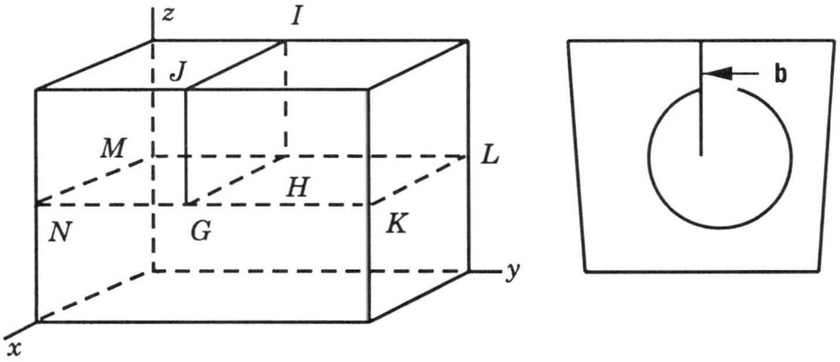

Figure 17.1 Schematic of edge dislocation.

tions are covered in a series of texts from the classical period of dislocation theory development (Cottrell, 1953; Read, 1953; Weertman and Weertman, 1992; Hull, 1965) and in later books (Nabarro, 1979–1989; Hirth and Lothe, 1982; Hull and Bacon, 1982).

17.2 DEFINITION OF A DISLOCATION

A dislocation is a line imperfection in a crystal with an associated misfit in the crystal. A dislocation is thus characterized by a description of the line (which need not be straight) and by the associated misfit vector, called the Burgers vector. This definition is best understood from examples.

The simplest example is a straight dislocation running through a crystal. Figure 17.1 shows an edge dislocation consisting of an extra half-plane of atoms *GHIJ*. The dislocation can be considered to have been produced by making a cut *GHIJ*, inserting a half-plane of atoms, and rejoining the crystal. The dislocation line *GH* is the edge of the extra half-plane.

Another example of a simple straight dislocation is the screw dislocation shown in Figure 17.2. Here the dislocation can be produced by making a cut

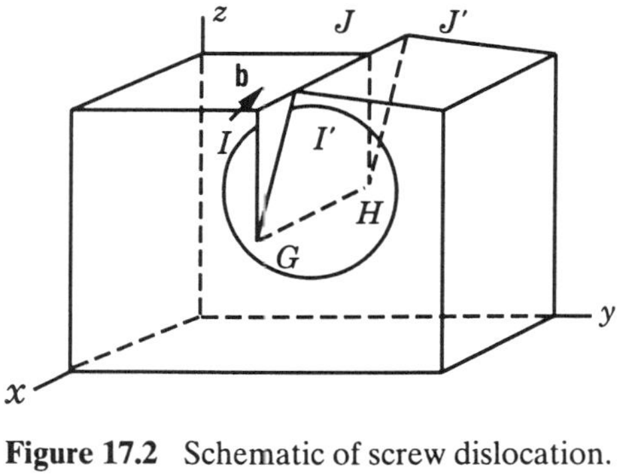

Figure 17.2 Schematic of screw dislocation.

GHIJ as before but instead of inserting an extra plane of atoms, one face of the cut is displaced parallel to *GH* through one atomic distance and rejoined. That is, *I* goes to *I*′ and *J* to *J*′. This is termed a screw dislocation. The dislocation line is again *GH*.

Dislocations are conveniently discussed using the concept of the Burgers circuit. Such a circuit is any path enclosing the dislocation line that would close in a perfect crystal. F. C. Frank considers a reference crystal of the material under consideration, which is perfect (i.e., free of imperfections), and draws a closed path in it. A path is then drawn in the real (imperfect) crystal with one-to-one correspondence between the atoms in the reference and the real crystal. The Burgers vector is the vector needed to close the Burgers circuit around the dislocation line from the point that the circuit would close if the crystal were perfect. The convention concerning Burgers vectors requires choosing a positive direction along a dislocation line such as *GH* (the opposite direction, *HG*, could equally well be chosen). A circuit is then drawn in a clockwise direction while looking along the positive direction of the dislocation line, as shown in Figure 17.1 **b**. An equivalent way to express this rule is to use the right-hand rule with thumb along the positive dislocation direction and fingers curled in the circuit direction. Unit cells are counted along the line *AB* until enough have been traversed to close the line in a perfect crystal. The distance required to close the line in the actual crystal is the Burgers vector. In Figure 17.1 the Burgers vector **b** for the edge dislocation *GH* is perpendicular to the plane *GHIJ*. In Figure 17.2 the Burgers vector **b** for the screw dislocation *GH* is parallel to the dislocation *GH*.

A dislocation can change character from edge to screw, as shown in Figure 17.3. Here a cut has been made on *ABC*, the point *B* is displaced to *B*′, and the crystal is rejoined. An extra row of atoms can be identified above point *C*. A Burgers circuit drawn in the plane about *C* gives a Burgers vector parallel to *BC* and hence to the segment of the dislocation line *AC* nearest to *C*. That is, the dislocation *AC* is an edge dislocation at *C*. A Burgers circuit drawn about *A* also gives a Burgers vector parallel to *BC*. However, the segment of the

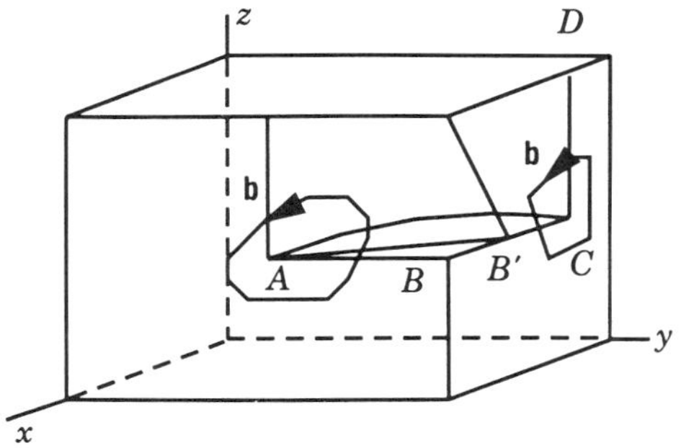

Figure 17.3 Burgers circuit for curved dislocation showing continuous change from screw to edge character.

dislocation line nearest A is parallel to **b**. That is, the dislocation AC is a screw dislocation at A. The character of the dislocation changes continuously from screw to edge in going from A to C. At any point the dislocation can be described as having a given amount of edge and screw character in terms of the components of the Burgers vector perpendicular and parallel to the dislocation at that point.

If a Burgers circuit does not close in the real crystal it encloses a dislocation. The vector from the end of the path to the beginning is the Burgers vector of the dislocation. Such a Burgers circuit can be moved along the dislocation and the Burgers vector will remain constant unless the circuit cuts through a dislocation as the circuit moves. Evidently dislocations cannot terminate within a crystal but only on the surface of the crystal. A dislocation line can close within a crystal by forming a loop, as shown in Figure 17.4. Here the dislocation can be considered as having been made by cutting the crystal in half, adding a disk of material one atomic layer thick, and rejoining the crystal. The Burgers vector is normal to the plane.

17.3 GLIDE AND CLIMB OF DISLOCATIONS

A dislocation can move by glide (also called slip) without needing the removal or addition of material. The slip surface at any point along a dislocation is limited to the plane at that point passing through the dislocation and the Burgers vector. In Figure 17.1 the slip plane for the edge dislocation GH is the plane $KLMN$ containing both GH and **b**. A screw dislocation has greater freedom of glide movement. In Figure 17.2 any plane through the screw dislocation GH will contain the Burgers vector so that the screw dislocation can glide on any plane through itself. In Figure 17.3 the plane ABC is the slip plane. In Figure 17.4 the slip surface is a cylinder, the glide cylinder, consisting of all the straight lines parallel to **b** and passing through the dislocation loop.

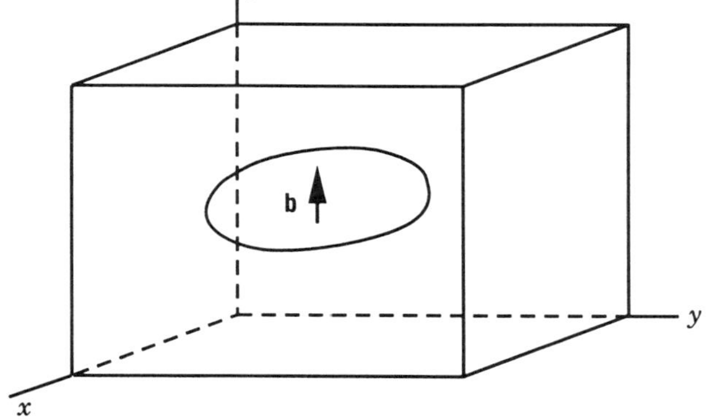

Figure 17.4 Closed dislocation loop in crystal.

Glide (slip) is a motion that conserves material, in contrast to climb, which requires the addition or removal of material. Suppose that in Figure 17.1 atoms diffuse to the line *GH* and add an extra row, thus moving the dislocation line one atomic spacing perpendicular to the plane *KLMN*. This is an example of climb. A general theorem is that if a closed loop of dislocation moves so as to change its projected area on a plane normal to the Burgers vector, then mass transport is required.

17.4 FORCE ON A DISLOCATION

A dislocation in a general stress field will experience a force component tending to cause glide and a force component tending to cause climb (Read, 1953; Weertman and Weertman, 1992).

To obtain the force causing glide, consider the effect of a shear stress $\tau = \sigma_{yz}$ on the edge dislocation shown in Figure 17.1. If the dislocation glides in the y direction the crystal will undergo a shear strain ϵ_{yz} and work will be done. This work is τb per unit area because the upper surface of the crystal is offset by b and the force on it is τ per unit area. The force per unit length of the dislocation is defined as the work done per unit length of dislocation and per unit of motion. The force per unit length on the dislocation caused by the stress τ is thus τb. The force is in the y direction; it tends to cause glide. It is noteworthy that the magnitude of this force does not depend on the direction of the dislocation line.

To obtain the force tending to cause climb, consider the effect of a tensile stress $\sigma = \sigma_{yy}$ on the edge dislocation of Figure 17.1. If enough atoms diffuse to the dislocation line *GH* to move the dislocation down one atomic spacing, the crystal will expand and work will be done by the tensile stress. The work is σb per unit area. The force per unit length on the dislocation caused by σ is σb. This force is in the z direction; it tends to cause climb.

An equation for the force per unit length on a dislocation segment of arbitrary character in an arbitrary direction in a general stress field can be written (Weertman and Weertman, 1992). The dislocation segment is defined by a unit vector **t** along the segment and the Burgers vector **b** which can be at any angle to **t**. The most general stress has components σ_{ij}. It is convenient to define a vector **G** in terms of the stress and the dislocation direction by

$$G_x = \sigma_{xx} b_x + \sigma_{xy} b_y + \sigma_{xz} b_z \tag{17.1a}$$

$$G_y = \sigma_{yx} b_x + \sigma_{yy} b_y + \sigma_{yz} b_z \tag{17.1b}$$

$$G_z = \sigma_{zx} b_x + \sigma_{zy} b_y + \sigma_{zz} b_z \tag{17.1c}$$

The vector **F** giving the force per unit length on the dislocation is given by the Peach–Koehler equation:

$$\boldsymbol{F} = \boldsymbol{t} \times \boldsymbol{G} \tag{17.2}$$

The total force on the dislocation line is always perpendicular to the line because of the vector cross product.

It is instructive to examine the force acting on the curved dislocation of Figure 17.3 under the stress $\tau = \sigma_{xz} = \sigma_{zx}$. The dislocation is shown in the xy plane in Figure 17.5. Consider the forces acting on the dislocation at the three points A (where the dislocation is pure screw), C (where the dislocation is pure edge), and E (where the dislocation is mixed with half edge and half screw character). t_z is zero everywhere so that only t_x and t_y need be considered. The components of the Burgers vector are $b_x = b$, $b_y = b_z = 0$. Since both σ_{ij} and b_i are constant along the dislocation, G_{ij} will be constant along the dislocation. Accordingly, $G_x = 0$, $G_y = 0$, and $G_z = b\,\sigma_{zx}$.

At the point A, the unit vector tangent to the dislocation is $t_x = -1$ and $t_y = 0$. The force components are $F_x = 0$, $F_y = -\tau b$, and $F_z = 0$.

At the point C, the unit vector tangent to the dislocation is $t_x = 0$ and $t_y = 1$. The force components are $F_x = \tau b$ and $F_y = 0$.

At the point E, the unit vector tangent to the dislocation is $t_x = \sqrt{2}/2$ and $t_y = \sqrt{2}/2$. The force components are $F_x = (\sqrt{2}/2)\,\tau b$ and $F_y = (\sqrt{2}/2)\,\tau b$. The magnitude of the total force at point E is τb and the direction is perpendicular to the dislocation at that point.

The result for the dislocation segment AEC is that the shear stress $\tau = \sigma_{xz} = \sigma_{zx}$ applies a constant force everywhere perpendicular to the curved dislocation line, tending to expand the area that has undergone slip.

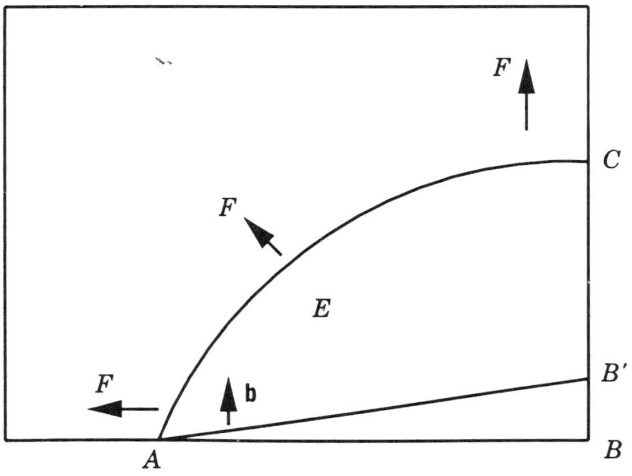

Figure 17.5 Force on curved dislocation.

17.5 STRESS FIELD AND ENERGY OF A DISLOCATION

The stresses associated with a dislocation in an otherwise perfect crystal free of stress from any external forces are of interest in determining the energy of the dislocation and its ability to interact through stress with other dislocations. The calculations are too lengthy to give here. The results for an edge dislocation lying along the z direction with the Burgers vector **b** in the x direction are

$$\sigma_{xx} = \frac{\mu b}{2\pi(1-\nu)} \frac{y(3x^2+y^2)}{(x^2+y^2)^2} \tag{17.3}$$

$$\sigma_{yy} = -\frac{\mu b}{2\pi(1-\nu)} \frac{y(x^2-y^2)}{(x^2+y^2)^2} \tag{17.4}$$

$$\sigma_{zz} = \nu(\sigma_{xx}+\sigma_{yy}) = \frac{\mu b \nu}{\pi(1-\nu)} \frac{y}{(x^2+y^2)} \tag{17.5}$$

$$\sigma_{xy} = -\frac{\mu b}{2\pi(1-\nu)} \frac{x(x^2-y^2)}{(x^2+y^2)^2} \tag{17.6}$$

$$\sigma_{xz} = \sigma_{yz} = 0 \tag{17.7}$$

The self-energy of a dislocation can be obtained by integrating the energy density expressed in terms of the stresses associated with the dislocation. The stresses are written in polar coordinates and integrated over the distance r from the dislocation. The integral diverges both at the origin and at infinity. Accordingly, the lower limit of integration is taken to be $5b$ and the energy from the material within this core is evaluated separately. The upper limit of integration is left as R, corresponding to the size of the crystal. The result for the energy per unit length of an edge dislocation is

$$U = \frac{\mu b^2}{4\pi(1-\nu)} \ln \frac{R}{5b} \tag{17.8}$$

An important feature of this result is the dependence on b^2. Evidently dislocations with the smallest Burgers vector possible are favored from the point of view of minimizing energy.

Weertman and Weertman (1992) estimate the core energy as being about $\mu b^2/10$ and conclude that the core energy is typically only 10–20% of the energy from the logarithmic formula above. They estimate the self-energy of a dislocation in aluminum to be about 3.3 eV per atomic length and contrast this with the energy of about 1 eV needed to produce a lattice vacancy in aluminum.

17.6 FORCE REQUIRED TO MOVE A DISLOCATION

The concept of dislocations was originally introduced to explain how crystals could undergo slip at shear stress values far below the theoretical shear stress of a perfect crystal. It is evident from qualitative arguments that a dislocation should be able to move at far lower stresses. For an edge dislocation, the atoms above the slip plane are displaced in one direction on one side of the dislocation and in the other direction on the other side. The component of the interatomic forces parallel to the slip plane will be in opposite directions for atoms an equal distance from the dislocation on either side. The force to slightly displace the dislocation appears to be zero from this qualitative argument. A more detailed treatment leads to the Peierls–Nabarro equation for the resistance, σ_P, to motion by a screw dislocation:

$$\sigma_P = \alpha\frac{\mu b}{2c}\exp(-2\pi w/c) \tag{17.9}$$

Here α is a constant around ½, c is the atomic spacing in the direction of motion of the dislocation, and w is the width of the dislocation. This equation is interesting in showing that the resistance to dislocation motion depends strongly on the width of the dislocation. In ceramics dislocations generally have small widths and high resistance to motion. More detailed calculations show that the resistance to motion depends very strongly on the interatomic forces. Accurate calculations of the resistance appear to await improved knowledge of interatomic force laws.

17.7 LINE TENSION OF A DISLOCATION

The self-energy per unit length of a dislocation loop of radius R can be shown to be approximately

$$U \approx \frac{\mu b^2}{4\pi}\ln\frac{R}{5b} \tag{17.10}$$

This self-energy per unit length is equivalent to a line tension T of the same value

$$T \approx \frac{\mu b^2}{4\pi}\ln\frac{R}{5b} \tag{17.11}$$

It is customary to take

$$\frac{1}{4\pi}\ln\frac{R}{5b} \approx 1 \tag{17.12}$$

giving

$$T \approx \mu b^2 \tag{17.13}$$

A curved dislocation line at a point r under tension T is subject to a force F per unit length perpendicular to the line given by (Weertman and Weertman, 1992)

$$F = T\frac{d^2\mathbf{r}}{dr^2} \text{ and } \frac{d^2\mathbf{r}}{dr^2} \approx \frac{1}{R} \tag{17.14}$$

This gives a force per unit length perpendicular to the dislocation due to line tension of

$$F \approx \frac{\mu b^2}{R} \tag{17.15}$$

17.8 DISLOCATION MULTIPLICATION

As noted above, a shear stress on a dislocation segment in a slip plane will tend to cause slip. The dislocation should move over the entire plane and be annihilated by forming a step at the surface of the crystal. Thus plastic deformation of a crystal would be expected on this simple picture to remove dislocations from crystals. Experiments showed the opposite. Plastic dislocation greatly increases the dislocation density. This puzzling discrepancy was solved by the discovery of mechanisms for dislocation multiplication.

Read's version of this multiplication mechanism is shown in Figure 17.6. A dislocation ABC is made up of two straight segments AB and BC. The segment BC lies in a slip plane $EFGH$ to which a shear stress is applied. The segment AB does not glide, but the segment BC rotates indefinitely about AB, causing an unlimited amount of slip.

The Frank–Read source is shown in Figure 17.7. Here the dislocation $ABCD$ is made up of three straight line segments as shown. The initially straight dislocation (labeled 1 in Figure 17.8) bows out under a shear stress until a position (labeled 2) is reached at which the restoring force (force due to line tension plus the Peierls force) equals the driving force caused by the applied stress. If the stress is increased the loop expands until position 3 is reached. Any further increase in stress causes the radius of curvature to decrease and the loop becomes unstable, sweeping around as shown in position

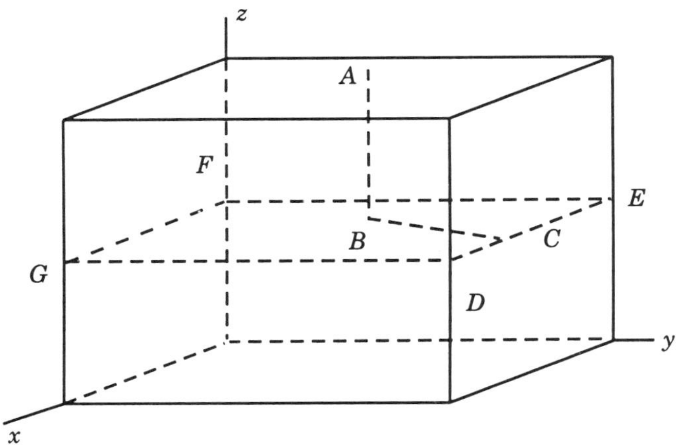

Figure 17.6 Read's dislocation source.

4. When the lobes of the loop meet below BC the sections of dislocation join, producing two separate dislocations. One of these is a free loop labeled 5, which can now expand throughout the crystal and the other forms a segment labeled 1′ that can repeat the multiplication process.

An extension of the Frank–Read source by Koehler (Weertman and Weertman, 1992) is shown in Figure 17.8. A portion JK of a curved dislocation in the slip plane $ABCD$ is parallel to the Burgers vector and so has screw character. This portion can slip on any slip plane containing the Burgers vector such as

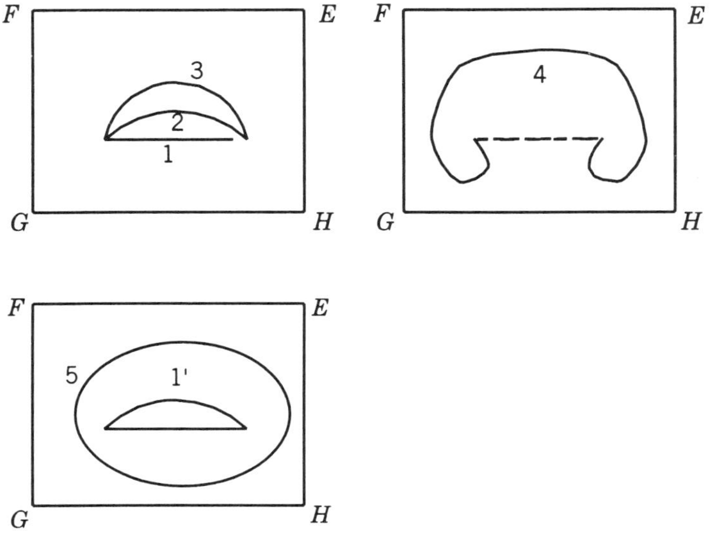

Figure 17.7 The Frank–Read dislocation source.

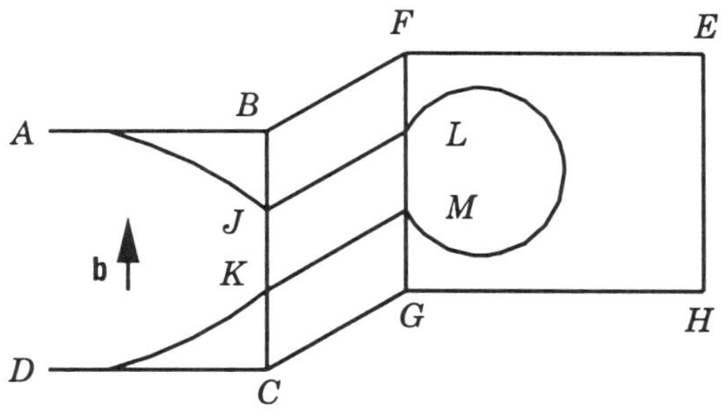

Figure 17.8 Cross slip leading to Frank–Read source.

the plane *FBCG*. After *JK* slips to the position *LM*, it then forms a Frank–Read source on the slip plane *EFGH*.

The stress needed to cause a Frank–Read source to spin off dislocations can now be calculated. If the pinning points of a Frank–Read source are a distance L apart in a slip plane and if a shear stress τ is applied, the dislocation will bow out and the radius of curvature will decrease until $R = L/2$. Since the force on a dislocation is τb the critical stress needed to bow the Frank–Read source into a semicircle is

$$\tau = \mu \frac{b}{R} \tag{17.16}$$

A slight increase in stress will push the dislocation further out, the radius of curvature will decrease, and the dislocation loop becomes unstable, causing the source to produce free dislocation loops.

17.9 KINKS AND JOGS

A typical crystal contains many dislocations, called a dislocation forest. When a moving dislocation crosses one of these forest dislocations (called a tree), the moving dislocation is sheared as it passes through the tree dislocation. The resulting step produced in the dislocation is called a kink if it lies on the slip plane and a jog if it lies perpendicular to the slip plane. The moving dislocation can rid itself of a kink by slip. Removal of a jog, however, requires diffusion of atoms.

17.10 FORCES BETWEEN DISLOCATIONS

Two dislocations will in general cause forces on each other because one produces a stress field that causes a force on the other. Two screw dislocations lying parallel to the z axis are shown in Figure 17.9 If they have Burgers vectors of magnitude b_1 and b_2 they produce a force along the line joining the dislocation center of magnitude

$$F = \frac{\mu b_1 b_2}{2\pi r} \tag{17.17}$$

The force is repulsive if the Burgers vectors are in the same direction and attractive if the two Burgers vectors are oppositely directed.

Two edge dislocations lying parallel to the z axis are shown in Figure 17.10. If they have Burgers vectors of magnitude b_1 and b_2 they create force components given by

$$F_x = \frac{\mu b_1 b_2}{2\pi r} \cos\theta\,(\cos^2\theta - \sin^2\theta) \tag{17.18}$$

$$F_y = \frac{\mu b_1 b_2}{2\pi r} \sin\theta\,(1 + 2\cos^2\theta) \tag{17.19}$$

If the angle θ from the slip plane to the line joining the dislocations is equal to 45°, the component of force between the dislocations parallel to the slip plane is zero. For $0 < \theta < 45°$ or $135° < \theta < 180°$ the force parallel to the slip plane is repulsive so that the dislocations will be driven apart. For $45° < \theta < 135°$ the force is attractive. In the latter range of θ the dislocations will be attracted to a position directly above one another. For a whole series of parallel edge dislocations the result is the formation of a small-angle grain boundary.

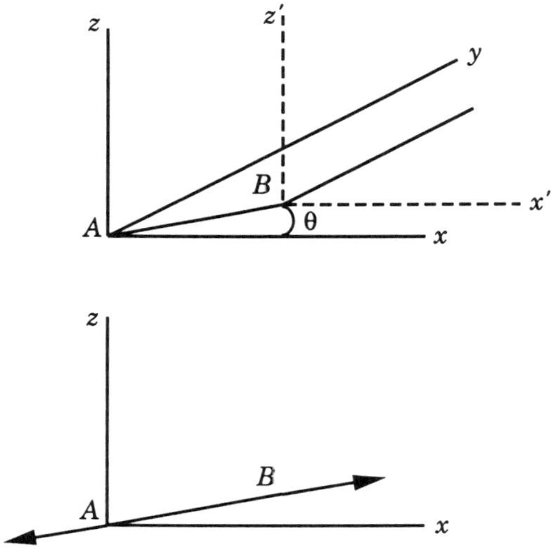

Figure 17.9 Force between parallel screw dislocations.

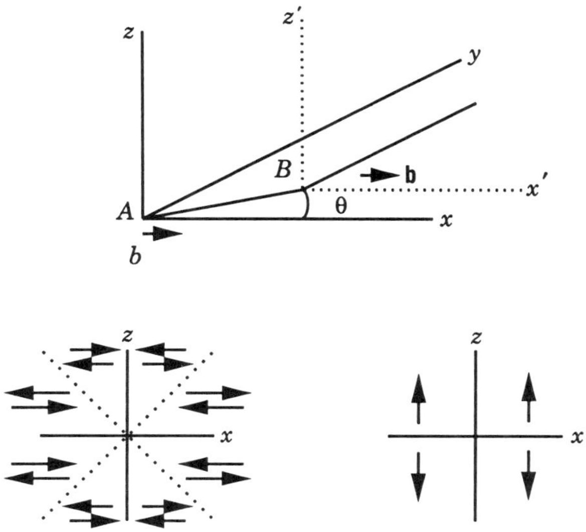

Figure 17.10 Force between parallel edge dislocations.

17.11 DISLOCATION PILEUPS

If a dislocation source operates under a shear stress in a plane containing an obstacle to dislocation motion, such as a grain boundary, dislocations pile up against the boundary and exert a back stress on subsequent dislocations emerging from the source. The result affects either the yield behavior or the fracture behavior depending on whether the boundary permits a dislocation to pass through or causes a crack to nucleate. The resulting equation for the yield stress or the fracture stress is called the Hall–Petch relation (Hall, 1951; Petch, 1953, 1968). We consider the case of fracture and follow the treatment given by Davidge (1979). Figure 17.11 shows a dislocation pileup against a boundary with the edge dislocations indicated schematically by an inverted T. A set of N dislocations under a shear stress τ produces a tensile stress $\sigma = N\tau$ at the head of the pileup. Each dislocation has to overcome the slip resistance τ_0 so that the fracture condition is

$$N(\tau - \tau_0) = \sigma_{theo} \tag{17.20}$$

where σ_{theor} is the theoretical strength. The number of dislocations in the pileup can be expressed in terms of the length of the pileup l by setting the elastic shear displacement written in terms of the dislocation parameters equal to its value written in terms of stress and shear modulus μ to give

$$Nb = \frac{(\tau - \tau_0)\, l}{\mu} \tag{17.21}$$

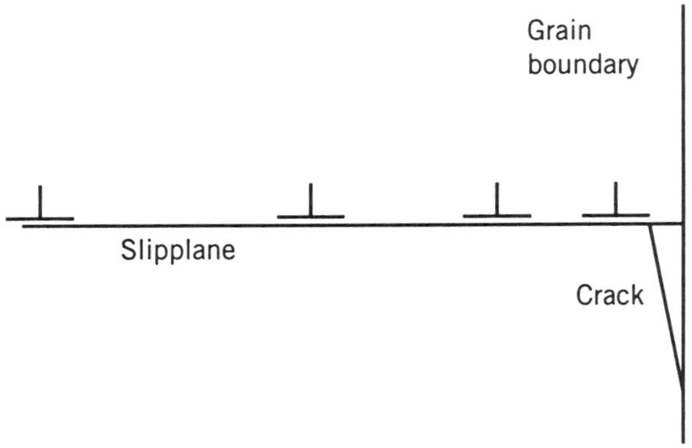

Figure 17.11 Dislocation pileup causing crack at grain boundary.

which leads to

$$(\tau - \tau_0)^2 = \frac{\mu b}{l} \sigma_{\text{theo}} \tag{17.22}$$

Recalling that $\mu = E/[2(1 + \nu)]$ and that for the most favorable orientation for slip $\sigma = 2\tau$ gives the tensile stress to open a crack at the grain boundary as

$$\sigma_f = \sigma_0 + \left(\frac{2E\, b\, \sigma_{\text{theo}}}{1 + \nu}\right)^{1/2} l^{-1/2} = \sigma_0 + k_f\, l^{-1/2} \tag{17.23}$$

The quantity k_f has the dimensions of a critical stress intensity factor but is associated with the length of the dislocation pileup rather than a crack length. This equation gives a dependence of strength upon the reciprocal square root of grain size, as is often observed in polycrystalline ceramics over some range of grain size. A dislocation mechanism appears to be responsible for fracture in polycrystalline MgO so that use of this simple model may be justified as an approximation for MgO. Dislocation mobility appears to be too low to justify use of this fracture model at or near room temperature in most other ceramics, however.

For materials with appreciable dislocation-based ductility, the Petch–Hall relationship is also used to describe the grain size dependence of the yield stress σ_y by writing

$$\sigma_y = \sigma_0 + k_y\, l^{-1/2} \tag{17.24}$$

17.12 OROWAN'S EQUATION FOR STRAIN RATE

Orowan's equation gives the strain rate produced by glide or climb of a dislocation. The treatment of Poirier (1985) is followed here for the case of glide of an edge dislocation but the resulting equation is general for climb or glide of an arbitrary dislocation. Consider a block of crystal with dimensions L, l, and h, respectively, in the x, y, and z directions. An edge dislocation running through the crystal in the y direction glides in the x direction, causing a displacement of the upper surface with respect to the lower by the Burgers vector, **b**. The shear strain is

$$\varepsilon = \frac{b}{h} \tag{17.25}$$

If the dislocation moves only a distance ΔL, the shear strain is

$$\varepsilon = \frac{b}{h}\frac{\Delta L}{L} \tag{17.26}$$

If N parallel dislocations each move through ΔL, the strain is

$$\varepsilon = \frac{N}{hL} b\Delta L = \rho\, b\, \Delta L \tag{17.27}$$

where $\rho = N/hL$ is the dislocation density. Taking ρ and ΔL as average values and differentiating with respect to time gives

$$\dot{\varepsilon} = b[\Delta L \frac{d\rho}{dt} + \rho \frac{d(\Delta L)}{dt}] \tag{17.28}$$

In some cases, such as steady-state creep, it is reasonable to assume that the dislocation density changes only slowly with time and that its time derivative can be taken as zero compared with the other term, giving Orowan's equation

$$\dot{\varepsilon} = \rho b v \tag{17.29}$$

where v is the average dislocation velocity.

17.13 DISLOCATION VELOCITY

The velocity of dislocations gliding on a slip plane under shear stress is the result of a balance of the driving force caused by the stress (discussed earlier) and the resistance force. The latter is the sum of the resistance of the otherwise perfect crystal (termed the Peierls force) and the resistance caused by other dislocations and other imperfections. Several studies of isolated dislocations (Johnston and Gilman, 1959; Gilman and Johnston, 1960; Stein and Low, 1960) indicate that the velocity of a dislocation at a given temperature in the absence of other imperfections is a strong function of the shear stress. For LiF, edge dislocations were found to move somewhat faster than screw dislocations, but over a range of about 6–20 MPa both ranged over about six orders of magnitude and followed the empirical equation

$$v = v_0 \left(\frac{\tau}{\tau_0} \right)^P \tag{17.30}$$

where v_0, τ_0, and P are experimentally determined constants. For LiF, $P \approx 16.5$. This equation cannot hold to very high stress levels because the limiting velocity of a dislocation is thought to be that of sound shear waves. The experimental data on LiF indicate a bending over of the velocity curve at high stresses consistent with this theoretical limit.

Dislocation velocity is usually limited by obstacles such as discrete particles and other dislocations so that the above equation for dislocation velocity is not generally used in the description of plasticity or creep. The process of dislocation motion is instead viewed as the overcoming of obstacles with the assistance of thermal activation. The mean velocity is written as

$$v_{mean} = \beta\, b\, \nu\, e^{-\left(\frac{\Delta G(\tau)}{kT} \right)} \tag{17.31}$$

where β is a numerical constant, b is the magnitude of the Burgers vector, ν is a frequency, and ΔG is the activation energy that depends on the shear stress, τ (Frost and Ashby, 1982). The form of the stress dependence of the activation energy depends on the type and geometrical arrangement of the obstacles. For example, for regular array of box-shaped obstacles the result is

$$\Delta G(\tau) = \Delta F \left(1 - \tau / \hat{\tau}\right) \tag{17.32}$$

where ΔF is the activation energy without applied stress and $\hat{\tau}$ is the shear strength in the absence of thermal energy and includes the effects of the obstacles. This method of treating dislocation velocity is the basis for constructing portions of the deformation maps developed by Frost and Ashby. We consider primarily plastic effects in the present chapter, take up creep and related dynamic effects in Chapter 19, briefly discuss the representation of plasticity and creep in deformation maps in Chapter 19, and consider fracture maps in Chapter 20.

17.14 HARDENING BY SOLID SOLUTION AND PRECIPITATION

Impurities and solid solution atoms can interact with dislocations through elastic, electrical, and chemical effects. Elastic interactions are particularly important and easy to visualize quantitatively. An edge dislocation has a region of compression on the side of the slip plane with the extra half-plane of atoms and a region of expansion on the other side of the slip plane. Evidently impurity atoms larger than the normal atoms of the crystal can minimize their elastic misfit energy by locating in the expanded region. Given time to diffuse to the dislocation they will tend to do so. Similarly, impurity atoms smaller than the normal atoms of the material will tend to diffuse to the compression side of the slip plane near an edge dislocation. The increased concentration of impurities that develops near a dislocation is termed a dislocation atmosphere. It requires energy to separate a dislocation from its atmosphere. If the dislocation is forced by stress to move faster than the atmosphere can diffuse to keep up with its motion an extra resistance must thus be overcome. Once the dislocation has moved some distance away from its atmosphere the force to continue motion is less. This is one effect that leads to a yield point in a constant strain rate experiment.

An alloy with a solid solution in equilibrium at a high temperature may no longer be in equilibrium if the temperature is lowered. Typically, the new equilibrium state is a mixture of precipitates of one composition in a more dilute solid solution. A dislocation moving on a slip plane containing a precipitate may be able to glide through the precipitate particle but more typically finds it difficult to do so. Instead, the dislocation can bend around the particle (much in the manner of a Frank–Read source), rejoin on the other side, and leave behind a loop of dislocation surrounding the precipitate.

Both dislocation atmospheres and precipitate particles are important sources of hardening in metals and alloys. Ceramics at temperatures and stresses that permit dislocation motion show similar effects.

For strong obstacles every obstacle will pin the dislocation so that the mean distance between pinning points will be the same as the mean spacing of obstacles, L. Then the shear stress necessary to break free a semicircular loop of dislocation is given by

$$\tau_{max} = \frac{\mu b}{L} \tag{17.33}$$

If the obstacles are not too strong the dislocation will pull free at a less than semicircular shape and the average spacing between pinning points will be larger. Both effects lower the hardening effect of obstacles (Courtney, 1990).

17.15 YIELD POINT BEHAVIOR

Yield points are often seen in constant strain rate measurements on ductile materials. A crystal in which all dislocations have impurity atmospheres will deform elastically as the stress increases until a stress sufficient to pull the dislocations free of the restraining atmosphere is reached. After the dislocation is pulled away from its atmosphere it can move under a lower stress, so that plastic strain occurs to accommodate the displacement imposed by the testing machine. The stress accordingly drops somewhat and the material is said to show a yield drop after the yield point. In addition, the dependence of dislocation velocity on stress can make the yield drop greater. This can be seen by substituting the velocity equation in Orowan's equation for strain rate and solving for stress to obtain

$$\tau = \tau_0 \left(\frac{\dot{\varepsilon}}{v_0 \rho\, b} \right)^{1/P} \tag{17.34}$$

Once the yield point is reached the dislocation density increases rapidly, causing the shear stress required to sustain an imposed plastic strain rate to decrease, so that the yield drop is increased. This effect is often observed in metals, where it is primarily due to impurity atmospheres, and also occurs in single crystal aluminum oxide at temperatures above about 1000 °C, where it is primarily due to the low value of the velocity exponent P.

17.16 WORK HARDENING

Once the yield point is exceeded and dislocation multiplication begins, the dislocation density increases rapidly with increasing strain. Interaction of dislocations produces work hardening; that is, as the strain increases greater stress is required to produce further strain. The models for work hardening are complex in detail but a common experimental result in ductile materials is that work hardening is proportional to the square root of the dislocation density (Courtney, 1990) according to

$$\tau = \tau_0 + \alpha\, \mu\, b\, \rho^{1/2} \tag{17.35}$$

Here α is a numerical constant equal to about 0.4 for body-centered cubic metals and 0.2 for face-centered metals.

17.17 HARDENING IN GENERAL

Ductile materials often yield at quite low stresses compared to stress levels used in engineering practice. The use of much higher stresses to carry load is possible because of hardening effects so that hardening is a key technology in the load-carrying use of metals as well as an interesting basic phenomenon. All the hardening effects mentioned above can act simultaneously. For example

(Hertzberg, 1989), an equation proposed to describe the yield strength of lath martensite is

$$\tau_{ys} = \tau_0 + k\, C^{1/2} + k_y\, d^{-1/2} + \alpha\, \mu\, b\, \rho^{1/2} \tag{17.36}$$

where k is a constant, C is the concentration of solute in solid solution, d is the size of packets of martensite, and the rest of the quantities have already been defined. The terms in sequence describe the effects of the intrinsic resistance to dislocation glide (the Peierls stress), solid solution strengthening, boundary hardening, and strain hardening.

17.18 THE PLASTIC STRESS–STRAIN CURVE FOR DUCTILE MATERIALS

We have noted that the behavior of dislocations is essentially kinetic. That is, a given shear stress produces a dislocation velocity. As dislocation motion and multiplication proceed, hardening takes place, typically in proportion to the square root of the dislocation density as stated above. After hardening has developed to the point where resistance to further dislocation motion is equal to the applied stress, dislocation motion ceases at a particular plastic strain. The plot of this value of plastic strain as a function of applied stress (taken as force divided by the initial cross-sectional area) is the engineering stress–strain curve. For ductile materials the strain before failure occurs may be large, typically over 1 (i.e., over 100%).

Stress–strain curves can also be determined by applying a strain and measuring the stress that results. That is, a stiff testing machine can be used to force the specimen to strain at a given rate. For strain rates too great to allow work hardening to develop fully at each strain, the resulting stress–strain curve depends on the strain rate. Below some strain rate for a given material and a given temperature the stress required for a given strain is independent of strain rate and the material can be described by a single stress–strain curve in this range of strain rates. This is the range of plastic behavior.

An effect sometimes noted in ductile materials is that the engineering stress–strain curve characterizing plastic behavior exhibits a maximum stress followed by a declining stress with a considerable period of continuing strain before the specimen fails. This effect is due to the fact that the cross section of the specimen decreases. Plastic deformation is a constant volume process to a good approximation. Taking A_0 and L_0 as the initial cross-sectional area and initial length the constant volume condition is

$$A\, L = A_0\, L_0 \tag{17.37}$$

Defining

$$\sigma_t = \frac{P}{A} \text{ and } \sigma_0 = \frac{P}{A_0} \tag{17.38}$$

leads to

$$\sigma_t = \frac{P}{A_0}\frac{A_0}{A} = \frac{P}{A_0}\frac{L}{L_0} = \frac{P}{A_0}\left(\frac{\Delta L + L_0}{L_0}\right) = \sigma_0(\varepsilon + 1) \tag{17.39}$$

The definition of engineering strain (i.e., the strain definition used throughout this book) uses the change of length, ΔL, divided by the initial length, L. An alternative definition is to consider the strain to occur in small increments, δL, and to divide each such increment by the length at the beginning of the increment, giving

$$\varepsilon_t = \frac{\delta L_1}{L_0} + \frac{\delta L_2}{L_1} + \cdots = \int_{L_0}^{L} \frac{dL}{L} = \ln\left(\frac{L}{L_0}\right) = \ln(\varepsilon + 1) \tag{17.40}$$

This quantity is usually called the "true" strain (Dieter, 1986). A plot of σ_t as a function of ϵ_t is called the true-stress–true-strain curve. Such true stress–strain curves typically show a curve of stress that continues to increase until the point of failure, reflecting the continuing strain hardening as strain increases all the way to the failure strain. However, this does not remove the fact that a maximum load (leading to a maximum engineering stress) usually occurs before the ultimate strain is reached. The reason for the decreasing load beyond the maximum load value is the occurrence of necking. The latter is the development of concentrated plastic deformation in a limited portion of the length of the specimen.

Necking is the result of plastic instability. The condition for necking can be derived by setting the change in load equal to zero, giving

$$dP = d(\sigma_t A) = \sigma_t \, dA + A \, d\sigma_t = 0 \tag{17.41}$$

Using the constant volume condition gives

$$\frac{d\sigma_t}{\sigma_t} = -\frac{dA}{A} = \frac{dL}{L} = d\varepsilon_t \tag{17.42}$$

so that

$$\frac{d\,\sigma_t}{d\,\varepsilon_t} = \sigma_t \tag{17.43}$$

Thus necking will occur in uniaxial tension at the strain at which the slope of the true stress–strain curve equals the true stress at that strain. The necking condition can also be written in terms of the engineering strain as

$$\frac{d\sigma_t}{d\varepsilon} = \frac{\sigma_t}{1+\varepsilon} \tag{17.44}$$

The strain at which necking begins, ϵ_u, can be expressed in terms of a strain hardening coefficient, n_s. This coefficient ranges from about 0.02 to 0.50 for metals near room temperature (Courtney, 1990). The true stress–strain curve is often represented by a power law

$$\sigma_t = K\,(\varepsilon_t)^{n_s} \tag{17.45}$$

Using equation (17.43) gives

$$\frac{d\,\sigma_t}{d\,\varepsilon_t} = n_s\,K\,(\varepsilon_t)^{(n_s-1)} = \sigma_t = K(\varepsilon_t)^{n_s} \tag{17.46}$$

or

$$\varepsilon_{t,\,\text{necking}} = n_s \tag{17.47}$$

Thus the true strain at which necking begins in conditions of plastic behavior is equal to the strain-hardening coefficient. Note that necking can also occur under creep conditions; a necking condition in terms of steady-state creep parameters then applies.

17.19 PARTIAL DISLOCATIONS

Up to this point we have considered dislocations whose Burgers vectors are a repeat distance in the crystal. Such a dislocation is termed a perfect dislocation and is a line defect. That is, large distortions of the crystal structure occur only in a small core region surrounding the mathematical line describing the position of the dislocation. Outside the core the crystal structure is perfect apart from some degree of elastic deformation. In some crystal structures a perfect dislocation can break up (dissociate) into two partial dislocations connected by a strip of material (a stacking fault) across which the normal interatomic distances and angles are not those of the perfect crystal. This process can be visualized for a face-centered cubic crystal having one atom per lattice point for which the atomic positions are shown in Figure 17.12. If the atoms are considered to be spheres, the (111) plane is a hexagonal array as shown in Figure 17.13. The initial (111) plane is labeled *A*. The next plane of atoms above the *A* plane is located above the site labeled *B* and the plane above that is located above the site labeled *C*. The stacking sequence of (111) planes in this face-centered cubic (fcc) structure is thus *ABCABC* . . . . A hexagonal close-packed (hcp) structure with a single atom per lattice point has (0001) plane with the same atomic arrangement as layer *A* of Figure 17.13. The next (0001) plane is located above the site labeled *B* but the plane above that is located above *A*. The stacking sequence of (0001) planes is thus *ABABAB* . . . . The nearest-neighbor relationships are the same in the fcc and hcp structures so that the energy difference results from differences in the relationships of the next-nearest and more distant neighbors.

A total dislocation (*td*) with Burgers vector $\mathbf{b}_1$ can dissociate into two partial dislocations (*pd*) and a stacking fault (*sf*) as indicated symbolically using boldface to indicate vectors by

$$td\,(\mathbf{b}_1) \rightarrow pd\,(\mathbf{c}_1) + pd\,(\mathbf{c}_2) + sf$$

with

$$\mathbf{b}_1 = \mathbf{c}_1 + \mathbf{c}_2 \tag{17.48}$$

so that the Burgers vector is conserved. From Figure 17.14, $\mathbf{c}_1$ is given by the vector sum

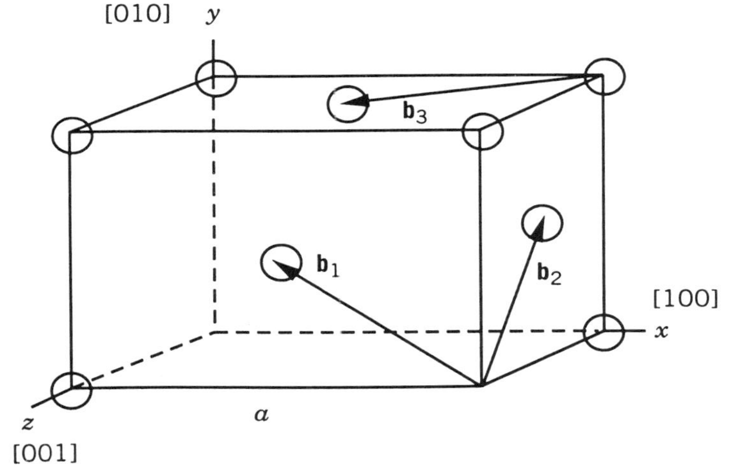

Figure 17.12 Face-centered cubic structure $\mathbf{b}_1 = (a/2)\,[1\bar{1}0]$, $\mathbf{b}_2 = (a/2)\,[01\bar{1}]$, and $\mathbf{b}_3 = (a/2)\,[\bar{1}01]$.

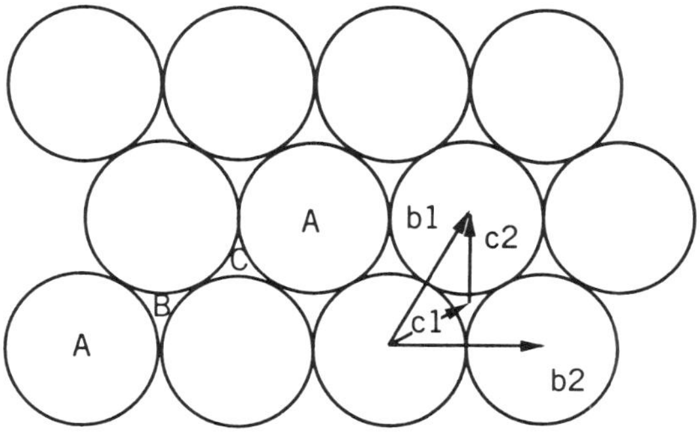

Figure 17.13 Atomic arrangement in (111) plane. Successive (111) planes are stacked in the sequence $ABCA\ldots$.

$$\boldsymbol{c}_1 = \frac{1}{3}(\boldsymbol{b}_1 + \boldsymbol{b}_2) = \frac{a}{6}[\bar{1}2\bar{1}] \tag{17.49}$$

and

$$\boldsymbol{c}_2 = \boldsymbol{b}_1 - \boldsymbol{c}_1 = \frac{a}{6}[\bar{2}11] \tag{17.50}$$

Taking the energy of a dislocation as μb^2 as discussed in Sections 17.5 and 17.7 it follows that the energy of the two partial dislocations is less than that of the total dislocation because

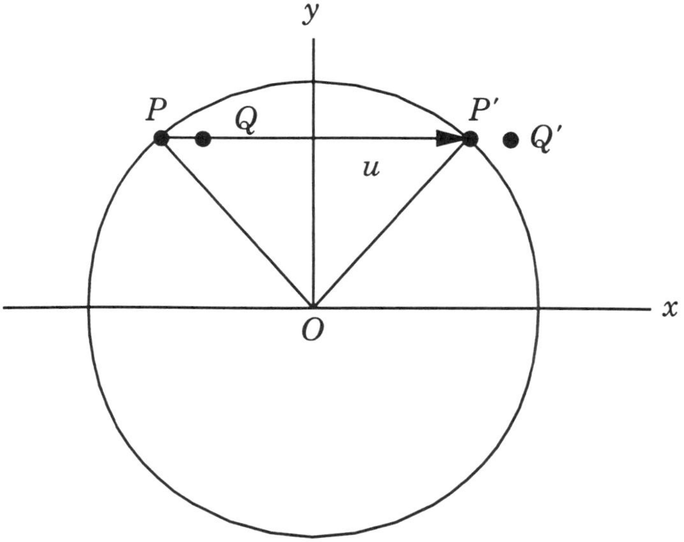

Figure 17.14 Twinning showing uniform shear.

$$b1^2 = a^2/2 \tag{17.51}$$

and

$$c1^2 = c2^2 = a^2/6 \tag{17.52}$$

so that the energy difference

$$\mu\,[(b1)^2 - 2\,(c1)^2] = \mu\, a^2/6 \tag{17.53}$$

is available to form a strip of stacking fault between the two partial dislocations. Another important feature of partial dislocations is that they can act to form a deformation twin, which is an important mechanism of deformation in some ceramic crystals including aluminum oxide and zirconium oxide.

17.20 DEFORMATION TWINNING

Deformation twinning is a process by which a portion of a crystal transforms into a different position while remaining joined to the rest of the crystal along a plane called the twin plane. The shifted (twinned) portion of the crystal cannot be brought into registry with the rest of the crystal by translation but can be made to coincide by a symmetry operation such as a mirror plane. The history of attempts to understand and correctly describe twinning is extensive (Cahn, 1954); we shall give only a brief summary of the deformation produced and the geometrical elements used to describe this deformation.

Twinning is produced by a uniform shear of a portion of a crystal as shown in Figure 17.14. Below $y = 0$ the crystal is undisturbed while the portion above undergoes a uniform shear s. The twin is described by two undistorted planes (K_1, K_2), two special directions (η_1, η_2), and s. The displacements u, v, w of a point initially at x, y, z are $u = sy$, $v = w = 0$. The plane $y = 0$ is undistorted by the shear in the x direction; this plane and direction x are termed K_1 and η_1, respectively. There is a second plane within which distances are unchanged by the shear; a point P is carried into a point P' such that $OP = OP'$. In contrast, the point Q is carried into Q' but $OQ \neq OQ'$. Distances in the plane through OP and the z axis are unchanged; this plane is termed K_2 and OP is termed η_2. The result of the motions is that K_1 is a mirror plane (Figure 17.15). Twins may be related by other symmetry elements such as a glide plane.

The above summary of the geometry of twinning describes the motion of lattice points; for a crystal with a single atom per lattice point as in a face-centered cubic metal the atomic motions in deformation twinning correspond to the lattice point motions. For a crystal with more than one atom per lattice point the atomic motions must correspond to the symmetry element, often a mirror plane, linking the twinned and untwinned portions of the crystal.

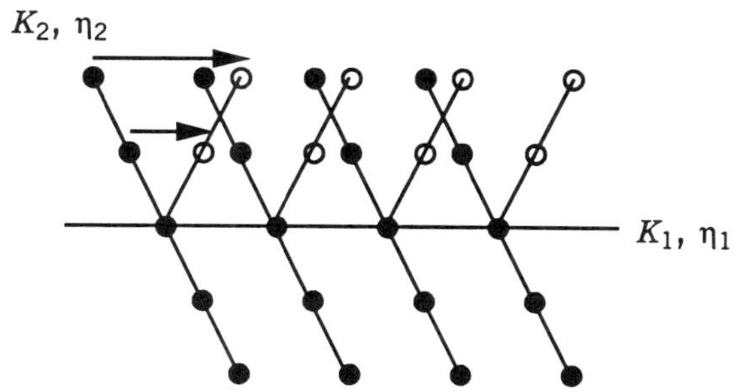

Figure 17.15 Twinning showing motion of lattice points.

The formation of a deformation twin can be described by the motion of a partial dislocation on each plane above the K_1 plane. Following Weertman and Weertman (1992), Figure 17.16 shows schematically the formation of a deformation twin in a fcc structure with $K_1 = (111)$. On each plane above K_1 a partial dislocation with $\mathbf{c}_1 = \frac{a}{6}[\bar{1}2\bar{1}]$ has entered the crystal from the left and moved halfway across. The first plane above K_1 is shifted $A \rightarrow$ B. The second plane above experiences the shifting effect of two partials and so is shifted $B \rightarrow$ C $\rightarrow$ A. Successive planes are shifted in this pattern, resulting in the stacking

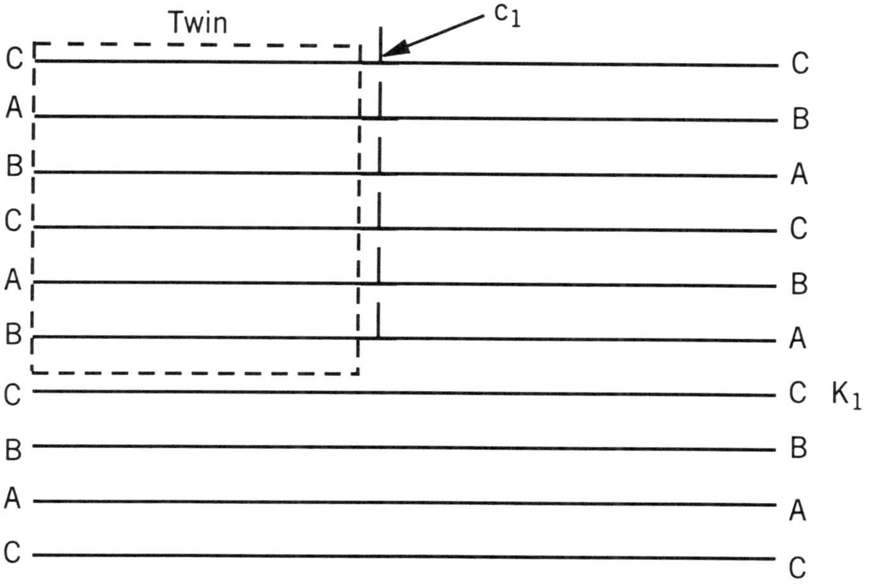

Figure 17.16 Deformation twin formed by motion of partial dislocations in fcc crystal.

sequence shown on the left. Plane K_1 is thus a mirror plane relating the twinned portion of the crystal to the untwinned portion. A stacking fault (incorrect next-nearest-neighbor stacking sequence between the planes immediately on either side of K_1) extends from the left edge of the crystal to the stack of partial dislocations.

CHAPTER 18

DISLOCATIONS AND PLASTIC DEFORMATION IN CERAMICS

18.1 INTRODUCTION

Polycrystalline ceramics are usually considered to be completely brittle at room temperature and to exhibit permanent deformation before failure only at elevated temperatures. This view is correct in an engineering sense for practical purposes, although considerable plastic deformation is possible in individual ceramic single crystals depending on the ceramic involved and the stress system. The lack of significant plasticity in polycrystalline ceramics is the result of limitations on the number of slip systems and on interactions between different slip systems leading to the opening of cracks. In this chapter we sketch the basic behavior of two ceramics that may serve as prototypes of the plastic behavior of ceramics: alumina and magnesia.

18.2 SLIP SYSTEMS IN CERAMIC CRYSTALS

Mobile dislocations in ceramic crystals permit the development of plastic shear strain under shear stress. In most cubic metals the crystallographically equivalent slip systems are sufficiently numerous to permit complete flexibility in the sense that glide of dislocations on slip planes can produce all the strain components and thus produce any change of shape. If the number of operative slip systems is limited, only certain components of strain can be produced. This has pertinence not only for allowing arbitrary deformation of single crystals but also for allowing the grains of a polycrystalline aggregate to conform as the sample as a whole is deformed. A minimum of five independent slip systems is needed to permit an arbitrary change of shape. A procedure for determining the number of independent slip systems was developed by Groves and Kelly (1963) and the systems have been tabulated for many crystals by Kelly and MacMillan (1986). The result is that, for sapphire, basal slip alone gives only two independent slip systems so that arbitrary deformation of a single crystal or any uniform plastic deformation of a polycrystalline body cannot be produced by basal slip alone.

For MgO below 650 K, slip on all the equivalent (110) planes in $\langle 1\bar{1}0 \rangle$ directions gives only two independent slip systems. Above 650 K slip is also possible on (001) planes in $\langle 110 \rangle$ directions, giving a total of five independent slip systems so that arbitrary deformation on a single crystal or uniform deformation of a polycrystalline material is possible in principle. However, stresses arising as a result of the operation of these slip systems unfortunately lead to fracture before significant bulk plasticity occurs in MgO (Kelly and MacMillan, 1986).

For zirconia stabilized in the cubic form the slip system is $\{001\}\langle 110 \rangle$ (Heuer, Lanteri, and Dominguez-Rodriguez, 1989).

18.3 PLASTIC DEFORMATION OF SINGLE CRYSTAL ALUMINA

The crystal structure of sapphire (single crystal aluminum oxide) is more complex than that of the fcc and hcp metals used for the initial development of dislocation theory. More complex dislocations and dislocation behavior are to be expected. We present a portion of the treatment of the structure and dislocations in sapphire given by Kronberg (1957). The crystal structure of sapphire corresponds closely (with small deviations in atomic positions) to oxygen ions in hexagonal close packing with aluminum ions in two out of three of the octahedral interstitial positions between the oxygen ions. Kronberg refers to the stacking sequence of the oxygen ions as 1,3,1,3 (rather than *ABAB* as in the Section 18.19) and reserves 2 to refer to the sheet of octahedral interstitial sites between the sheets of oxygen ions. The basic stacking sequence is thus 1,2,3,1,2,3 before the differences in position of the holes (interstitial octahedral sites unoccupied by aluminum ions) is taken into account. He uses the notation *ABCABC* to refer to the stacking sequence of the holes and aluminum ions as shown in Figure 18.1. Thus the complete description of the stacking sequence in Kronberg's notation is 1,2*A*,3,2*B*,1,2*C*,3,2*A*,1,2*B*,3,2*C*,1, . . . Accordingly, six sheets of oxygens must be stacked above the first one to return the sequence of oxygen and aluminum ions to coincidence with the starting ones; that is, the translation vector perpendicular to the sheets of oxygen atoms is $c = 1.297$ nm. The symmetry of the

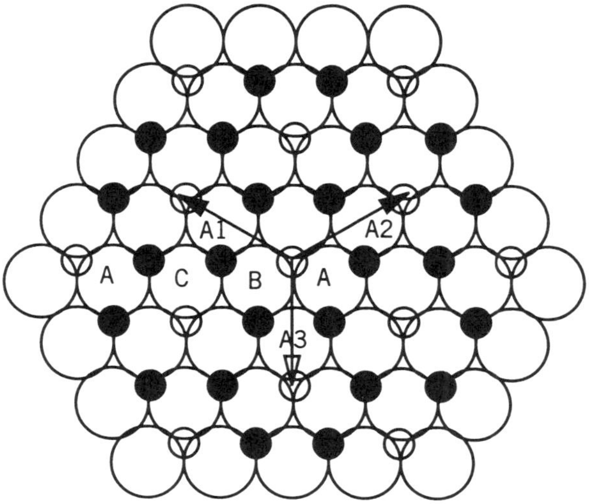

Figure 18.1 Oxygen ions (large open circles), aluminum ions (small solid circles), and unoccupied octahedral interstitial sites = holes (small open circles).

crystal structure is trigonal rather than hexagonal, but it is convenient and conventional to use a hexagonal rather than a rhombohedral unit cell. Moreover, a morphological hexagonal cell with $c = 0.6485$ and $c/a = 1.365$ is generally used rather than a structural cell with $c/a = 2.730$. The hexagonal base plane axes are shown in Figure 18.1 as A_1, A_2, and A_3, which are, respectively, the $[2\bar{1}\bar{1}0]$, $[\bar{1}2\bar{1}0]$, and $[\bar{1}\bar{1}20]$ directions. The directions of the rows of oxygen atoms in the morphological hexagonal cell are $[\bar{1}0\bar{1}0]$ and equivalent directions.

Moderately high temperatures are required for significant macroscopic plastic deformation of nominally pure alumina single crystals (synthetic sapphire rods) by slip in tension at atmospheric pressure and the stress required for deformation increases with the deformation rate imposed. In static tests this effect is manifested as an incubation time required for measurable creep in a fixed time of observation. Wachtman and Maxwell (1954) investigated the plastic deformation of sapphire in the temperature range 900–1300 °C and found that slip took place on the (0001) plane in the $[11\bar{2}0]$ direction. They used a 100-hour period and considered that creep occurred if more than 0.01% strain was measured over this time after the initial elastic deformation. In practice, either less than 0.01% deformation took place before a given stress, termed the creep yield stress for 100 hours, or much more deformation occurred if the stress was higher. The controlling stress was found to be the shear stress acting on the slip plane in the slip direction (i.e., the critical resolved shear stress) given for a tensile specimen under tensile stress σ by

$$\tau = \sigma \sin\chi \cos\lambda$$

where χ is the angle between the rod axis and the projection of the rod axis on the slip plane and λ is the angle between the rod axis and the slip direction. Typical creep curves for a sapphire rod in tension at 1100 °C are shown in Figure 18.2 from Wachtman and Maxwell (1954). Under 690 kg/cm^2 resolved shear stress, the creep rate is at first immeasurably slow, then rises to a maximum, and decreases again to an immeasurably slow rate. Subsequent application of a somewhat higher stress causes a second, similar curve. Further application of a still higher stress causes a much faster initial creep rate. Two explanations (or a combination of both) appear possible for the initial period of slowly increasing creep rate: a yield point due to dislocation breakaway from impurities or slow multiplication of dislocations to give a great enough number to support a measurable creep rate. Experiments on sapphire containing chromium (ruby) indicate an increase in creep yield point. However, dynamic yield points measured in constant strain rate experiments by Kronberg are consistent with the dislocation multiplication hypothesis. An indication that some type of yielding has taken place is provided by Figure 18.3 from Wachtman and Maxwell (1957), showing that the stage of relatively rapid creep is reached under 450 kg/cm^2 at 1000 °C and that it will continue if the stress is reduced to 350 kg/cm^2, which is below the creep yield stress for a previously untested crystal at this temperature.

The 100-hour creep yield stress was determined as a function of temperature by Wachtman and Maxwell (1957) with the result shown in Figure 18.4. The creep yield stress rises rapidly as the temperature is decreased and becomes immeasurably high below 900 °C for the experimental conditions used. One sees in this a reason for the brittleness of aluminum oxide at lower temperatures. Even in the temperature range for which observable creep is found in alumina single

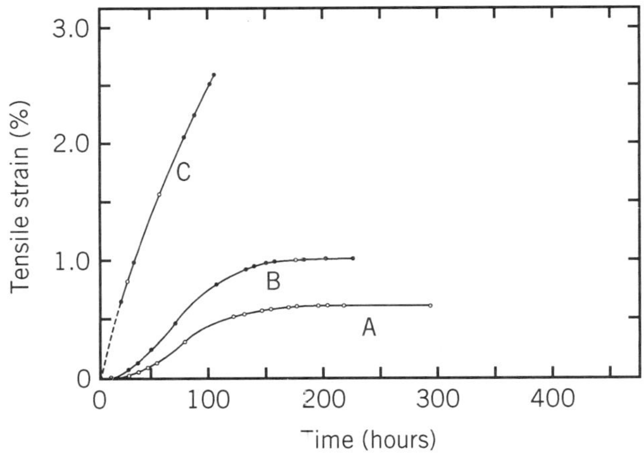

Figure 18.2 Creep of sapphire at 1100 °C under resolved shear stress values in units of 100 kg/cm^2 of (A) 0.69, (B) 0.93, and (C) 1.23. (From J. B. Wachtman, Jr. and J. H. Maxwell, "Plastic Deformation of Ceramic-Oxide Single Crystals," *J. Am Ceram. Soc.*, Vol. 37, 291–299, 1954. Reprinted by permission of the American Ceramic Society.)

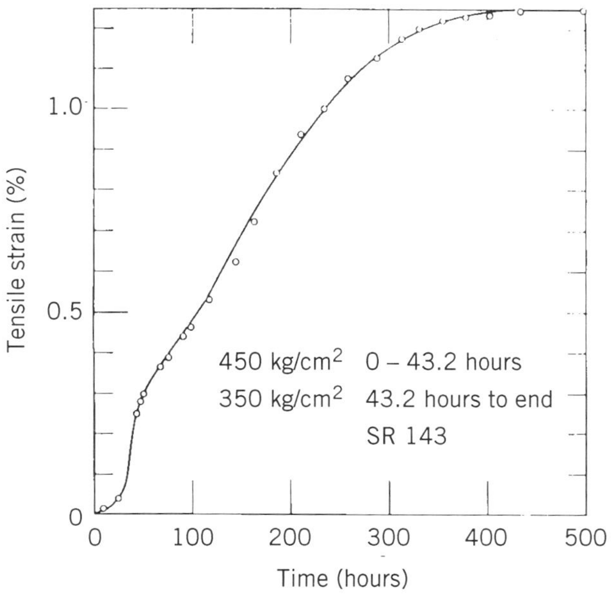

Figure 18.3 Creep curve of sapphire at 1000 °C showing creep continuing below the creep yield stress after being started under higher stress. (From J. B. Wachtman, Jr. and J. H. Maxwell, "Plastic Deformation of Ceramic-Oxide Single Crystals II," *J. Am Ceram. Soc.*, Vol. 40, 377–385, 1957. Reprinted by permission of the American Ceramic Society.)

crystals, there is still reason for brittle behavior of polycrystalline alumina in the fact that the (0001) $[11\bar{2}0]$ slip system does not provide the five degrees of freedom needed for the most general change of shape. This slip system alone can accommodate shear in any direction only on the hexagonal (strictly the trigonal) base plane. Attempts to deform sapphire in tension with the base plane parallel or perpendicular to the tensile axis gave no measurable plastic deformation until a temperature of about 1600 °C was used.

Kronberg (1962) used constant strain rate testing to show that a dynamic yield point that is the analog of the creep yield point occurred in sapphire oriented for basal slip. He developed (Kronberg, 1957) detailed models for dislocations in the corundum structure and discussed their role in slip and deformation twinning. Firestone and Heuer (1976) studied the yield point in constant strain rate experiments and concluded that the yield point was caused by a dislocation multiplication mechanism rather than an impurity atmosphere.

Creep of sapphire single crystals with the tensile axis parallel to the *c* axis has been studied at temperatures above 1600 °C by Tressler and Barber (1974) and Firestone and Heuer (1976). The latter concluded that slip on rhombohedral planes was involved. The crystallography of slip systems in sapphire including their description in alternative unit cells was reviewed by Snow and Heuer (1973).

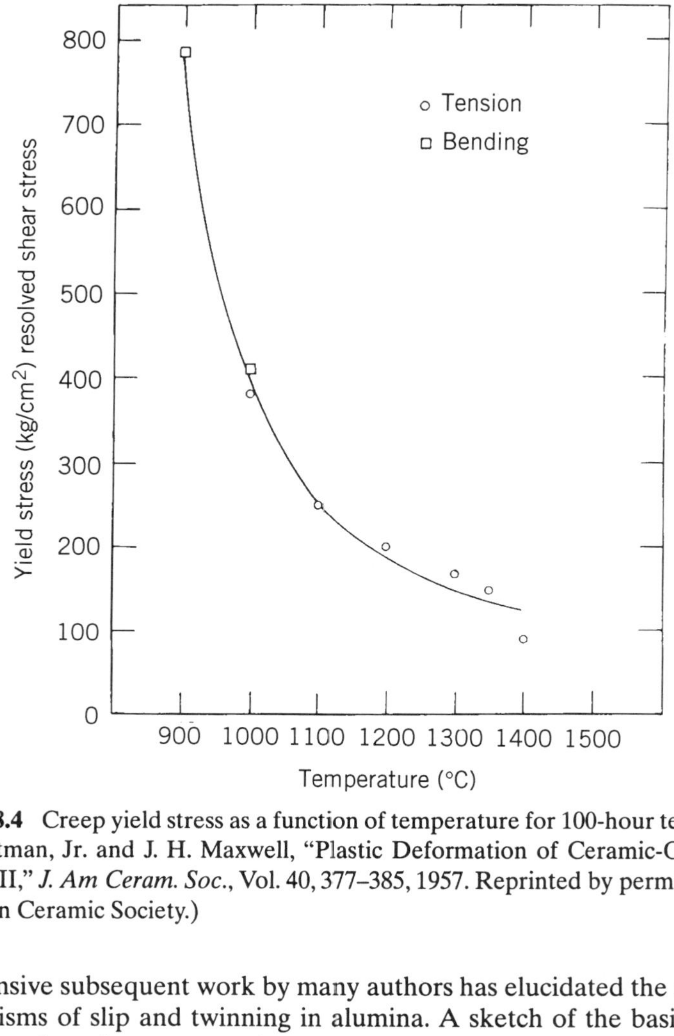

Figure 18.4 Creep yield stress as a function of temperature for 100-hour tests. (From J. B. Wachtman, Jr. and J. H. Maxwell, "Plastic Deformation of Ceramic-Oxide Single Crystals II," *J. Am Ceram. Soc.*, Vol. 40, 377–385, 1957. Reprinted by permission of the American Ceramic Society.)

Extensive subsequent work by many authors has elucidated the dislocation mechanisms of slip and twinning in alumina. A sketch of the basic results is given here based on a general summary presented by Lagerlof, Heuer, Castaing, Riviere, and Mitchell (1994).

Single crystals of alumina in the stable corundum structure show no macroscopic deformation below about 900 °C under 1-atmosphere pressure in macroscopic tensile tests. Hockey (1971) has shown that the complex stress system existing under a hardness indentation does produce local deformation (microplasticity) and that dislocations are generated. At higher temperatures three slip systems are known to operate in simple tension under 1-atmosphere as listed in Table 18.1.

Lagerlof et al. conducted a series of experiments by superimposing uniaxial compressive stress on silver-sheathed sapphire single crystals under isostatic pressure of 1.5 GPa applied by NaCl. One set of specimens, termed type I, was favorably oriented for prismatic slip and the other, termed type II, was favor-

TABLE 18.1 Slip Systems in α-Alumina (Corundum = Sapphire)

System Name	Slip System	Remarks
Basal	$(0001)\ \frac{1}{3} < 2\bar{1}\bar{1}0 >$	Dominant system under shear superimposed on 1-atm pressure
Prismatic	$\{\bar{1}2\bar{1}0\} < 10\bar{1}0 >$	Occurs above 1600 °C under shear superimposed on 1-atm pressure
Pyramidal	$\{10\bar{1}1\}\ \frac{1}{3} < \bar{1}101 >$ or possibly $\{\bar{1}012\}\ \frac{1}{3} < \bar{1}101 >$ or possibly $\{2\bar{1}\bar{1}3\}\ \frac{1}{3} < \bar{1}101 >$	Occurs above 1600 °C under shear superimposed on 1-atm pressure

ably oriented for basal slip. Their experiments were carried out at a constant strain rate of 2×10^{-5}/s. Under these conditions it was found that type I specimens could be plastically deformed at temperatures as low as 200 °C while 400 °C was required for type I specimens. Typical stress–strain curves are shown in Figure 18.5 from Lagerlof et al. An interesting point is that the initial portion of the stress–strain curve had a slope less than that for elastic deformation so that some type of plastic deformation was occurring even in this initial portion of the curve, perhaps twinning. The accuracy of measurement was not sufficient for detailed analysis of this slope (A. H. Heuer, personal communication, 1994). However, the authors took the point of pronounced change in slope to indicate yielding for slip under these experimental conditions. Type I specimens showed no work hardening after yield, but type II specimens showed appreciable work hardening after yielding. Plots of the yield stress as a function of temperature are shown in Figure 18.6. Under these experimental conditions prismatic slip predominates below about 700 °C while basal slip predominates at higher temperatures. The critical resolved yield stress, τ_{crs}, was found to vary with temperature according to

$$\ln \tau_{crs} = \ln \tau_0 - BT \tag{18.1}$$

where τ_0 and B are constants. This form of temperature dependence cannot be explained on the basis of the usual picture of thermally activated slip with a constant activation energy. Lagerlof et al. propose a model of stress-dependent activation energy involving dissociation of dislocations into partial dislocations that leads to the observed form of temperature dependence.

An earlier summary of the available data on the temperature dependence of the tensile stress required for flow (the creep yield stress) on sapphire single crystals oriented for slip on the basal, prismatic, and pyramidal systems is

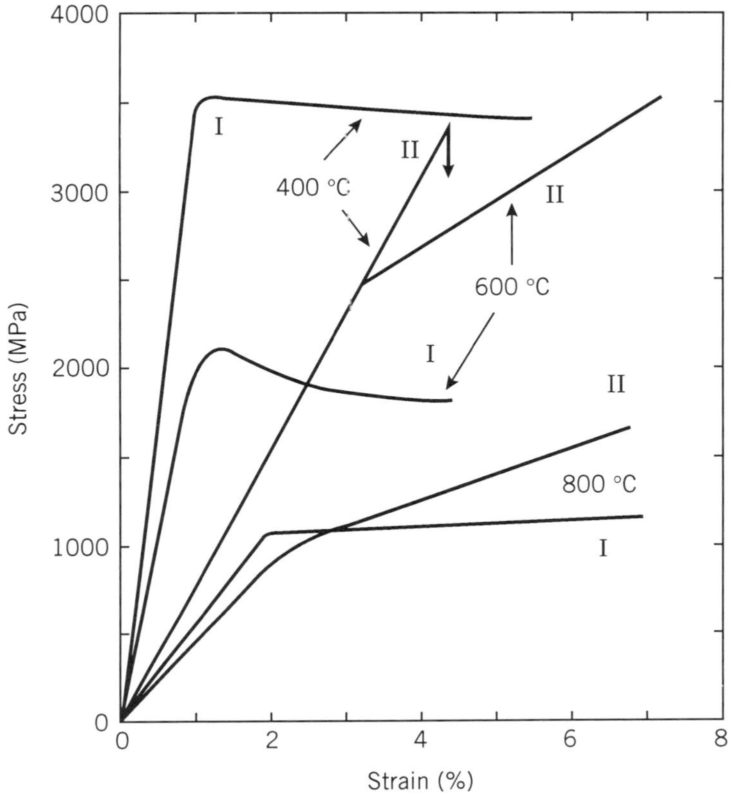

Figure 18.5 Stress–strain curves for samples oriented for prism plane slip (type I) and basal slip (type II). (From K. P. D. Lagerlof, A. H. Heuer, J. Castaing, J. P. Riviere, and T. E. Mitchell, "Slip and Twinning in Sapphire (α-Al_2O_3)," *J. Am. Ceram. Soc.*, Vol. 77, 385–397, 1994. Reprinted with permission of the American Ceramic Society.)

shown in Figure 18.7 (Heuer, Tighe, and Cannon, 1980). The polycrystalline data were taken at a strain rate of 0.000001/s.

The atomic mechanisms of slip and twinning in sapphire have recently been reviewed in detail, including consideration of the real structure rather than the slightly idealized structure used by Kronberg (Bilde-Sorensen, Lawlor, Geipel, Pirouz, Heuer, and Lagerlof, 1995; Pirouz, Lawlor, Geipel, Bilde-Sorenson, Heuer, and Lagerlof, 1995; Geipel, Bilde-Sorenson, Lawlor, Pirouz, Lagerlof, and Heuer, 1995).

18.4 TWINNING IN ALUMINUM OXIDE

Deformation twinning has long been known to occur in alumina under atmospheric pressure at temperatures as low as 77 K (Heuer, 1966; Scott and Orr, 1983). Both basal and rhombohedral twins occur.

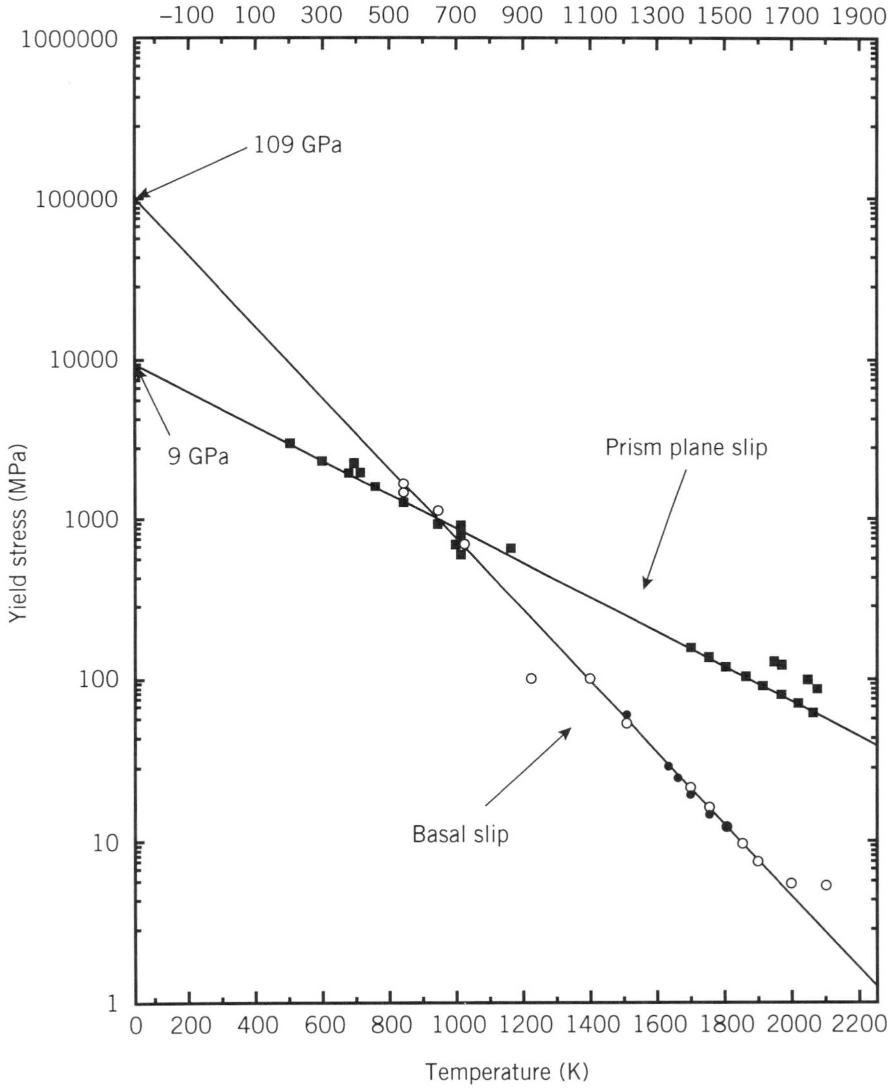

Figure 18.6 Yield stress as a function of temperature for prism plane slip and basal plane slip. (From K. P. D. Lagerlof, A. H. Heuer, J. Castaing, J. P. Riviere, and T. E. Mitchell, "Slip and Twinning in Sapphire (α-Al_2O_3)," *J. Am. Ceram. Soc.*, Vol. 77, 385–397, 1994. Reprinted with permission of the American Ceramic Society.)

The elements of basal twinning are K_1 = (0001), η_1 = $<10\bar{1}0>$, K_2 = $\{10\bar{1}1\}$, η_2 = $<\bar{1}012>$ using structural indices with c/a = 2.73, and s = 0.635 (Lagerlof, Lawlor, Geipel, Pirouz, Heuer, and Bilde-Sorenson, 1994). The authors propose a model for basal twinning based on an expanding partial dislocation loop similar to a Frank–Read source on the basal plane with a Burgers vector of

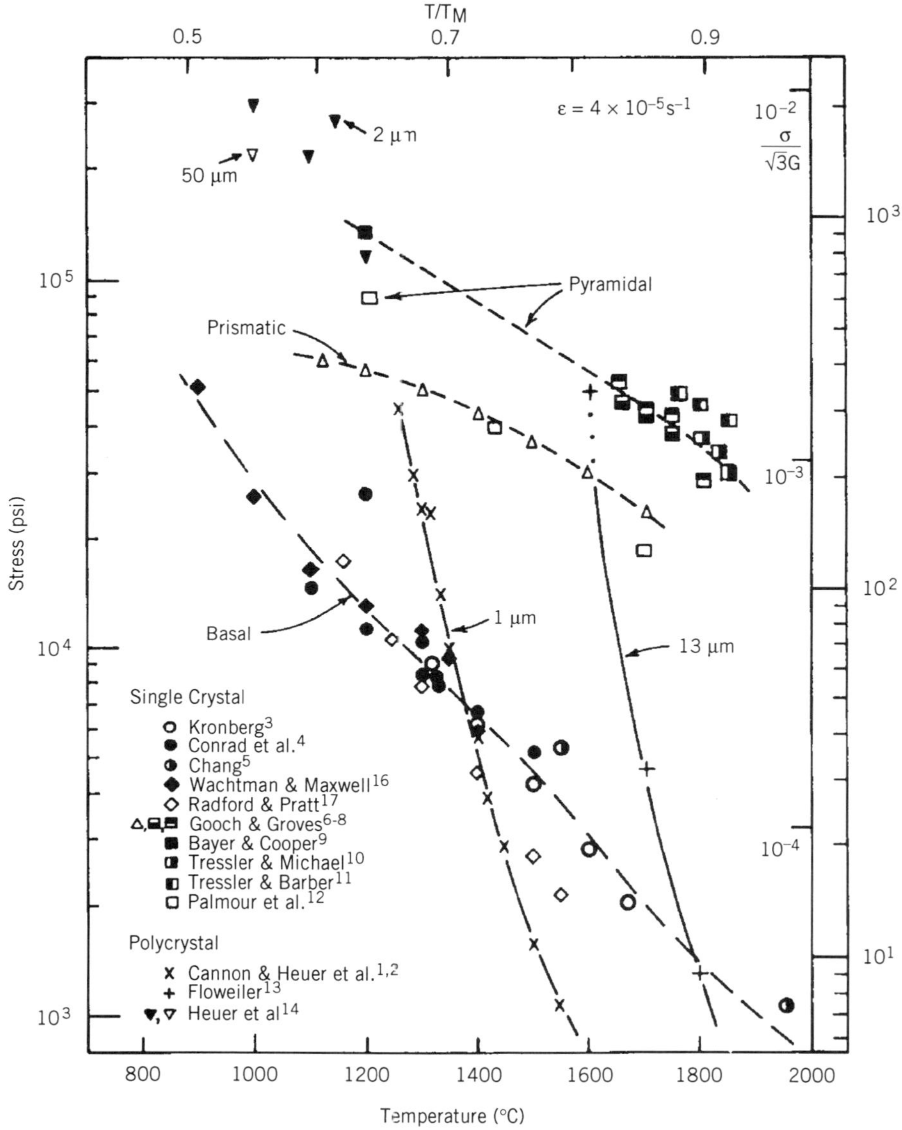

Figure 18.7 Tensile flow stress in single crystal and polycrystalline alumina. (From A. H. Heuer, N. J. Tighe, and R. M. Cannon, "Plastic Deformation of Fine-Grained Alumina (Al_2O_3): II, Basal Slip and Nonaccommodated Grain-Boundary Sliding," *J. Am. Ceram. Soc.*, Vol. 63, 53–58, 1980. Reprinted by permission of the American Ceramic Society.)

⅓ <$10\bar{1}0$>. The loop is assumed to expand once and to be blocked from a second expansion because the stacking fault energy formed on the second expansion is of a different type and has higher energy. Instead, the loop is assumed to cross slip and to expand again on a new plane. Repetition of this process forms the basal twin.

The elements of rhombohedral twinning are $K_1 = \{01\bar{1}2\}$, $\eta_1 = <0\bar{1}11>$, $K_2 = \{0\bar{1}14\}$, $\eta_2 = <022\bar{2}1>$, and $s = 0.202$, where again structural indices are used (Heuer, 1966). The symmetry element relating the twinned portion to the untwinned portion of the crystal is a screw axis consisting of a twofold rotation about η_1 accompanied by a translation along the rotation axis of ⅙ [$0\bar{1}11$]. A zonal dislocation mechanism has been proposed by Geipel, Lagerlof, Pirouz, and Heuer (1994).

18.5 PLASTIC DEFORMATION OF SINGLE CRYSTAL MAGNESIUM OXIDE

Magnesium oxide has the rock salt structure and its mechanical properties parallel those of softer materials with the rock salt structure (such as NaCl) to some degree, but the behavior of dislocations in MgO is sufficiently different in detail to make it a unique material. Magnesium oxide single crystals were extensively studied in the late 1950s and in the 1960s with the hope that the degree of dislocation mobility they possessed could be manipulated to produce ductile polycrystalline MgO materials. This was not achieved and MgO is today considered to be a semibrittle material rather than a ductile material. Many people contributed to the understanding of the mechanical properties of MgO, especially R. J. Stokes and his collaborators (Stokes, 1962,1965,1966a, 1966b, 1968, 1972; Stokes, Johnston, and Li, 1960; Stokes and Li, 1962, 1963, 1964). The following discussion is based primarily on an informal, unpublished summary by R. J. Stokes (personal communication, 1994).

Stokes considers the mechanical properties of MgO and related materials in two categories: (1) dislocation behavior and the mechanical yield strength of materials and (2) slip band characteristics and the fracture behavior of semibrittle materials.

From the Orowan equation for strain rate

$$\dot{\varepsilon} = \rho\, b\, v$$

it is evident that the strain rate depends on the density of mobile dislocations and their velocity. The distinction of the density of mobile dislocations rather than total density of dislocations is crucial in MgO. A striking fact about MgO is that the dislocations produced by crystal growth do not move under stresses up to values approaching the theoretical shear strength. They are very effectively pinned. However, fresh dislocations can very easily be introduced by the slightest contact due to the concentrated Hertzian stresses upon contact. The fresh dislocations do move under relatively low stresses. The problem in studying "grown-in" dislocations is to avoid introducing fresh dislocations during the preparation of the sample. Careful chemical polishing combined with annealing to pin the remaining fresh dislocations can produce samples with no fresh dislocations. Such samples are completely elastic up to very high stresses.

The velocity is very sensitive to stress, typically varying as stress to a power such as 10. The velocity is also affected by any feature of the crystal that retards the slip of the dislocation. These include interactions between the stress field of the mobile dislocation and the internal stress fields due to impurity particles (solid solution strengthening and vacancy cluster hardening), dislocation entanglement with precipitate particles (precipitation hardening), and dislocation entanglement with other dislocations (strain hardening). Thus hardening can be considered in two categories: static hardening (stationary dislocations are pinned by heat treatment) and dynamic hardening (moving dislocations are impeded by interaction with other defects) (Sprackling, 1976; R. J. Stokes, personal communication, 1994).

The density of mobile dislocations in MgO grows rapidly as slip proceeds due to the double cross slip multiplication mechanism. Extensive slip takes place in a narrow band from a single fresh, mobile dislocation as it moves, cross slips, and multiplies under stress. This produces a highly deformed region (the slip band) adjacent to a region of crystal that is purely elastically deformed. Evidently large stresses can develop at the junction of these two regions as the elastic strain increases so that the two regions remain compatible.

Cracks in MgO initiate (nucleate) because of the inability of the crystal to respond to the demands of anisotropic shear such as that at the junction of a slip band with a region with no plastic deformation. The theoretical stress can be exceeded locally and a small crack suddenly forms. This occurs in MgO at room temperature and has been seen in LiF at liquid nitrogen temperature. These small cracks can then undergo subcritical crack growth as the crystal undergoes further plastic strain and eventually a length is reached for catastrophic crack growth.

The strength of MgO containing mobile dislocations is thus at or near the yield stress since the process of yielding leads to crack nucleation and growth. MgO single crystals with no mobile dislocations can have much higher strength.

19.5 PLASTIC DEFORMATION IN SINGLE CRYSTAL CUBIC ZIRCONIA

Cubic yttria-stabilized zirconia single crystals containing tetragonal zirconia precipitates were studied by Heuer, Lanteri, and Dominguez-Rodriguez (1989), Martinez-Fernandez, Jimenez-Melendo, Dominguez-Rodriguez, Lagerlof, and Heuer (1993), and Martinez-Fernandez, Jimenez-Melendo, Dominguez-Rodriguez, Cordier, Lagerlof, and Heuer (1995). Stress–strain curves were measured at 1400 °C after various annealing times at 1600 °C. The annealing causes precipitation of lamellae whose spacing increases with time due to Ostwald ripening. The initial yield stress for 0.2% strain offset was about 400 MPa but increased in proportion to lamella spacing and was about 500 MPa after 150 hours. Annealing increased the spacing from the initial value of 20 nm to 60 nm.

CHAPTER 19

CREEP IN CERAMICS

19.1 INTRODUCTION

Creep can most simply be described as the phenomenon of continuing plastic deformation under constant stress. Creep effects may be studied with other loading schedules, but the same basic processes are involved. Typically, creep in ceramics takes place at high temperatures under modest stress levels and with low strain rates. Creep is sometimes discussed as if it were a different phenomenon than plastic deformation and the latter is treated as if it occurred instantaneously when stress is applied. Fundamentally, however, plastic deformation is not qualitatively different from creep.

In this chapter the phenomenological description of creep in terms of stages is first given and then several important mechanisms of creep in ceramics are discussed. An excellent general reference on creep is Poirer (1985).

Creep in ceramics has been reviewed thoroughly by Cannon and Langdon (1983, 1988). These authors classify creep in ceramics into two categories depending on the way in which the creep rate depends on the stress. In the first category the creep rate depends on the stress to a power close to 1. In the second category the creep rate depends on stress to a higher power, typically approximately 3 or 5. Creep with a stress exponent of 5 is believed to take place as in metals by a dislocation motion with five independent slip systems operating. Diffusion plays a role associated with the freeing of dislocations rather than with the primary transport process. Creep with a stress exponent of about 3 is taken to result from dislocation climb under conditions such that five independent slip systems cannot interpenetrate or when there are fewer than five independent slip systems operating.

The creep strain ϵ resulting from application of constant stress is often written as

$$\varepsilon=\varepsilon_0+\beta\, t^{1/3}+kt \tag{19.1}$$

Here ϵ_0 is the elastic strain and β and k are constants for a given material with a given microstructure (including defect structure) at a particular temperature. The second and third terms describe what is called, respectively, transient and steady-state creep (also sometimes called primary and secondary stages of creep). The justification for this equation is empirical; it fits much of the experimental creep data. If creep is continued long enough a tertiary stage is encountered, especially in metals. In tension this tertiary stage is an accelerating creep rate leading to failure.

Most of the attention in creep studies is focused on steady-state creep or at least what is taken as steady-state creep. The assumption is made here that a sufficient period of time giving nearly constant creep rate is studied experimentally to justify the fitting of models for steady-state creep and these models are presented.

19.2 NABARRO–HERRING CREEP

Nabarro (1948) considered a small cube of material under a pure shear stress resulting from tension in one direction (say, the z direction) and sufficient compression in the x and y directions to give zero pressure. He showed that vacancies would flow from the face under tension to the faces under compression, causing a shear of the material so as to elongate it in the z direction and cause it to shrink in the x and y directions. The equilibrium atomic fraction of vacancies at temperature T is given by

$$N_v = \exp\left(-\frac{\Delta G_f}{kT}\right) \tag{19.2}$$

where ΔG_f is the Gibbs free energy of formation of one vacancy and k is Boltzmann's constant. The equilibrium concentration is then

$$C_0 = N_v/\Omega \tag{19.3}$$

where Ω is the atomic volume. Creation of a vacancy at the face under tension σ is assisted by a contribution $\sigma\Omega$ to the free energy, while extra free energy of the same amount is needed to create a vacancy at the face under compression.

The local equilibrium concentrations at the tension and compression faces, respectively, are then

$$C^{+} = C_0 \exp\left(\frac{\sigma\,\Omega}{kT}\right) \tag{19.4}$$

$$C^{-} = C_0 \exp\left(\frac{-\sigma\,\Omega}{kT}\right) \tag{19.5}$$

There will be a flux of vacancies from the higher concentration face to the lower concentration face given by Fick's equation for diffusion

$$J = -D_v \nabla C \approx \alpha D_v \frac{C^{+} - C^{-}}{d} \tag{19.6}$$

where α is a numerical constant and d is the edge of the cube. The number of vacancies passing through an area d^2 per second, φ, is

$$\varphi = Jd^2 \tag{19.7}$$

The volume arriving per second is $\varphi\Omega$. Dividing by the area of a face, d^2, gives the elongation per second and dividing this in turn by the length, d, gives the strain rate

$$\dot{\varepsilon} = \varphi\frac{\Omega}{d^3} \tag{19.8}$$

$$\dot{\varepsilon} = \frac{\Omega}{d^3} Jd^2 = \frac{\Omega}{d}\,\alpha D_v \frac{C^{+} - C^{-}}{d} = \frac{\alpha\,\Omega\, D_v\, C_0}{d^2}\left(\exp\frac{\sigma\,\Omega}{kT} - \exp\frac{-\sigma\,\Omega}{kT}\right) \tag{19.9}$$

For $\sigma\Omega/kT < 1$

$$\dot{\varepsilon} = 2\alpha \frac{D_v\,\Omega^2\,\sigma\, c_0}{d^2} \tag{19.10}$$

Recognizing that the self-diffusion constant, D_l, is given by

$$D_l = D_v\,\Omega\, c_0 \tag{19.11}$$

leads to

$$\dot{\varepsilon} = 2\alpha \frac{D_l \Omega \sigma}{d^2 kT} \tag{19.12}$$

Herring (1950) gave a thermodynamic treatment and considered the case of grains in a solid and found $\alpha = 16$ when there is no grain boundary sliding or $\alpha = 40$ when there is sufficient grain boundary sliding to completely relax shear stresses on the boundary. The usual creep experiment is done in simple tension. When Herring's result and the relationship of simple tensile stress to the stress system used above to give simple shear are taken into account, the result for complete grain boundary sliding is (Poirier, 1985; Cannon and Langdon, 1988)

$$\dot{\varepsilon} = 13.3 \frac{D_l \Omega \sigma}{d^2 kT} \tag{19.13}$$

where σ is now the tensile stress in a simple tensile test.

19.3 COMBINED DIFFUSIONAL CREEP MECHANISMS

The above derivation of Nabarro–Herring creep assumed only one diffusion path (through the solid grains) and only one diffusing species. In ceramics there are often two possible diffusion paths (through the grains or in the grain boundaries) and two diffusing species (e.g., oxygen and aluminum in aluminum oxide). Cannon and Langdon (1988) have summarized the equations for these situations as presented in the following.

For diffusion through the grains and with complete grain boundary relaxation, Herring (1950) obtained a constant with value of 13.3 in Nabarro's equation. Taking the atomic volume as $0.7b^3$, where b is the magnitude of the Burgers vector, leads to

$$\frac{d\varepsilon}{dt} = 9.3 \frac{D_l G b}{kT} \left(\frac{b}{d}\right)^2 \left(\frac{\sigma}{G}\right) \tag{19.14}$$

where D_l is the diffusion constant through the grain, G is the shear modulus, k is Boltzmann's constant, d is the grain diameter, and σ is the tensile stress.

If the transport of matter is through the grain boundaries, the process is termed Coble creep (Coble, 1963). The diffusional creep equation then is

$$\frac{d\varepsilon}{dt} = \frac{150 \Omega \, \delta \, D_{gb} \, \sigma}{\pi d^3 kT} \tag{19.15}$$

where D_{gb} is the diffusion constant for transport in the grain boundaries and δ is the grain boundary width. If the atomic volume is taken as before as $\Omega = 0.7b^3$, the creep rate is

$$\frac{d\varepsilon}{dt} = 33.4 \frac{D_{gb}\, Gb}{kT}\left(\frac{\delta}{b}\right)\left(\frac{b}{d}\right)^3\left(\frac{\sigma}{G}\right) \tag{19.16}$$

Because the activation energy for grain boundary diffusion is less than that for diffusion through the grains, Coble creep is favored over Nabarro–Herring creep at lower temperature. Because of the inverse cubic dependence on grain size, Coble creep is favored over Nabarro–Herring creep at very small grain sizes.

Actually, Nabarro–Herring creep and Coble creep take place in parallel so that the total diffusional creep rate is given by

$$\frac{d\varepsilon}{dt} = 9.3 \frac{D_l Gb}{kT}\left(\frac{b}{d}\right)^2\left(\frac{\sigma}{G}\right)\left[1 + 3.6\frac{D_{gb}}{D_l}\left(\frac{\delta}{d}\right)\right] \tag{19.17}$$

In ceramics both anions and cations can diffuse, leading to (Gordon, 1973, 1975)

$$\frac{d\varepsilon}{dt} = 9.3 \left(\frac{(1/\alpha)[D_{c,l} + 3.6D_{c,gb}(\delta_c/d)]}{1 + (\beta/\alpha)\,[D_{a,l} + 3.6D_{a,gb}\left(\delta_a/d\right)]}\right)\frac{DGb}{kT}\left(\frac{b}{d}\right)^2\left(\frac{\sigma}{G}\right) \tag{19.18}$$

where subscripts a and c refer to the anion and cation, respectively, and α and β are the valences of the anion and cation.

Evans and Langdon (1976) considered the possibilities for the usual cases in which there are large differences in the diffusion rates. Their general conclusion is that the creep rate is controlled by the slower diffusing species along the faster diffusion path.

19.4 POWER LAW CREEP IN CERAMICS

In this section the concept of creep by slip with the rate-controlling process being the thermally activated freeing of dislocations is summarized. The succession of ideas is that of long-range and short-range stress, thermal activation of dislocations over barriers, the balancing of hardening and recovery and the resulting Bailey–Orowan equation, and the Weertman model of creep by slip with rate control by climb.

Metals often creep with a steady-state creep rate varying with approximately the fifth power of the stress. Weertman (1957) developed a theory for glide of piled-up dislocations from an operating source with climb being the rate-controlling process. He obtained

$$\frac{d\varepsilon}{dt} = \frac{B_2\,\Omega D_l\,\sigma^{4.5}}{G^{3.5}\,M^{0.5}\,b^{3.5}\,kT} \tag{19.19}$$

where B_2 is a constant and M is the concentration of the active dislocation sources. Taking $B_2 = 0.2$ (Hazzledine, 1967; Cannon and Langdon, 1988) gives

$$\frac{d\varepsilon}{dt} = \frac{0.14}{b^{1.5}\,M^{.5}}\left(\frac{D_l\,G\,b}{kT}\right)\left(\frac{\sigma}{G}\right)^{4.5} \tag{19.20}$$

Some ceramics exhibit stress dependence with a power near 5 and are assumed to creep by a dislocation mechanism of the Weertman type.

Metals or ceramics undergoing power law creep form a substructure within grains with the subgrains bounded by small-angle grain boundaries with an average size λ. The dislocation density within the grains is ρ. Cannon and Langdon (1988) conclude that ceramics show the same behavior as metals in that the average subgrain size is inversely related to the stress

$$\frac{\lambda}{b} = \text{constant} \times \frac{G}{\sigma} \tag{19.21}$$

The constant is in the range of 20–30 for ceramics and is typically 20 for metals. The normalized dislocation density within subgrains for ceramics was found to be given by

$$b\rho^{1/2} = \text{constant} \times \frac{\sigma}{G} \tag{19.22}$$

Here the constant ranges from 1 to 2 for ceramics and is around 1 for metals.

Other ceramics exhibit a stress dependence with a power near 3. Cannon and Langdon (1988) examined theories giving this type of stress dependence and concluded that the most likely was a mechanism of creep through climb of dislocations from Bardeen–Herring sources. The theory predicts

$$\frac{d\varepsilon}{dt} = 0.22\frac{D_l G b}{kt}\left(\frac{\sigma}{G}\right)^3 \tag{19.23}$$

Chokshi and Langdon (1991) examined the power law creep of ceramics and concluded that there are two broad categories of behavior: a stress exponent close to 5 represents control by dislocation climb and fully ductile behavior whereas a stress exponent close to 3 arises from control by climb from Bardeen–Herring sources under conditions where there are less than five interpenetrating independent slip systems.

19.5 THE ROLE OF GRAIN BOUNDARIES IN HIGH-TEMPERATURE DEFORMATION AND FAILURE

Grain boundaries often play a crucial role in deformation and fracture of ceramics. We follow Langdon (1982, 1991a, 1991b, 1993a, 1993b, 1994a, 1994b) in presenting a summary considering behavior of metals and ceramics. Grain boundaries play two important but conflicting roles in deformation: contributing to the deformation and promoting local cracking and ultimate failure. The grain size determines the degree of grain boundary contribution to deformation and failure. Langdon defines four grain sizes in this regard: macroscopic, mesoscopic, microscopic, and nanoscopic.

Macroscopic grains are large enough, typically with grain diameter $d > 1000$ μm, so that grain boundaries play no significant role in deformation.

Mesoscopic grains are smaller but still large enough that the grain boundaries make a relatively minor contribution to deformation (such as in grain boundary sliding as a secondary process) but the boundaries may be dominant in controlling fracture. In metals typical mesoscopic grains range from 10 to 1000 μm in diameter. For materials deforming by dislocation slip, a substructure forms within mesoscopic grains consisting of subgrains of average diameter λ with small misorientations between adjacent subgrains.

Microscopic grains are still smaller (typically 10–0.01 μm) and grain boundaries play a dominant role in flow and often also in fracture.

Nanoscopic grains are still smaller and are beginning to be investigated as processing methods for making polycrystalline materials with grains in this range are developed. It is anticipated that grain boundaries will contribute most of the deformation in these materials.

Grain boundaries are important for ultimate failure in tension because they provide sites at which cavities can be nucleated and paths for the growth of cavities into intergranular cracks. For ceramics these processes are significant in damage-assisted creep and subsequent failure. However, under suitable conditions, both metals and ceramics with microscopic grain size can undergo large tensile deformations before failure, as discussed in a later section on superplasticity.

19.6 DAMAGE-ENHANCED CREEP

An additional feature of creep in ceramics that has not been considered so far is damage-enhanced creep. It is often observed, especially at higher stresses in ceramics with a viscous phase in the grain boundaries, that voids form at the grain boundaries as creep continues. Results of Wiederhorn, Roberts, Chuang, and Chuck (1988) and Chuang and Wiederhorn (1988) on siliconized silicon carbide are reviewed here to illustrate the phenomenon. Damage-enhanced creep has also been reported by other investigators in various materials including lithium zinc silicate glass ceramics (Morrell and Ashbee, 1973), reaction-bonded silicon nitride (Birch, Wilshire, and Godfrey, 1978), siliconized silicon carbide (Carroll and Tressler, 1988), and SiAlON–YAG ceramics (Chen, Wiederhorn and Chuang, 1991).

The experiments of Wiederhorn, Roberts, Chuang, and Chuck (1988) were done with a siliconized silicon carbide, KX01, made by SOHIO Corp. This material is approximately 33% silicon by volume and contains grains of silicon carbide ranging from 2 to 5 μm in diameter. These authors measured creep at 1300 °C in tension and compression and found that the creep rate in tension was 30 times or more greater than the creep rate in compression at the same stress. Furthermore, the dependence of creep rate on stress was bimodal in both cases. Their results are summarized in Figure 19.1, a log–log plot of creep rate as a function of stress. The results can be described by a power law equation of the form

$$\frac{d\varepsilon}{dt} = A\sigma^n \qquad (19.24)$$

where A and n are constants. For creep in tension $n = 3.5$ at stresses below 100 MPa and $n = 11$ for higher stresses. For creep in compression $n = 3.5$ for stress below 200 MPa and $n = 14$ for higher stresses. The authors attribute the creep in the range of stresses corresponding to the lower n values to power law deformation in the silicon phase. They find some formation of voids (cavita-

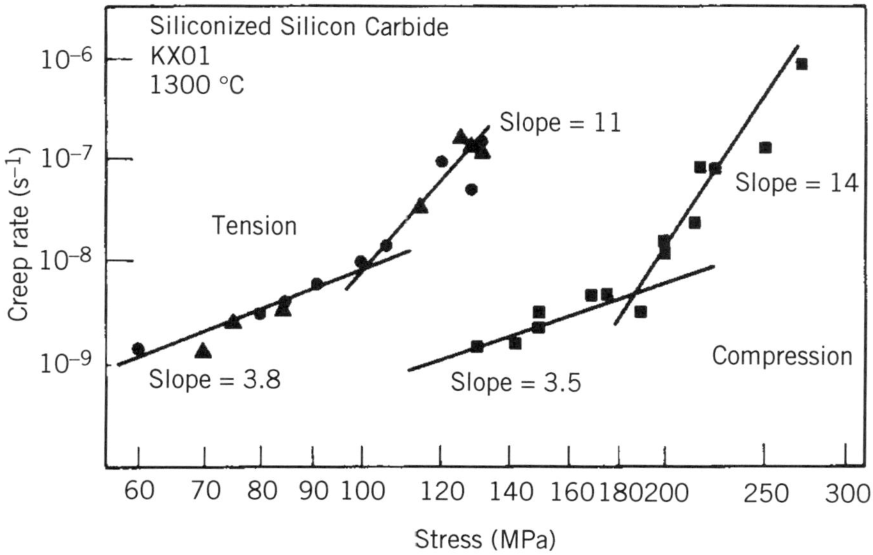

Figure 19.1 Stress dependence of the creep rate. (From S. M. Wiederhorn, D. E. Roberts, T.-J, Chuang, and L. Chuck, "Damage-Enhanced Creep in a Siliconized Silicon Carbide: Phenomenology," *J. Am. Ceram. Soc.*, Vol. 71, 602–608, 1988. Reprinted by permission of the American Ceramic Society.)

tion) at tensile stresses as low as 75 MPa, but very much more cavitation in experiments at 100 MPa for 100 hours. They attribute the higher creep rates above 100 MPa in tension to the effects of cavitation. For creep in compression in the stress range corresponding to the lower n values, power law creep is also assumed. The authors suggest that friction between silicon carbide grains affects the stress on the silicon phase and hence the creep rate. It is plausible that these frictional forces would be greater in compressive creep and so increase the stress required to achieve the same creep rate. The authors searched for cavitation in the compressive creep specimens but did not find evidence of it. The cause for the change to $n = 14$ at higher stresses in compression was not determined.

Bend testing leads to special problems when creep is involved. As noted earlier, the creep behavior may not be the same in tension as in compression; an important reason leading to this type of difference in ceramics is the opening of voids during tension. Creep results obtained in bending thus require special analysis if indeed they are analyzable at all. Analysis procedures have been presented by Hollenberg, Terwilliger, and Gordon (1971) and Chen and Chuang (1990).

The data of Figure 19.1 on creep of siliconized silicon carbide were used by Chuang and Wiederhorn (1988) to model the behavior in flexural creep. They used the bimodal results for the portion of the bar in tension but only the $n = 3.5$ behavior for the portion in compression. Thus the creep rate in tension was taken as

$$\dot{\varepsilon} = \dot{\varepsilon}_0 \, (\sigma/\sigma_0)^n \quad \text{for } \sigma < \sigma_0 \tag{19.25}$$

They used $\dot{\varepsilon}_0 \approx 5 \times 10^{-9} \text{s}^{-1}$, $\sigma_0 = 100$ MPa at 1300 °C and $\dot{\varepsilon}_0 \approx 10^{-10} \text{s}^{-1}$, $\sigma_0 = 125$ MPa at 1100 °C and $n = 4$–5. For creep in tension above σ_0 they used

$$\dot{\varepsilon} = \dot{\varepsilon}_0 \, (\sigma/\sigma_0)^n \quad \text{for } \sigma > \sigma_0 \tag{19.26}$$

with n 10. For creep in compression they took a fraction λ of the creep rate in tension; that is,

$$|\dot{\varepsilon}| = \lambda \, \dot{\varepsilon}_0 \, (|\sigma|/\sigma_0)^n \tag{19.27}$$

with $\lambda = 0.1$ typically.

They carried out an analysis of both transient and steady-state creep in flexure based on Bernoulli's hypothesis that plane sections remain plane, which they verified experimentally. The result is that there is a critical bending moment for cavity formation for a given temperature. Below this bending moment the neutral axis will shift toward the compressive side. The strain varies linearly with distance from the maximum compressive value at the compressive face to the maximum tensile value at the tensile face. The stress varies in a nonlinear manner with distance. The compressive stress is concentrated to a higher value at the compressive face than that for elastic bending and the tensile stress has a lower value at the tensile face than that for elastic bending.

At bending moment values above the critical value, a cavitation zone develops near the tension face.

A study showing the combined effect of a basic creep mechanism and damage-assisted creep in siliconized silicon carbide was reported by Carroll and Tressler (1989). They found a threshold for creep damage of 132 MPa at 1100 °C. Below this stress the stress exponent was about 4. At stresses above the creep damage threshold the stress exponent increased with increasing stress to around 10 at 172 MPa. The activation energy for creep also increased with increasing stress above the threshold. Creep below the stress threshold was attributed to dislocation activity in the silicon phase. Above the stress threshold for damage, cavities formed and contributed to the creep strain. Models developed by Raj (1982) and Hasselman and Venkateswaran (1984) were used to separate the damage component, leading to the conclusion that the remaining component of the creep strain is still well described by a stress exponent of 4 to the highest stresses used.

19.7 SUPERPLASTICITY

Many polycrystalline materials including some ceramics can sustain very large creep deformations (several hundred percent) in uniaxial tension at high temperature (typically above one-half the absolute melting temperature) before failing (Nieh and Wadsworth, 1990b). Fine grain size (a few micrometers or less) and an equiaxed structure are required. A striking feature of superplastic behavior is that grain shape does not change appreciably during deformation in contrast to normal plastic deformation in ductile polycrystalline materials. In the latter, the grains typically elongate to the same extent as the specimen as a whole.

One aspect of superplasticity is that large strains occur without necking. This is usually considered in terms of an empirical equation for true stress as a function of true strain rate

$$\sigma_t = K'(\dot{\varepsilon}_t)^{m_s} \tag{19.28}$$

where m_s is a number in the range 0.01–1. From constant volume

$$\dot{\varepsilon}_t = \frac{1}{L}\frac{dL}{dt} = -\frac{1}{A}\frac{dA}{dt} = \left(\frac{\sigma_t}{K'}\right)^{\frac{1}{m_s}} \tag{19.29}$$

so that

$$\frac{dA}{dt} = -\left(\frac{F}{K'}\right)^{\frac{1}{m_s}} A^{\left(1 - \frac{1}{m_s}\right)} \tag{19.30}$$

One can now consider what would happen under constant force creep conditions to a portion of the specimen in which the cross-sectional area, A_2, is smaller than the rest of the specimen, A_1, so that $A_2/A_1 < 1$. If $m_s = 1$ the rate of decrease of the cross-sectional area is constant everywhere. If $m_s < 1$ so that $1 - 1/m_s = -\delta$ with δ positive, then

$$\frac{d(A_2/A_1)}{dt} = (A_1/A_2)^{\delta} \tag{19.31}$$

The right-hand side is a positive number greater than 1 so that the smaller cross-sectional area decreases faster than the larger cross-sectional area. Thus as m_s approaches 1 the tendency to necking decreases. Langdon (1982) has shown that the total creep strain before failure correlates well with m_s for many materials.

The observed constancy of grain size and shape in superplastic deformation has been assumed to result from a compound mechanism. This mechanism has been considered to be grain boundary sliding with the resulting misfit relieved by a secondary mechanism—either diffusional creep or dislocation creep or possibly both acting in tandem. The approximate constancy of grain size and shape has been proposed (Ashby and Verrall, 1973) to result from a grain-switching mechanism, illustrated schematically in Figures 19.2 to 19.4, that gives a qualitative picture of how approximate constancy of grain size and shape might be accomplished. Langdon (1991a) has discussed difficulties with this model, including the fact that it does not provide for an increase in surface area that must take place. He proposed an alternative model in which grains slide apart, facets become rounded, and grains in a parallel plane move in to fill the openings and to increase the surface area. The issue of how much of the total strain is caused by grain boundary sliding is difficult to answer. Langdon (1994a) has discussed the measurement techniques and evaluated data on superplasticity in many metallic alloys with the conclusion that for these materials grain boundary sliding accounts for all the superplastic deformation.

Maehara and Langdon (1990) have considered superplastic deformation in more detail. They define two principal types: transformational and structural superplasticity. The former involves superplastic behavior in materials that are temperature-cycled through a crystallographic transformation. Some of the experiments on ceramics were done in compression, but true superplasticity in tension has been demonstrated with elongations of 800% in 3Y–TZP (tetragonal ZrO_2 stabilized with 3 mol % of Y_2O_3) (Nieh and Wadsworth, 1990a) and

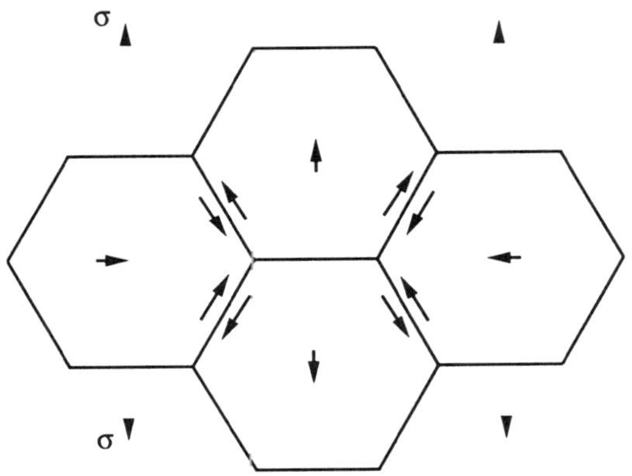

Figure 19.2 Initial grain boundary sliding. (Redrawn from Ashby and Verrall, 1973.)

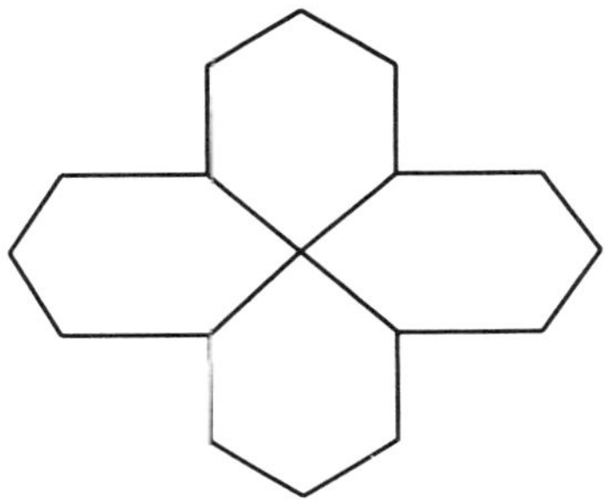

Figure 19.3 Intermediate grain positions. (Redrawn from Ashby and Verrall, 1973.)

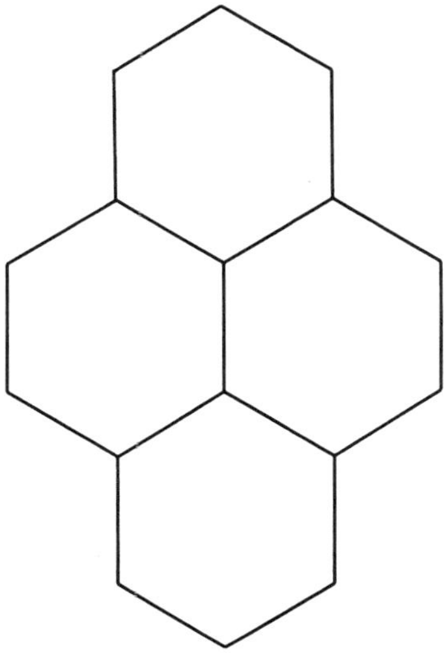

Figure 19.4 Positions after grain switching. (Redrawn from Ashby and Verrall, 1973.)

1038% in 2.5 Y–TZP containing 5 wt. % of SiO_2 (Kajihara, Yoshizawa, and Sakuma, 1990). In general, superplastic ceramics tend to exhibit a stress exponent for flow of about 2–3, but the precise value depends critically on the impurity level of the material (Langdon, 1990). A stress exponent of 2 is consistent with a theoretical model in which grain boundary sliding occurs by the movement of dislocations along the boundaries with accommodation by some limited intragranular slip (Langdon, 1994b).

Structural superplasticity in metals shows several regions of strain rate in each of which the equation

$$\sigma_t = K'(\dot{\varepsilon}_t)^{m_s} \tag{19.32}$$

holds with a transition between the regions. m_s has a small value in regions I and III corresponding to low and high strain rates, respectively, but has a higher value (typically in the range 0.5–1) in region II corresponding to intermediate strain rates. The maximum strain before failure occurs in region II since there is a correlation between strain at failure and m_s as noted earlier.

Structural superplasticity has been found in many ceramics at high temperatures (above half the absolute melting point) and with small grain size (typically 2–0.2 μm) (Maehara and Langdon, 1990). Some of the experiments on ceramics were done in compression but true superplasticity in tension has been demonstrated in Y–TZP (Langdon, 1993a).

Superplasticity in ceramics has been considered as a forming mechanism. A potential problem is the development of internal voids. Ma and Langdon (1993) have studied superplastic deformation at 1723 K of dense tetragonal zirconia stabilized with 3 mol % yttria with a grain size of 0.5 μm. For strain rates of 0.0028, 0.00028, and 0.000028/s, the engineering strain at failure was 85%, 195%, and 355%, respectively. Cavitation took place in each case, but the amount at failure increased with increasing strain rate. They concluded that cavitation is likely to be a problem in materials with very small grain sizes because of the large number of grain boundaries providing fast diffusion paths to feed cavity growth.

A comparison of superplasticity in metals and ceramics from Langdon (1991b) is given in Table 19.1.

19.8 DEFORMATION MECHANISM MAPS

A very useful way of summarizing the dominant deformation mechanisms in a single drawing was apparently first suggested by Weertman and Weertman (1965) and has been developed by Ashby and his collaborators (Ashby, 1972; Frost and Ashby, 1982). The basic concept is to show, for a given grain size, the regions of stress and temperature in which each deformation mechanism predominates. The border between two regions is the line along which the neigh-

TABLE 19.1 Comparison of Features of Superplasticity in Metals and Ceramics

Feature	Metals	Ceramics
Shape of the logarithmic plot of flow stress, σ, as a function of strain rate, $\dot{\varepsilon}$	Sigmoidal with three regions (I, II, and III)	Linear with only one region
Effect of purity on the slope, $1/m_s$, of logarithmic $\dot{\varepsilon}$ versus σ	The value of $1/m_s$ is independent of purity in regions II and III; the advent of region I depends on purity	The value of $1/m_s$ increases with increasing purity
Effect of grain size on the value of $1/m_s$	No effect	The value of $1/m_s$ decreases as the grain size increases
Dependence of elongation to failure on strain rate	Maximum elongations occur at intermediate strain rates (region II)	Maximum elongation occurs at the lowest strain rate
Variation of cavitation with strain rate	Maximum cavitation occurs at the slowest strain rate	Maximum cavitation occurs at the fastest strain rate

boring mechanisms have equal strain rates. Several types of plot are possible, but the most useful is a plot of log homologous stress (stress divided by the shear modulus) as a function of homologous temperature (absolute temperature divided by the melting point). A schematic plot for MgO with 10-μm grain size is shown in Figure 19.5 based on a more detailed plot by Frost and Ashby

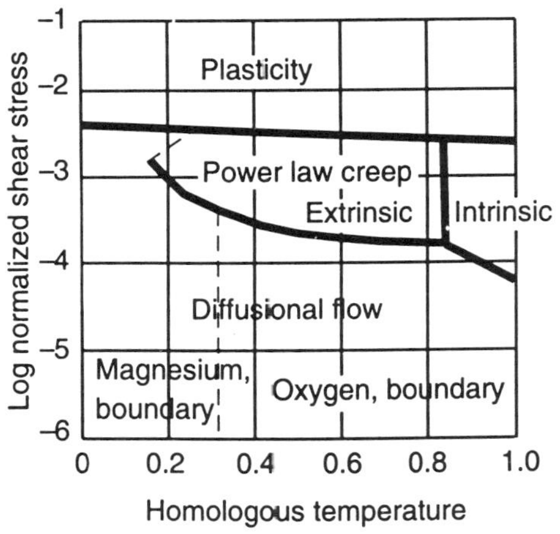

Figure 19.5 Deformation mechanism map for MgO with 10-μm grain size. (Adapted from Frost and Ashby, 1982.)

(1982). Additional information is usually provided on these diagrams by overlaying lines of constant strain rate. Note that a new plot must be made for each grain size and that new features may become evident with a considerable change in grain size. Thus for MgO of 1-μm grain size Frost and Ashby indicate a region where the dominant diffusional flow mechanism is controlled by magnesium diffusion in the lattice and another where magnesium diffusion in the boundaries is controlling.

A second type of deformation map can be constructed at constant temperature (Langdon and Mohamed, 1976, 1978; Langdon, 1980). The axes are homologous grain size (grain size divided by magnitude of Burgers vector) and homologous stress (stress divided by the shear modulus). Lines on this map indicate the locus where two types of deformation give equal rates. This type of deformation map clearly shows the regions in which the predominant deformation mechanism is Nabarro–Herring creep, or Coble creep, or dislocation creep. Lines of constant strain rate can be overlaid.

CHAPTER 20

CREEP RUPTURE IN CERAMICS AT HIGH TEMPERATURES AND SAFE-LIFE DESIGN

20.1 INTRODUCTION

All ceramic materials undergo some degree of creep at sufficiently high temperatures. Polycrystalline ceramics in tension at high temperatures typically develop small internal flaws that eventually lead to fracture. Design for safe use of ceramics under load at high temperature seeks to avoid both excessive deflection (e.g., avoidance of sufficient creep of turbine blades to cause contact between the blade tips and the shroud) and fracture. It is thus important to know the creep strain as a function of stress, temperature, and time and to have a failure criterion to predict the time to failure at a given stress and temperature.

20.2 THE GENERAL PROCESS OF CREEP DAMAGE AND FAILURE IN CERAMICS

We follow Wilkinson (1992) in summarizing creep damage mechanisms in ceramics. He classifies the microstructures of ceramic materials for high-temperature use into three categories. The first category is single-phase materials processed to give clean grain boundaries such as pure alumina. This class of materials constitutes a good system for model studies but such materials are not generally regarded as useful as high-temperature structural materials. The second category is materials with a bonding phase such as sintered silicon nitride. This boundary phase is typically liquid at the sintering temperature and facilitates sintering to full density. In older silicon nitrides the grain boundary phase usually became a glass at lower temperatures, leading to a microstructure of crystalline grains embedded in a continuous glass matrix distributed as a thin layer between grains and larger pockets. In modern silicon nitrides intended for load bearing at high temperatures, combinations of composition and heat treatment are used to form a crystalline grain boundary phase. The third category is multi-phase crystalline ceramics, often containing discontinuous reinforcements such as particles, platelets, or whiskers.

All these types of ceramic materials exhibit cavity formation under appropriate circumstances. Ultimate failure at high temperatures typically involves cavity growth at grain boundary triple junctions and linking along grain boundaries to form a crack that ultimately propagates catastrophically. The nucleation of these cavities is apparently not fully understood but often seems to occur at inhomogeneities caused by processing. In particular, some small pores can generally be found even in material nominally pore-free such as hot-pressed alumina. Other stress-concentrating inhomogeneities may also be involved in cavity nucleation. Cavities often nucleate at three-grain junctions. Thus microstructure always plays a key role in failure but the flaws responsible for high-temperature failure may be quite different from those responsible for failure at low temperatures. The process leading to failure can be complicated by additional, parallel processes such as grain growth or devitrification of an initially glassy grain boundary phase.

The status of knowledge of creep damage development in structural ceramics was reviewed by Chan and Page (1993), who also presented new data on cavity development in alumina and silicon nitride. They distinguish two extremes of cavity damage accumulation and ultimate failure. In one extreme, cavities nucleate and grow throughout the specimen. Failure occurs when enough cavities coalesce to form a microcrack that can propagate by fast fracture. In the other extreme, cavities nucleate and grow in the concentrated stress field associated with some microstructural inhomogeneity. These cavities coalesce to form a microcrack that then grows subcritically by means of cavitation. Actual failure may involve a combination of these two extremes. Bulk damage is thought to be favored at low stresses and long times while high stresses and short failure times favor localized damage.

They obtained data on cavity number and size by the technique of small-angle neutron scattering (SANS). This technique presents average data and so does not give a direct picture of coalescence. For polycrystalline alumina without a glassy phase (Lucalox), their results are shown in Figure 20.1. The density of cavities increases with increasing creep strain, but the size does not increase, indicating nucleation-dominated behavior. For polycrystalline alumina with a glassy silica phase (Coors AD99), their results are shown in Figure 20.2. Here the density of cavities and their size both increase, indicating con-

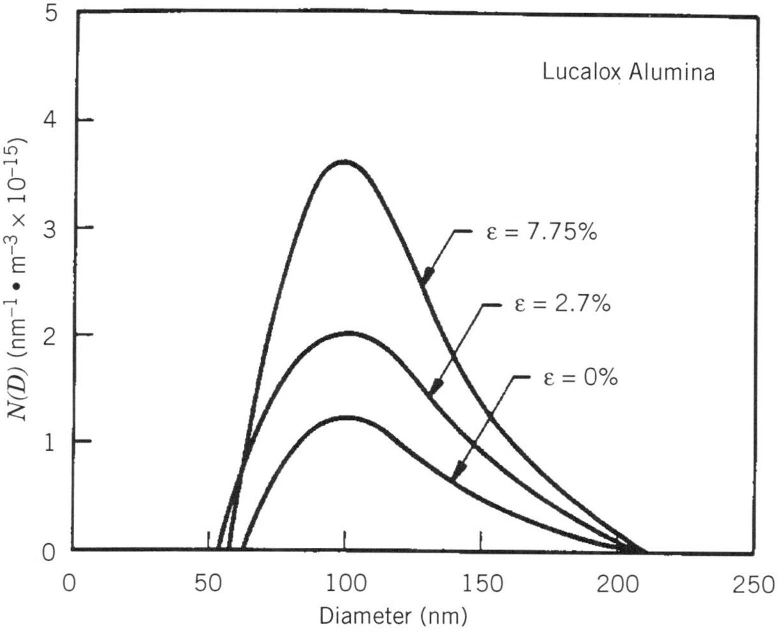

Figure 20.1 Cavity-size distribution as a function of creep strain in alumina without a glassy phase. (From K. S. Chan and R. A. Page, "Creep Damage Development in Structural Ceramics," *J. Am. Ceram. Soc.*, Vol. 76, 803–826, 1993. Reprinted by permission of the American Ceramic Society.)

current nucleation and growth. For polycrystalline SiC (Norton NC203), the peak density of cavities changes little, as shown in Figure 20.3, but the size at the peak increases with increasing creep strain, indicating growth-dominated behavior.

Direct observations of creep-crack growth in many ceramics suggest the schematic crack growth curve shown in Figure 20.4. First is the near-threshold region although thresholds are not always found. Second is a power law growth region in which the crack velocity is characterized empirically by

$$v = v_0\left(\frac{K}{K_C}\right)^{\alpha} \tag{20.1}$$

Third is the fast fracture region. Chan and Page have compiled the available data on the crack growth exponent and find values ranging from 1.88 to 270 for silicon nitride. Values in the range 1.88–13 appear to be characteristic of region II and the mechanism of creep-crack growth is thought to be viscous or diffusive flow in this region. An example of this type of behavior for MgO-doped alumina is given in Section 23.4 (Blumenthal and Evans, 1984; Johnson, Dalgleish, and Evans, 1984).

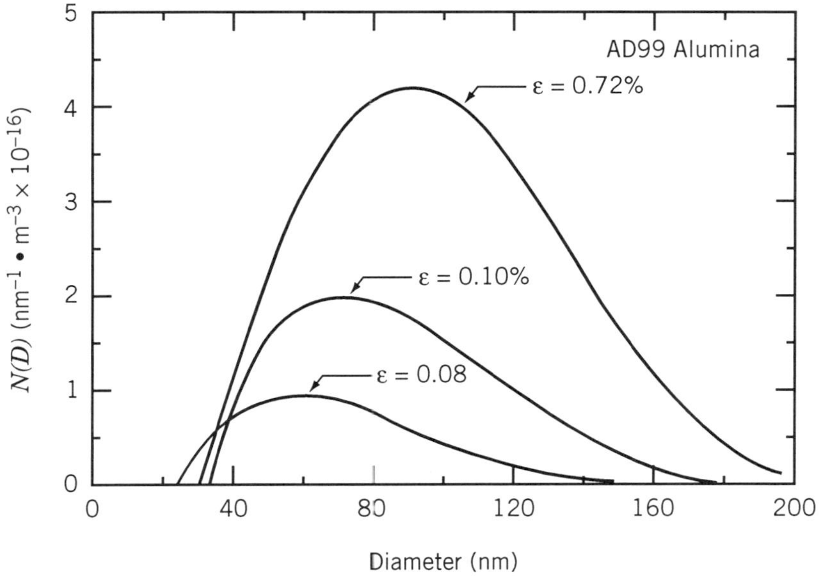

Figure 20.2 Cavity-size distribution as a function of creep strain in alumina with a glassy phase. (From K. S. Chan and R. A. Page, "Creep Damage Development in Structural Ceramics," *J. Am. Ceram. Soc.*, Vol. 76, 803–826, 1993. Reprinted by permission of the American Ceramic Society.)

20.3 THE MONKMAN–GRANT TECHNIQUE OF LIFE PREDICTION

Most ceramic materials at high temperature follow the Monkman–Grant type of empirical creep–rupture relationship, so that use of this relationship provides a good way to summarize data and also leads to one means of lifetime prediction under creep conditions. An empirical correlation of steady-state creep rate with time to failure was suggested by Monkman and Grant (1956). For many materials a plot of the logarithm of the steady-state creep rate, $\dot{\varepsilon}_s$, and the logarithm of the time to failure, t_f, can be well represented by a straight line leading to the empirical relationship

$$t_f = C\,\dot{\varepsilon}_s^{-M} \tag{20.2}$$

where M is a dimensionless constant close to 1. The important point for the use of this relationship for prediction of lifetime under stress is that this single curve holds for data for a given material over a wide range of stresses and temperatures. Monkman and Grant found that $M = 1$ fit the data on their materials well. This led to the use of a constant total strain as a criterion for failure caused by creep. More recent work on ceramics creeping in tension shows that M cannot be taken as 1; values of 1.45 and 2.39 were found for two

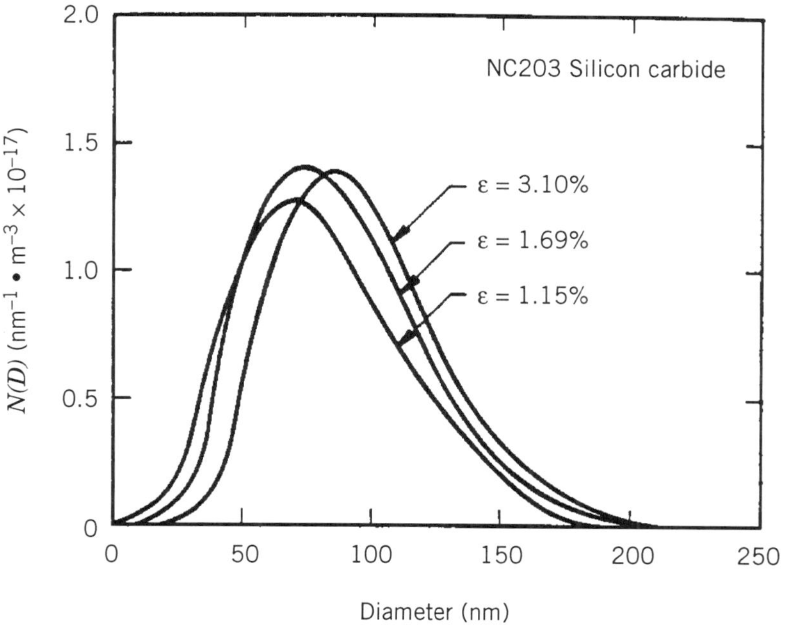

Figure 20.3 Cavity-size distribution as a function of creep strain for polycrystalline SiC. (From K. S. Chan and R. A. Page, "Creep Damage Development in Structural Ceramics," *J. Am. Ceram. Soc.*, Vol. 76, 803–826, 1993. Reprinted by permission of the American Ceramic Society.)

different types of siliconized carbide and a value of 1.69 was found for silicon nitride reinforced with silicon carbide whiskers (Wiederhorn and Hockey, 1991; Wiederhorn, Hockey, and Chuang, 1991). The total strain for failure thus decreases with increasing strain rate for materials with $M > 1$. The Monkman–Grant relation can still be used as a failure criterion with the proper M value for the material. In developing lifetime prediction equations, Wiederhorn, Hockey, and Chuang presented derivations of the Monkman–Grant equation for either of two extreme types of failure from cavity formation: (1) creep-crack nucleation and propagation and (2) cavity coalescence to form a critical size crack.

For the first type of failure they assumed that the time to nucleate cracks was short compared to the time for crack growth. They assumed a power law crack velocity equation like (8.1) with a thermally activated factor:

$$v = v_0\,(K_{\mathrm{I}} / K_0)^p \exp(-Q / RT) \tag{20.3}$$

The time to failure from crack propagation under constant stress is then given by substitution in (8.15) and is

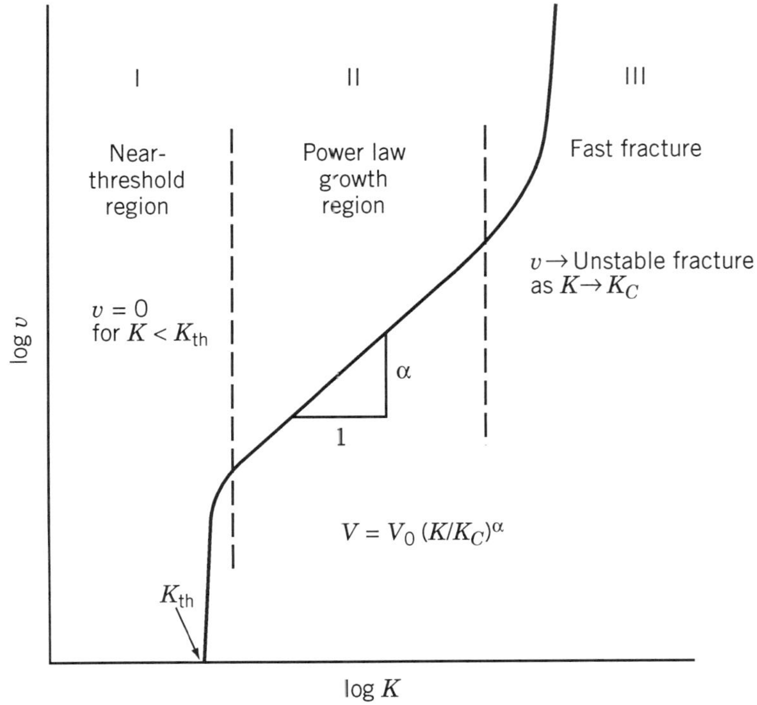

Figure 20.4 Schematic of creep-crack growth cruve for ceramics showing three characteristic regions. (From K. S. Chan and R. A. Page, "Creep Damage Development in Structural Ceramics," *J. Am. Ceram. Soc.*, Vol. 76, 803–826, 1993. Reprinted by permission of the American Ceramic Society.)

$$t_{f,\sigma,\mathrm{cp}} = \frac{B\ (\sigma_{f,t=0})^{p-2} \exp\,(Q\,/\,RT)}{\sigma^{p}} \tag{20.4}$$

Here $B = 2_0^{p}/[Y\nu_0(p - 2)K_I C^{p-2}]$ is a material constant and $\sigma_{f,\,t=0}$ is the strength at temperature set by the initial flaw size. They then assumed that the steady-state creep rate was given by a power law in stress

$$\dot{\varepsilon}_s = \dot{\varepsilon}_{s0}\,(T)\,(\sigma\,/\,\sigma_0)^{n_{\mathrm{cr}}} \tag{20.5}$$

and assumed a thermal activation factor to obtain the Norton equation for creep rate

$$\dot{\varepsilon}_s = \dot{\varepsilon}_{s0}\ \ (\sigma\,/\,\sigma_0)^{n_{\mathrm{cr}}} \exp\,(-H\,/\,RT) \tag{20.6}$$

Solving this for stress and substituting in (20.40) gives the failure time from crack growth in terms of the steady-state strain rate as

$$t_{f,\sigma,cp} = \dot{\varepsilon}^{-p/n_{cr}} B\,[exp\,([(-p/n_{cr})H + Q]/RT)]\,(\sigma_{f,t=0})^{p-2}\,(\sigma_0^{n_{cr}}/\dot{\varepsilon}_0)^{-p/n_{cr}} \quad (20.7)$$

From this equation the Monkman–Grant parameter is $M = p/n_{cr}$. The Monkman–Grant equation will be temperature independent only if $Q - pH/n_{cr} = 0$. Otherwise a series of parallel lines for different temperatures will be obtained on a Monkman–Grant plot.

For the case of cavity coalescence they eventually obtain

$$t_{f,\sigma,cc} = [(\pi/2b)^{1/2}\,(K_{th}/\sigma_0)\,\dot{\varepsilon}_{s0}^{1/n_{cr}}\;)]^{5.3}\,\dot{\varepsilon}_s^{-(1+5.3/n_{cr})} \quad (20.8)$$

where $b = 1 \times 10^{-4}$ μm and K_{th} is the threshold stress intensity factor for propagation of a crack formed by cavity coalescence. For a siliconized silicon carbide the value required to fit the data on lifetime was $K_{th} \approx 0.4$ MPa · m$^{1/2}$ in contrast to the value of critical stress intensity factor measured in a chevron notch experiment at high temperature of about 15 MPa · m$^{1/2}$. This high critical stress intensity value is thought to be due to the development of a substantial cavitation zone that develops around a deliberately introduced crack. In contrast, cracks that are nucleated by creep do not develop large cavitated zones. Thus self-nucleated cracks apparently propagate differently from those introduced by indentation or other external means.

They note that if Monkman–Grant data and data on stress dependence of creep rate are collected separately, only a relatively small number of experiments are need as a basis for creep time to failure. They suggest that only a small number, such as six, creep–rupture tests can be used to determine the Monkman–Grant parameters and that as few as two specimens (with multiple creep experiments on each) can be used to determine the stress and temperature dependence of the creep rate, although the use of more specimens is desirable. Wilkinson (1992) discusses models for the power law stress dependence of creep life and concludes that an exponent of 2.4 is predicted and that this is in good agreement with experiments.

20.4 THE TWO-STAGE STRAIN PROJECTION TECHNIQUE

As discussed previously, the traditional convention for metals and the continuing convention of most workers on ceramics is to regard the curve of strain as a function of time under creep conditions as divided into three stages: primary, steady state, and tertiary. A competing viewpoint developed more recently by some workers on metals is that creep curves in metals and alloys can generally be accurately described mathematically in terms of just two stages: primary and tertiary. The intervening portion previously termed "steady-state creep" is regarded as just a transition between the primary and tertiary stages. The transition can require a long time and produce a creep curve that gives a long interval of nearly constant creep rate. A good discussion of this two-stage representation including some theoretical justification in terms of creep mechanisms and practical examples of creep prediction for several metals is given by Evans and Wilshire (1993). We present a summary of this two-stage representation and refer the reader to Evans and Wilshire for fuller treatment.

The creep strain is written as the sum of a part due to the primary stage, ϵ_p, and a part, ϵ_t, due to the tertiary stage:

$$\varepsilon = \varepsilon_p + \varepsilon_t \tag{20.9}$$

The contributions from the two stages are written in terms of four experimental parameters, θ_1, θ_2, θ_3, and θ_4, as

$$\varepsilon_p = \theta_1 \left(1 - \exp\left(-\theta_2 t\right)\right) \tag{20.10}$$

$$\varepsilon_t = \theta_3 \left(\exp\left(\theta_4 t\right) - 1\right) \tag{20.11}$$

giving strain rates

$$\varepsilon_p = \theta_1 \theta_2 \exp\left(-\theta_2 t\right) \tag{20.12}$$

$$\dot{\varepsilon}_t = \theta_3 \theta_4 \exp\left(\theta_4 t\right) \tag{20.13}$$

The first equation describes a primary creep rate that decreases exponentially with time and approaches zero (i.e., the primary creep strain approaches a constant value) at long times. The second equation describes an accelerating process producing a tertiary strain and tertiary strain rate that both increase indefinitely with time. In practice the second process is terminated by failure of the specimen if creep continues for sufficient time.

The transition from predominantly primary to predominantly tertiary creep leads to a minimum creep rate at time

$$t_{\min} = \frac{1}{\theta_2 + \theta_4} \ln \frac{\theta_1 \theta_2^2}{\theta_3 \theta_4^2} \tag{20.14}$$

with value

$$\dot{\varepsilon}_{\min} = \theta_1 \theta_2 \left(\frac{\theta_1 \theta_2^2}{\theta_3 \theta_4^2}\right)^{-\left(\frac{\theta_2}{\theta_1+\theta_2}\right)} + \theta_3 \theta_4 \left(\frac{\theta_1 \theta_2^2}{\theta_3 \theta_4^2}\right)^{+\left(\frac{\theta_4}{\theta_2+\theta_4}\right)} \qquad (20.15)$$

From an engineering point of view these equations have proved very useful for metals and alloys in accurately describing creep curves. The four parameters are functions of temperature and stress and must be determined experimentally over a range of both variables and their dependence on these variables must be described by empirical equations. Once this information is available, the strain rate equations can be used to predict behavior at intermediate temperatures and stresses or even used to forecast behavior outside the range of measurement, although the latter is, of course, risky.

Two practical applications are to calculate the combination of stress and time that will lead to failure either through the occurrence of strain larger than the maximum tolerable or to the occurrence of stress rupture. The former condition simply requires integration of the strain rate equations to give strain as a function of time followed by determination of the time required to reach the maximum strain permitted by the limits of deflection of the part in question. The latter condition requires a fracture criterion. The fracture criterion commonly assumed for stress rupture under constant stress is that the total strain reaches a value, ϵ_f, termed the creep ductility. For a Cr–Mo ferritic steel, ϵ_f typically increases slightly with increasing stress at constant temperature and increases with temperature at constant stress (Evans and Wilshire, 1993). Values of ϵ_f for this steel range from 0.13 to 0.23 over a stress range of 150–350 Mpa and a temperature range of 808–868 K. The failure time, however, ranges over orders of magnitude over the same range of stress and temperature. Use of the creep strain equation including the stress and temperature dependence of the θ parameters combined with a value of the creep ductility allows the time to failure in stress rupture to be estimated.

20.5 FRACTURE MECHANISM MAPS

We are now in a position to consider fracture over a range of temperatures and stresses drawing on the background developed for different types of failure mechanisms. A useful device is the fracture mechanism map, which is analogous to the deformation mechanism maps discussed in Chapter 19. Quinn and Matsui were apparently the first to construct fracture mechanism maps for specific engineering ceramics (Quinn, 1984a; 1986; Matsui, Ishida, Soma, and Oda, 1986; Quinn and Braue, 1990), although conceptual work was also done by others (Evans and Blumenthal, 1983; Evans and Dalgleish, 1987; and Grathwohl, 1984). In our summary we follow Quinn (1990), who gives the history including the first suggestion by Wray and subsequent early work by Ashby and his collaborators (Ashby, Gandhi, and Taplin, 1979; Gandhi and Ashby,

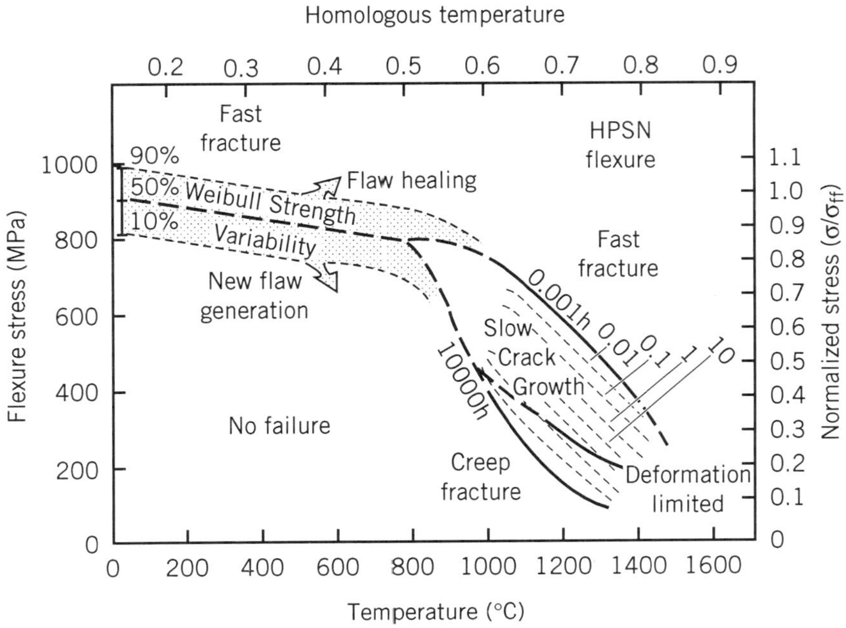

Figure 20.5 Fracture map for Mg-doped, hot-pressed silicon nitride in flexure in air. (From Quinn, 1990.)

1979). Three fracture mechanisms are found in silicon nitride (and in polycrystalline ceramics in general). These are fast fracture (treated in Chapter 5), slow crack growth (treated in Chapter 8), and creep fracture (treated in Chapter 20). The stress required for fracture from either of the last two mechanisms depends on the time allowed for fracture so that lines of constant time to fracture are plotted. Figure 20.5 shows a fracture map constructed for Mg-doped, hot-pressed silicon nitride (Norton NC 132) based on measurements in flexure in air by Quinn. The stress values are those calculated assuming elastic deformation and the creep strains were calculated assuming equal strain in tension and compression. As noted earlier, these approximations are progressively less good for increasing creep strain. Also, as discussed in Chapter 7, statistical effects lead to differences in strength in tension and bending in the fast fracture regime. Despite these cautions, this fracture map is useful in displaying at a glance the approximate maximum permissible stress without failure for given times under load as a function of temperature. The map is plotted as fracture stress as a function of temperature (in °C). A region of 90% strength variability in the fast fracture regime is shown based on the value of 16.6 for the Weibull modulus of this material measured in flexure as shown in Figure 20.6. The measured values of flexure failure stress as a function of time upon which the flexural failure map is based are shown in Figure 20.7.

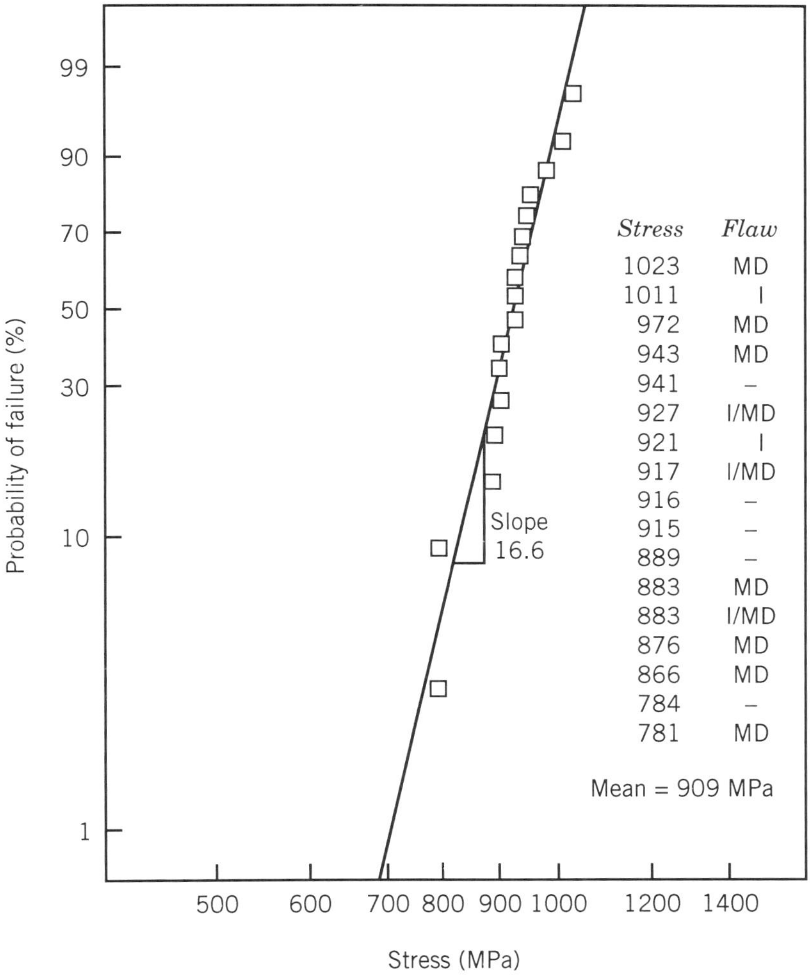

Figure 20.6 Weibull plot of flexure strength of MgO-doped silicon nitride. (From Quinn, 1990.)

Quinn used Weibull calculations and a comparison of failure times in tension with failure times in flexure to calculate a fracture map for the same silicon nitride in tension. For the flexure experiments, stresses were calculated by assuming elastic behavior; creep strains were calculated by assuming equal strain in tension and compression. His comparison of stress–failure lines in tension measured by Govila to those of his own in flexure is shown in Figure 20.8. A considerably lower allowable stress in tension at 1300 °C for a given failure time is evident. Based on these data, Quinn constructed the tension fracture map shown in Figure 20.9. Scales of normalized stress (fracture stress divided by fast fracture stress at room temperature) as a function of homolo-

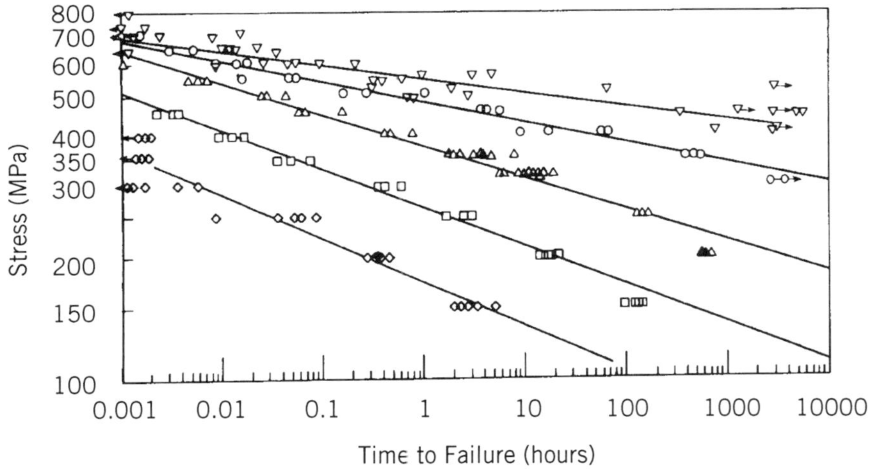

Figure 20.7 Flexural stress rupture results for MgO-doped, hot-pressed silicon nitride. (From Quinn, 1990.)

gous temperature (absolute temperature divided by the melting temperature with the latter taken as 2123 K, the decomposition temperature) are used.

This type of map is evidently of great value in design with ceramics. Indeed, its attractiveness is so great that caution is urged in the use of such maps. First, the map is for a particular silicon nitride with a specific microstructure. Even

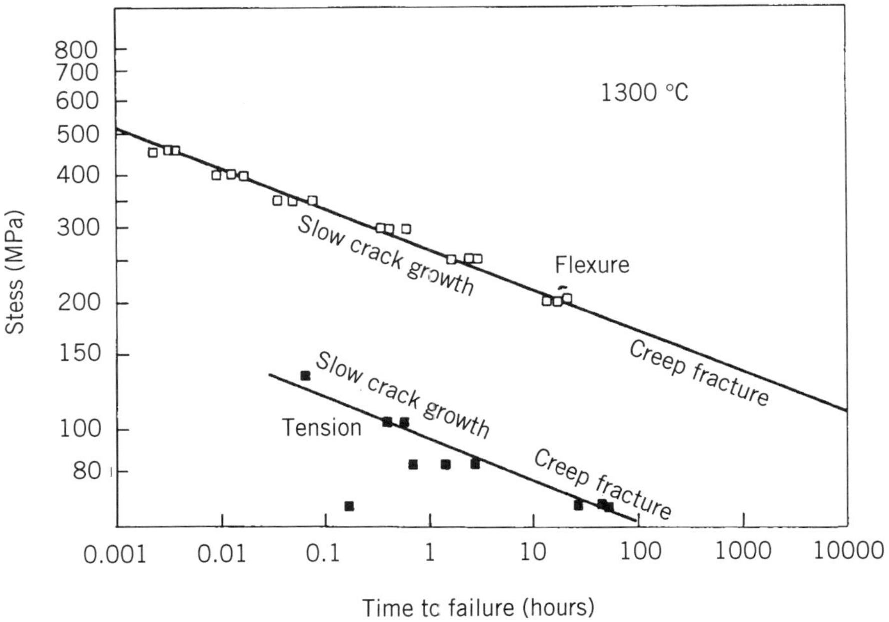

Figure 20.8 Comparison of tensile and flexural stress rupture data on MgO-doped, hot-pressed silicon nitride. (From Quinn, 1990.)

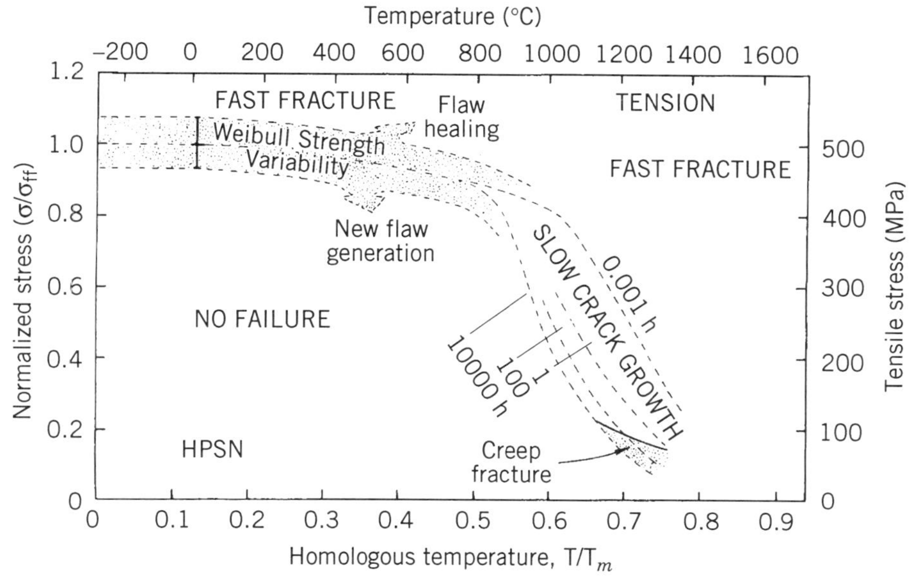

Figure 20.9 Fracture map for MgO-doped silicon nitride in air in tension. (From Quinn, 1990.)

apparently minor variations in sintering aids can alter creep behavior, as we shall show in Chapter 24. Second, the map is based on data taken in air. A change in environmental conditions can affect slow crack growth and perhaps creep rate. Third, this particular map is based in part on flexure data so that the calculated tensile behavior may not be accurate. Finally, the map takes no account of toughness and so gives no indication of degree of flaw sensitivity. The latter may be an important consideration in some applications. Also, in some applications deformation rather than failure may be the effective limitation. In summary, fracture mechanism maps provide an important and useful tool to be used as part of an overall consideration of materials behavior as part of design. They have been used in practice to design commercial turbocharger rotors (Matsui, Ishida, Soma, and Oda, 1986). An excellent discussion of the construction of fracture mechanism maps and their use in design is given by Wiederhorn, Quinn, and Krause (1994).

CHAPTER 21

MECHANICAL PROPERTIES OF GLASS AND GLASS-CERAMICS

21.1 INTRODUCTION

Inorganic glasses make up a major category of materials from the point of view of applications (both traditional and advanced) and from the point of view of materials science. In American usage, glasses are considered a category of ceramic materials so that a chapter on their mechanical properties is included in the present book. Also, glassy phases are present in many nominally polycrystalline ceramics and have major effects on the high-temperature mechanical properties of these ceramics. The present discussion of mechanical properties of glass is therefore presented not only for its direct importance to glasses but also as a prelude to discussion of polycrystalline ceramics.

Good general references include Doremus (1982), Varshneya (1994), and Kurkjian (1983).

21.2 TYPICAL INORGANIC GLASSES

A number of oxides, sulfides, tellurides, and halides form glasses under appropriate conditions. However, most of the glass used in practice is based on silica as the glass former and modified by the addition of other materials to give desired ranges of properties such as viscosity and elastic moduli. A summary of some of the most important categories of commercial silica glasses from Matthewson (1991) is given in Table 22.1.

TABLE 22.1 Important Categories of Silicate-Based Glasses

Type	Typical Composition	Typical Uses	Important Properties
Soda-lime	70–75% SiO_2, 12–16% Na_2O, 10–15% CaO	Bottles, glasses, windows	Optically clear, durable
Lead glasses	55–65% SiO_2, 18–38% PbO, 13–15% Na_2O or K_2O	Decorative items	High refractive index
Borosilicate	70–80% SiO_2, 7–13% Ba_2O_3, 4–8% Na_2O or K_2O, 2–7% Al_2O_3	Chemical apparatus, lamps and tubes	Chemical durability, low thermal expansion
Fused silica ("quartz")	100% SiO_2	High-temperature uses	High softening temperature, low thermal expansion

21.3 VISCOSITY OF GLASS

Glass in a liquid condition responds to a constant shear stress, σ_{xy}, by deforming with a constant shear strain rate, $\dot{\varepsilon}_{xy}$, given by

$$\sigma_{xy} = \eta\, \dot{\varepsilon}_{xy} \tag{21.1}$$

where η is the viscosity. When the viscosity is independent of the strain rate, the deformation is said to be Newtonian. The accepted unit of viscosity is the poise, which equals 1 dyn · s/cm^2 = 0.1 Pa · s. Viscosity in oxide glasses is usually taken to be Newtonian and this appears to be true for relatively low strain rates. At high strain rates deformation can depart from Newtonian behavior (Simmons and Simmons, 1989). The rate at which the departure becomes measurable depends on composition and microstructure of the glass including the possibility of phase separation.

A useful analogy exists between purely Newtonian viscous deformation and purely linear elastic deformation. The equations developed from elasticity to describe *strain* resulting from stress for an isotropic elastic body can be used to calculate *strain rate* produced by stress in a viscous body. To do this the elastic equations are written in terms of the shear modulus and Poisson's ratio and the shear modulus is replaced by the viscosity, Poisson's ratio is replaced by 0.5, and the strain components are replaced by the corresponding strain rates.

As a typical inorganic glass is cooled, the viscosity increases from values as low as 10^2 poise to as high as 10^{14} in the measurable range and presumably continues to increase as the temperature is further decreased. Several temperatures important in the technology of making glass and glass articles are defined in terms of viscosity values and given names. These names and the corresponding viscosities are: working point 10^4 poise, softening point $10^{7.65}$ poise, annealing point $10^{13.0}$ poise, and strain point $10^{14.5}$ poise. All of the foregoing are ASTM definitions except the working point.

The temperature dependence of the viscosity of inorganic glasses (and many other viscous materials) is accurately described by the Vogel–Fulcher–Tammann equation, usually termed the Fulcher equation in the ceramic community (Fulcher, 1925). This equation is

$$\ln \eta = A + \frac{B}{T - T_0} \tag{21.2}$$

where η is viscosity, T is temperature, and A, B, and T_0 are constants for a given glass. This equation can be derived from free-volume theory as summarized by Scherer (1992).

21.4 ELASTICITY OF INORGANIC GLASSES

At temperatures below the strain point glasses typically are elastic until brittle failure, except for limited inelastic deformation under pointed indenters as discussed later. The high strength of glass with a pristine surface causes the elastic strain to failure to be as high as 17% for fused silica. Above about 15% strain deviations from linear elasticity become evident and can be detected at lower strains with careful measurements. Silica is found to have a higher elastic modulus at higher strains while vitreous germanium oxide and vitreous boric oxide have a lower modulus at higher strain than at low strain (Doremus, 1982). For silica the strain dependence of Young's modulus is given approximately by $E(\epsilon) \approx E(\epsilon = 0)[1 + 3\epsilon]$ (M. J. Matthewson, personal communication, 1995); a more accurate polynomial approximation was given by Krause, Testardi, and Thurston (1979).

21.5 STRENGTH AND FRACTURE SURFACE ENERGY OF INORGANIC GLASSES

Silicate glass was used by Griffith in his classic work on strength of brittle materials and today is generally considered to typify brittle behavior, although some degree of local inelastic behavior can occur under concentrated loads (Marsh, 1964; Holloway, 1983). This inelastic deformation in high silica glasses appears to consist of two different types: (1) a densification that recovers completely upon subsequent heating and (2) shear deformation along fault lines. Doremus (1982) summarizes work by Hagan on this type of shear deformation of glass under a concentrated load and describes it as consisting of fault or flow lines separated by regions in which there is only elastic strain. He points out that it resembles geological faults and is quite different from plastic deformation in ductile crystals. Additional studies summarized by Kurkjian (1988) show that under pointed indenters a different type of behavior occurs in "anomalous" glasses (i.e., high silica glasses) and "normal" glasses (i.e., glasses having greater than 15% network modifier). In the latter a true volume conserving shear flow is possible. Mechanisms and theories of inelastic deformation are discussed further in a review by Argon (1980).

Many experiments indicate that glass with a "normal" surface usually fails from the surface and that flaws are responsible for its ordinary strength. The contemporary view is that Griffith's theory correctly describes the strength of glass having significant surface flaws. However, glass can be prepared without surface flaws (having a so-called "pristine" surface) and then has very high strength. Carefully prepared silica fibers have strengths as high as 14 GPa (2 million psi), which is about $E/5$ with $E = 70$ GPa (Kurkjian, 1988). Applying the Griffith–Irwin fracture mechanics with $K_IC = 0.81$ MPa $\cdot$ m$^{1/2}$ as determined from experiments with long cracks requires a crack length of 1 nm to account for a strength of 14 GPa. Such a short crack is probably outside the range of validity of the Griffith–Irwin theory and so 14 GPa is thought to be the theoretical strength of silica.

Carefully prepared silica fibers can have not only a high average strength but a small scatter. Schonhorn, Kurkjian, Jaeger, Vazirani, Albarino, and Di-Marcello (1976) drew silica fibers about 110 μm in diameter and coated them with an epoxy-acrylate film. The resulting fibers had an average strength of 3.5 GPa and a Weibull modulus m of about 25.

Several questions arise. What is the origin of flaws in normal glass surfaces? How can glass with a pristine surface be protected from strength degradation? What can be done to strengthen glasses with normal surfaces? Some of the surface flaws are evidently caused by mechanical damage. Studies using blunt indenters show a reduction of post-indentation strength with increasing load on the indenter (Kurkjian, 1988). The processes taking place depend on both the cleanness of the indenter and the initial state of the surface. Thus dust particles of small radius on the blunt indenter can be the actual indenter so that the stresses acting are greater than those calculated from the equation for stress under a ball

$$\sigma_t = \frac{1-2\nu}{2}\frac{P}{\pi a^2} \tag{21.3}$$

where

$$a = 1.1\left(\frac{Pr}{E}\right)^{1/3} \tag{21.4}$$

Here P is the load on the ball and r is the radius. A question is whether the concentrated local stress is initiating a surface flaw or causing growth of a small preexisting flaw. For glasses with nonpristine surfaces (including chemically polished glass) the latter is probably the case. The tensile stress causes small preexisting flaws in nonpristine surfaces to grow stably once a threshold stress is exceeded. This threshold stress is independent of flaw size and distribution unless the number of flaws is very small. Details of the process of mechanical damage of glass surfaces are complex and not yet completely understood but certainly involve local stresses from contact sometimes combined with chemical reactions. The tensile stress caused by a ball contact on a surface can be considerably larger than that given by the above equation if there is a component of force on the ball parallel to the surface (Hamilton and Goodman, 1966).

21.6 ACHIEVING HIGH STRENGTH IN BULK GLASS

The strength of bulk glass is typically far below the theoretical strength (usually a factor of 100 or more) unless special treatment has been used to remove surface flaws or render them less effective in initiating fracture. Removal of flaws can be done to a large extent by careful flame polishing or chemical polishing. Carefully polished rods can have high strengths, as high as 1400 MPa at liquid nitrogen temperature. However, such glass surfaces typically degrade by chemical or mechanical action and the high strength is then lost. A more useful approach in practice is to induce a compressive stress in a surface layer by thermal tempering, ion exchange, or other methods (Doremus, 1973; Kirchner, 1979; Schaeffer, 1983; Matthewson, 1991). The maximum strengthening factors achievable by various techniques have been summarized by Doremus and are given in Table 21.2.

The term tempered glass derives from the process of tempering steel in that a suitable thermal treatment can produce strengthening of either (Gardon, 1980). However, the process is quite different in terms of the basic mechanisms involved. In steel, carbides are precipitated and both hardness and toughness are changed. Thermal tempering leaves glass a single-phase material with no significant change in hardness. The increase in strength of thermally tempered glass results from a compressive stress at and near the surface balanced by a tensile stress within the body. Failure of glass typically takes place by propagation of surface cracks so that compressive prestressing produced by thermal tempering must be overcome by applied stress before cracks can propagate. The process of thermal tempering evidently causes an internal strain energy to be stored in the glass. When failure does occur in tempered glass this energy is released, resulting in the shattering of tempered glass into many small frag-

TABLE 21.2 Factors by Which Glasses Can Be Strengthened by Various Treatments

Treatment	Approximate Maximum Strengthening Factor
Quench hardening (thermal tempering)	6
Ion exchange	10
Surface crystallization	17
Ion exchange and surface crystallization	22
Etching	30
Fire-polishing	200
Second-phase particles	2

Source: Doremus (1973).

ments whose size and shape make them less dangerous than typical broken fragments of untempered glass.

The origin of a surface compressive stress caused by thermal tempering is qualitatively easy to understand. When glass is held for sufficient time above the annealing temperature, the stresses anneal out. If the glass is subsequently cooled rapidly, the temperature of the center lags behind that of the surface. As the surface drops below the annealing temperature it can no longer undergo viscous flow but the hotter center still can flow so that it adjusts to the thermal contraction of the surface until the center also passes through the annealing temperature. Thereafter center and surface contract together until the surface reaches room temperature. As the center continues to cool toward room temperature it attempts to contract further, causing the surface to go into compression. Stated differently, a nonuniform temperature distribution with a uniform (zero) stress distribution is converted to a nonuniform stress distribution when the temperature distribution becomes uniform. The surface compression of fully tempered glass is typically about 100 MPa (Gardon, 1980). The actual process of tempering is more complicated because the spatial and temporal distribution of temperature and stress must be treated quantitatively and because some degree of structural compaction as well as viscous flow must be taken into account. The theoretical situation is surveyed by Gardon. Bartholomew and Garfinkel (1980) state that the minimum sample thickness for effective thermal tempering is about 0.3 cm and that the achievable maximum strength is typically 170 MPa, while annealed glass typically has a strength of about 55 MPa.

Chemical strengthening can produce somewhat greater strengthening. Bartholomew and Garfinkel classify chemical strengthening into three types: compound glass, ion exchange, and surface crystallization. A compound glass made of surface layers of low thermal expansion on a core of high thermal expansion can produce high compressive surface stresses but suffers from edge effects. Ion exchange of Li for Na at high temperature has produced strength as high as 482 MPa. Surface crystallization of a low expansion phase at high temperature has produced strength in excess of 690 MPa.

21.7 GLASS-CERAMICS

The controlled crystallization of certain glasses can be used to produce fine grain sized polycrystalline ceramics with some amount of residual glassy phase that are termed glass-ceramics (McMillan, 1979; Beall, 1992). High strength can be achieved if the final grain size is small. Achieving fine grain size requires forming many nuclei from which the individual crystalline grains grow. Multiple nucleation is achieved by using a suitable nucleating agent combined with a heat treatment in an appropriate temperature range. Crystal growth is then accomplished by holding at a higher temperature.

McMillan gives as typical ranges of modulus of rupture values 55–70 MPa for glasses and 70–350 MPa for glass-ceramics. For a Li_2-SiO_2 glass he gives 30–50 MPa. The same base glass made into a glass-ceramic by using P_2O_5 as a nucleating agent gave 110–398 MPa.

CHAPTER 22

MECHANICAL PROPERTIES OF POLYCRYSTALLINE CERAMICS IN GENERAL AND DESIGN CONSIDERATIONS

22.1 INTRODUCTION

Polycrystalline ceramics comprise an extensive variety of materials with a wide range of mechanical properties. In the present chapter we present a brief overview to give perspective on their broad categories and corresponding mechanical properties. In subsequent chapters we present more detailed discussions of selected important families of ceramics.

Ceramics are conveniently considered as traditional and advanced (Phelps and Wachtman, 1986). There is no absolute distinction between these two categories; a particular ceramic such as an aluminum oxide with a silicate phase may appropriately be considered as one of the other depending on the degree of property optimization that has been accomplished through control of composition and processing to achieve a special microstructure. Nevertheless, the distinction is useful up to a point. Traditional ceramics are usually made in bulk from relatively inexpensive (usually natural) starting materials and tend to have low to modest strength values and creep resistance. Advanced ceramics are usually made from more highly refined or synthetic starting materials and often have microstructures designed to optimize mechanical or other functional properties.

22.2 MECHANICAL PROPERTIES OF POLYCRYSTALLINE CERAMICS IN GENERAL

Research intended to improve understanding of basic factors controlling mechanical properties is usually done on specimens with carefully controlled compositions and microstructures. One or at most a few features are systematically varied while other features are kept as constant as possible. The composition and microstructure are usually chosen to be as simple as possible consistent with the effect being studied. The result is that most of the basic research results available are on advanced ceramics or model materials rather than on traditional ceramics. Basic data on many advanced ceramics are presented in subsequent chapters. Engineering data on mechanical behavior of conventional ceramics are available, often from manufacturers, and a sample is given here to indicate a range of properties. Table 22.1 summarizes typical ranges of elastic moduli and strength of several classes of traditional ceramics

TABLE 22.1 Some Mechanical Properties of Ceramic Materials in Common Use

Material	Young's Modulus (GPa)	Flexural Strength (MPa)	Compressive Strength (MPa)
Brick	5–20	5–10	10–25
Roof tile	5–20	8–15	10–25
Steatite	1–3	140–160	850–1000
Silica refractories	96–97% SiO_2	8–14	30–80
Fireclay refractories, 10–44% Al_2O_3	20–45	5–15	10–80
Corundum refractories, 75–90% Al_2O_3	30–120	10–150	40–200
Forsterite refractories	25–30	5–10	20–40
Magnesia refractories	30–35	8–200	40–100
Zircon refractories	35–40	80–200	30–60
Whiteware	10–20	20–25	30–40
Stoneware	30–70	20–40	40–100
Electrical porcelain	55–100	90–145	55–100
Capacitor ceramics	—	90–160	300–1000

and two (electrical porcelain and capacitor electronic ceramics) that can be considered advanced.

The mechanical property data that one would like to have available for a particular ceramic ideally include a considerable list as follows: (1) elastic moduli as a function of temperature; (2) average strength and Weibull parameters, both as a function of temperature; (3) values of toughness as a function of temperature (actually, knowledge of the full R-curve is desirable at each temperature); (4) data characterizing slow crack propagation for each temperature of possible use (these should include curves of slow crack propagation rate as a function of applied stress intensity factor for each combination of environment and temperature of interest); (5) data on cyclic fatigue behavior for a range of amplitudes of the time-varying applied stress intensity factor and the static component; and (6) creep curves and stress rupture data at high temperatures.

Apart from the elastic moduli all these data typically depend on composition and microstructure and in some cases depend strongly on it. Evidently any sensible body of mechanical property data should be accompanied by a degree of characterization for each material sufficient to determine the mechanical properties being listed. Also, some mechanical property values depend on the test method so that accepted and specified test methods should be used. Both adequate characterization and good testing are much harder to do than is generally realized. A standard system of classifying ceramics and accepted standards for the measurement of mechanical properties and behavior are thus important. The general problem was recognized (Schneider, 1989) by the Versailles Project on Advanced Materials and Standards (VAMAS), and a system of classifying advanced ceramics was recently established in ASTM standard C 1286-95. Activities on standardization in Europe, Japan, and the United States were summarized by Padgett (1989) and the following list of organizations developing standards and recommendations for fine ceramics was given by the ISO (1994):

1. Divisional Council on Ceramics of Japanese Industrial Standards (JIS).
2. Committee C-28 on Advanced Ceramics of the American Society for Testing and Materials (ASTM Committee C-28).
3. Technical Committee on Advanced Technical Ceramics of European Committee for Standards (CEN/TC 184).
4. Versailles Project on Advanced Materials and Standards (VAMAS).
5. Cooperative Programme on Ceramics for Advanced Engines and Other Conservation Applications (Annex II) of International Energy Agency (IEA).

The Versailles Project on Advanced Materials and Standards has defined advanced ceramics as "a highly engineered, high performance, predominately non-metallic, inorganic, ceramic material having specific functional attributes" and has proposed a classification system for advanced technical ceramics in terms of four independent fields: application, chemical character (including form), processing, and property data (CIAC, 1994).

ASTM Committee C-28 on Advanced Ceramics has completed the standards listed in Table 22.2. Many additional standards for advanced ceramics are in the balloting process by this committee (G. D. Quinn, personal communication, 1995).

The Japan Institute for Standards has established standards for measurement of bending strength at room temperature, modulus of elasticity, bending strength at high temperatures, and elastic modulus at high temperature and is at work on many other tests (Japan Fine Ceramics Association, 1987).

A strong note of caution is therefore appropriate in the use of any mechanical property data from tables. With this caution in mind we mention some of the existing tabulations. An early and still useful series of reports was produced by Battelle Memorial Institute for the Air Force (Battelle, 1979). A report on structural ceramics produced in 1984 includes data on the strength and creep rate of many then-current materials as a function of temperature (Larson and Adams, 1984). A handbook series begun by the National Physical Laboratory provides an introduction and surveys data on high-alumina ceramics (Morrell, 1985, 1987). A survey of the principal families of structural ceramics as of 1989 including silicon nitride, silicon carbide, and zirconia is available (Wachtman, 1989).

Several computerized databases on ceramics exist and continue to be developed (Munro, 1991). A database on mechanical properties of ceramics studied in the Ceramic Technology Project of the Department of Energy is maintained by Oak Ridge National Laboratory (Keyes, 1994). Emphasis is on silicon-based ceramics and transformation-toughened ceramics. Data include computerized plots and scanned microstructures.

A computerized database on structural ceramics can be obtained from the National Institutes of Standards and Technology (Munro and Hubbard, 1989; Munro, Hwang, and Hubbard, 1989; R. G. Munro, personal communication, 1994). Both the database and a set of programs to use it are available on a computer disk. Both research materials and commercial materials are included when either meets stringent standards of quality regarding characterization and test methods. This database includes carbides (SiC, TiC, B_4C, etc.), nitrides (Si_3N_4, AlN, BN, etc.), oxides (Al_2O_3, ZrO_2, BeO, mullite, etc.), oxynitrides (sialon, silicon oxynitride, etc.), and borides (TiB_2, ZrB_2, etc.). The property data include a full range of thermal and mechanical properties and crystallography and corrosion property data are being added.

Another computerized database on ceramics contains only data on commercial ceramics as supplied by manufacturers. This is available from the Ceramic Information Center of the American Ceramic Society (C. Schnitzer, personal communication, 1994).

TABLE 22.2 ASTM Standards Completed by Committee C-28

Termminology	
C 1145-89	Standard Definition of Terms Relating to Advanced Ceramics
C 1286-95	Standard System for Classification of Advanced Ceramics
Properties and Performance (Monolithic)	
C 1161-90	Standard Test Method for Flexural Strength of Advanced Ceramics at Ambient Temperature
C1211-92	Standard Test Method for Flexural Strength of Advanced Ceramics at Elevated Temperatures
C 1259-94	Standard Test Method for Dynamic Young's Modulus
C 1273-94	Standard Practice for Tensile Strength of Monolithic Advanced Ceramics at Ambient Temperature
Design and Evaluation	
C 1175-91	Standard Guide to Test Methods for Nondestructive Testing of Advanced Ceramics
C 1198-91	Standard Test Method for Dynamic Young's Modulus
C 1212-92	Standard Practice for Fabricating Ceramic Reference Specimens Containing Seeded Voids
C 1239-93	Standard Practice for Reporting Uniaxial Strength Data and Estimating Weibull Distribution Parameters for Advanced Ceramics
Characterization and Processing	
C 1251-93	Standard Guide for Determination of Specific Surface Area of Advanced Ceramics by Gas Adsorption
C 1274-94	Standard Test Method for Advanced Ceramic Specific Surface Area by Physical Adsorption
C 1282-95	Standard Test Method for Determination of the Particle Size Distribution of Advanced Ceramics by Centrifugal Photosedimentation
Ceramic Composites	
C 1275-94	Standard Practice for Monotonic Tensile Strength Testing of Continuous Fiber-Reinforced Advanced Ceramics with Solid Rectangular Cross Section at Ambient Temperatures

22.3 DESIGN INVOLVING MECHANICAL PROPERTIES

The brittle failure typical of ceramics requires special care in design of structures or devices using ceramic components. If ceramics and their properties are taken as given, the problem becomes one of designing the device to survive within the properties available and under the service conditions. If modification of the ceramic to better serve the function required is included in consideration, the question of design of the microstructure to provide optimum properties, especially mechanical properties, becomes of interest. In either situation much of the material of the present book is pertinent to design. In a still broader consideration design can be taken to include economic factors, in which case consideration of mechanical properties still plays a role.

In the present section we briefly discuss the use of mathematical techniques for the analysis of stress and how the considerations of several earlier chapters come into play. Again we emphasize that mathematical sophistication in formal modeling should be supplemented by commonsense consideration of practical experience.

Within the general question of design with ceramics some particular situations have been developed into a theoretical model with a mathematical framework, which is complete within its assumptions. Examples include safe-life design for subcritical crack propagation and design for a desired probability of survival. How well the assumptions underlying these theoretical models fit the real behavior is always an important consideration. Knowledge of actual behavior of ceramics in comparable practical situations and judgment in application of formal treatments are evidently very important to design with ceramics.

Systematic qualitative discussions of design with ceramics including consideration of processing have been given by Creyke, Sainsbury, and Morrell (1982), Morrell (1987), and Richerson (1992). Design for survival from a stress distribution and statistical point of view has been reviewed in full mathematical detail by McLean and Hartsock (1989, 1991), who provide a completely worked out example. These excellent papers present an extensive discussion that is too long and complex to summarize here; they are highly recommended to the reader concerned with design for reliability with ceramics.

Design of ceramics with improved properties through control of composition and microstructure has been mentioned by various authors but is often focused only on some particular aspect of microstructure or some particular aspect of mechanical survival. Treatments of design of ceramic microstructures from a mechanical point of view with particular consideration of toughness, strength, and reliability include Becher (1991) and Lawn (1993).

Probabilistic design of ceramic components with the assistance of computer programs is discussed by Nemeth and Gyekenyesi (1991). Their basic model deals with overall fast fracture failure but is being modified to include subcritical crack growth. The initial step in their procedure is to use a computer program based on a finite element model. This model uses boundary conditions including loading and thermal conditions to obtain the stress distribution throughout the specimen or structure under consideration. The results from this model are input to a program called CARES (Ceramics Analysis and Reliability Evaluation of Structures) that combines three major elements: (1) linear elastic fracture mechanics, (2) extreme value fracture statistics to obtain the characteristic flaw size distribution function, and (3) material microstructure. The CARES program first analyzes statistically the data obtained from the fracture of simple specimens such as uniaxial tensile specimens or flexural specimens. The Weibull and Batdorf material parameters are then estimated from these data. A probability of failure by fast fracture is then done by taking into account the probability of failure of each volume element (or surface element) of the specimen or structure. Nemeth and Gyekenyesi give an example of a rotating annular disk and calculate the probability of failure as a function of the angular speed in revolutions per minute. Life prediction methodologies and data for ceramic materials have recently been reviewed in a conference proceedings book (Brinkman and Duffy, 1994).

CHAPTER 23

MECHANICAL BEHAVIOR OF ALUMINUM OXIDE

23.1 INTRODUCTION

Alumina (Al_2O_3) and the broad range of ceramics containing a major proportion of alumina typify ceramics in many ways. Alumina is available in single and polycrystalline form. It has traditional uses ranging from refractories to electrical porcelains and is available in various advanced forms such as whisker-reinforced and transformation-toughened alumina. The latter two special types of ceramics involving alumina are covered in other chapters: whisker-reinforced alumina is discussed in Chapter 13 and composites of alumina and zirconia are discussed in Chapter 26. High-alumina ceramics with more than 95% alumina have been surveyed by Dörre and Hubner (1984) and those with more than 80% alumina by Morrell (1985, 1987). Basic aspects of the plastic deformation, creep, and creep-rupture behavior of alumina have already been described in Chapters 18, 19, and 20. In the present chapter a brief summary of the range of values of strength and creep rate characteristic of alumina is given.

23.2 STRENGTH AS A FUNCTION OF GRAIN SIZE

Rice (1994) has recently surveyed the effect of temperature and environment on the dependence of strength on grain size of alumina, beryllia, magnesia, and zirconia. The general result found by many authors is that strength is proportional to $GS^{-1/2}$ over some range of grain size (GS) with two branches. The constant of proportionality is relatively small for large grain sizes and is larger for small grain size. Rice interprets the result in terms of two competing fracture mechanisms: microplasticity-controlled fracture (i.e., local plasticity nucleating cracks, which then cause failure) and brittle fracture (i.e., fracture from preexisting flaws without plastic effects). For microplasticity-controlled behavior, the branch for larger grain size extrapolates approximately to the single crystal value, reflecting the easiest mode of microplasticity activation. For brittle fracture the large grain size branch often extends below, sometimes well below, the lowest single crystal value (as a function

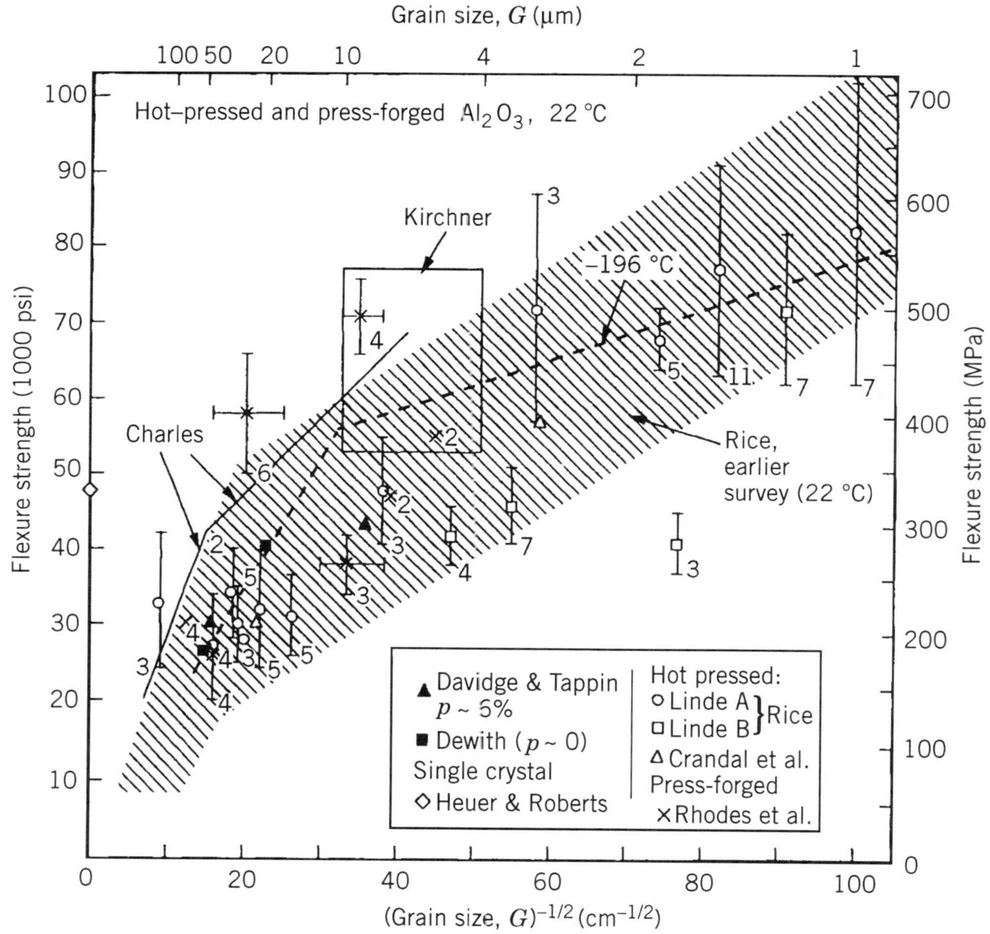

Figure 23.1 Strength of polycrystalline alumina at room temperature. (From Rice, 1996.)

of orientation) for comparable surface finish. When microplasticity-controlled fracture occurs it is in competition with fracture from preexisting flaws.

A compilation of flexural strength results for polycrystalline alumina at room temperature plotted as a function of reciprocal square root of grain size is shown in Figure 23.1 from Rice (1996). The data suggest a linear dependence with different slopes at small and large grain size. Similar data on alumina at liquid nitrogen temperature are shown in Figure 23.2, which also indicates a two-branch linear dependence. Finally, data taken in the range 1200–1315 °C show similar dependence on grain size as seen in Figure 23.3.

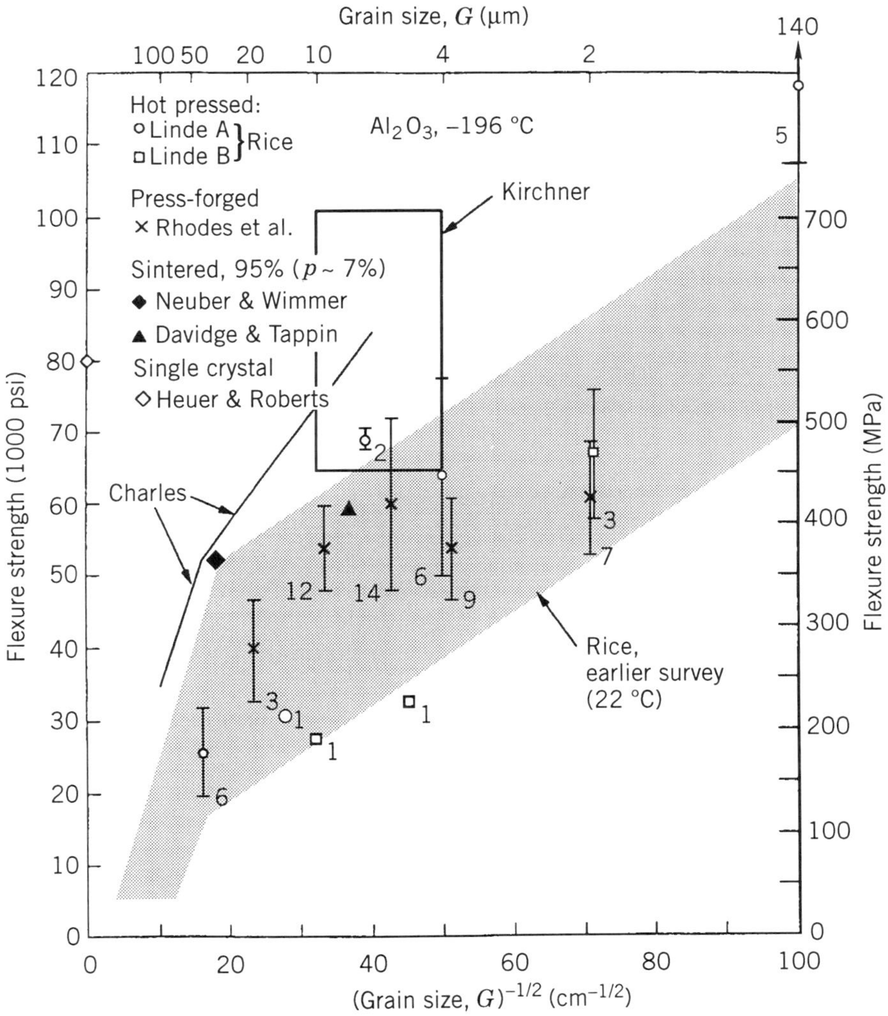

Figure 23.2 Strength of polycrystalline alumina at liquid nitrogen temperature. (From Rice, 1996.)

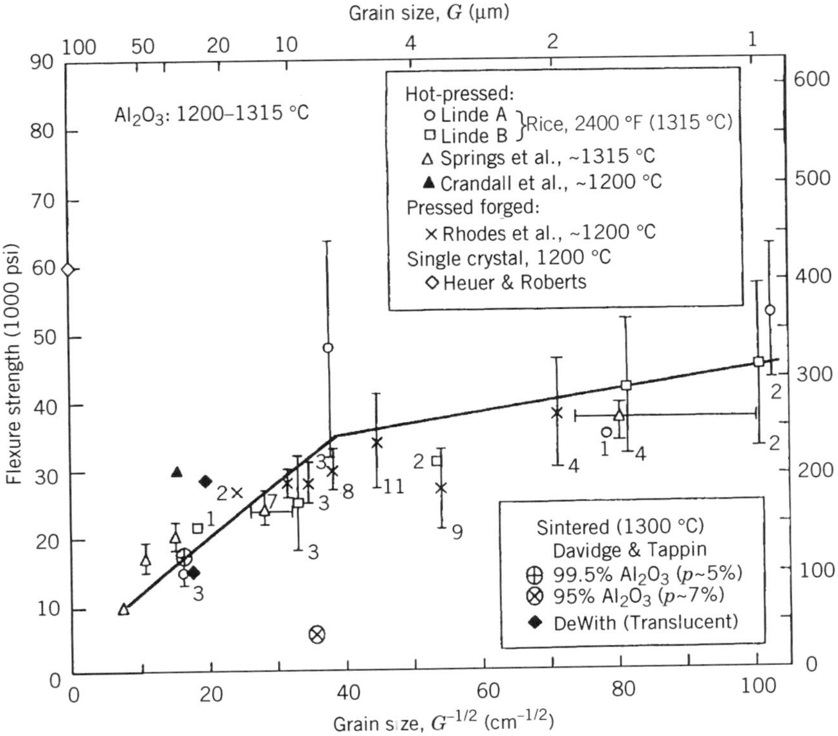

Figure 23.3 Strength of alumina as a function of grain size at 1200–1315 °C. (From Rice, 1996.)

23.3 STRENGTH AS A FUNCTION OF TEMPERATURE

The strength of polycrystalline alumina was found by Jackman and Roberts (1955) to show little change from room temperature up to about 700 °C and thereafter to decrease. In contrast, they found great scatter in the strength of single crystals but an indication of a minimum in the range 300–600 °C followed by an increase in strength with further increase in temperature. They speculated that the increase was associated with local plastic deformation to relieve stress concentrations. Wachtman and Maxwell (1959) measured the strength of a set of sapphire rods with the tensile axis at 45° to the hexagonal axis and another set with the tensile axis parallel to the hexagonal axis. The first set was thus favorably oriented for slip on the (0001) $[11\bar{2}0]$ system and the latter was unfavorably oriented. The results are shown in Figure 23.4 and provide some support for the plasticity hypothesis but no real proof. It is now thought that several effects may be involved in the strength dependence on temperature of single and polycrystalline alumina from liquid nitrogen temperature to 1000 °C: dislocation microplasticity, twinning, stresses due to thermal expansion anisotropy, and slow crack growth.

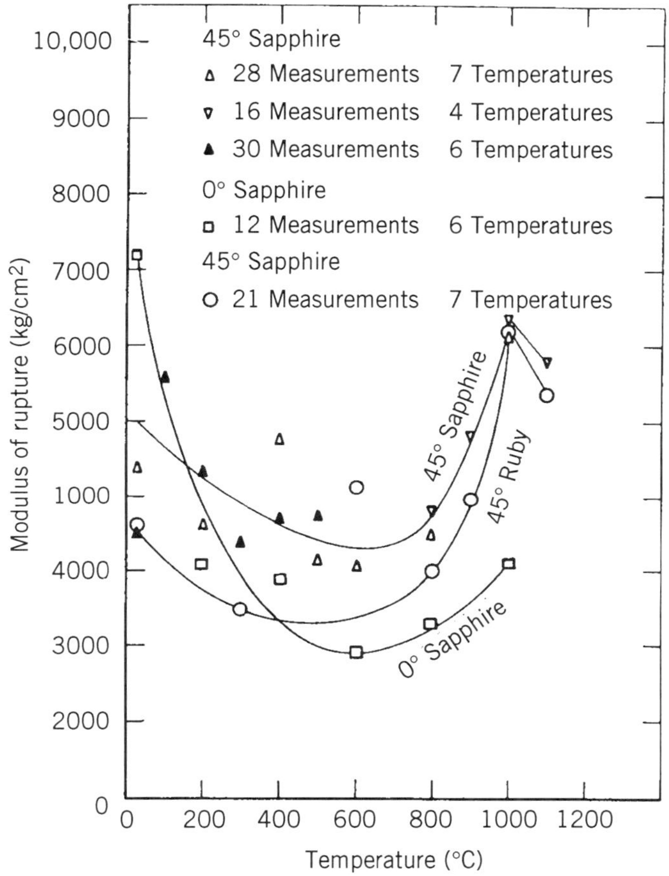

Figure 23.4 Strength of sapphire and ruby as a function of temperature. (From J. B. Wachtman, Jr. and L. H. Maxwell, "Strength of Synthetic Single Crystal Sapphire and Ruby as a Function of Temperature and Orientation," *J. Am. Ceram. Soc.*, Vol. 42, 432–433, 1959. Reprinted by permission of the American Ceramic Society.)

Rice (1994) compiled data on strength of polycrystalline alumina and found a wide variation, as shown in Figure 23.5. Some of the differences at a given temperature are evidently due to differences in grain size. Rice gives a qualitative discussion of the effects that may be dominant in different grain size and temperature regions but concludes that more experimental work is needed to sort out the effects of the different factors.

23.4 HIGH-TEMPERATURE CREEP AND STRESS RUPTURE

Deformation of pure alumina at high temperatures is complicated by the simultaneous process of grain growth (Xu and Solomon, 1992), unless special

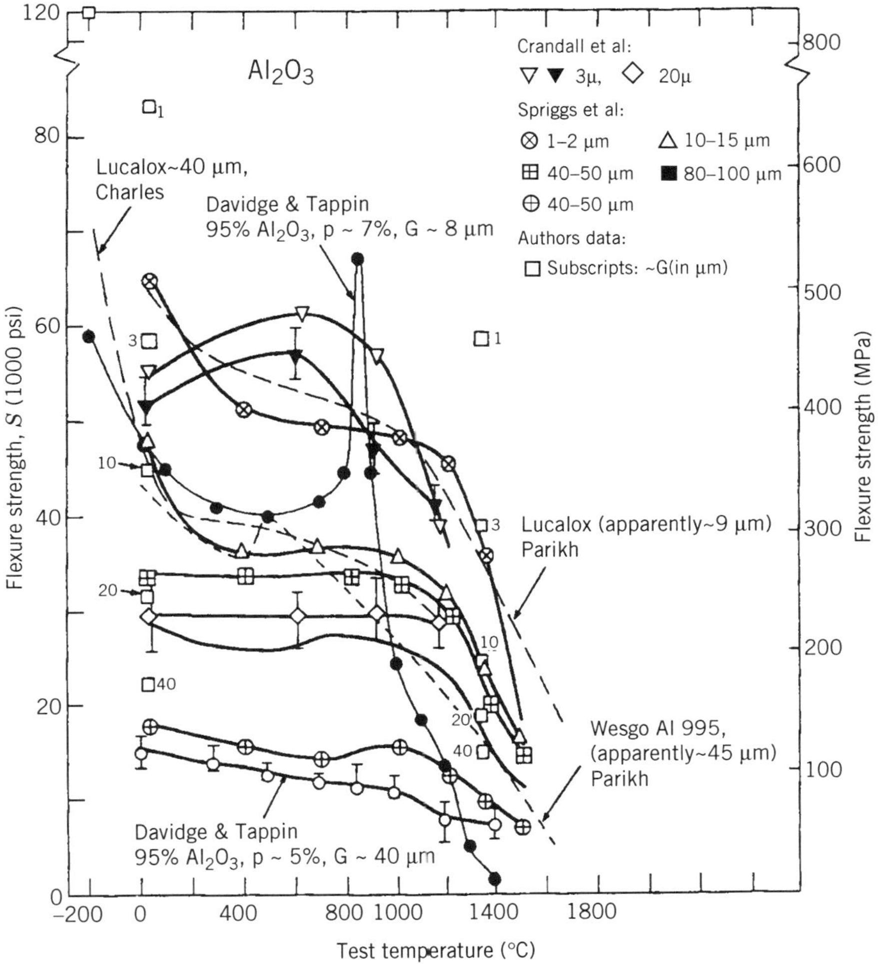

Figure 23.5 Strength of polycrystalline alumina as a function of temperature. (From Rice, 1996.)

means are taken to suppress grain growth such as doping with MgO. Deformation of fine-grained, MgO-doped polycrystalline alumina involves three deformation mechanisms: diffusional creep, basal slip, and unaccommodated grain boundary sliding. Over the temperature range of 1200–1750 °C, stress range of 1.4–310 MPa, and grain size range of 1–15 μm, the dominant mechanism is diffusional creep. However, Newtonian-viscous creep with stress exponent of 1 is not always found. Instead exponents of 0.5–1 were found. The departure from an exponent of 1 was attributed to (1) interface control of the rate of accommodation of diffusing material and (2) some degree of basal slip and unaccommodated grain boundary sliding (Cannon, Rhodes, and Heuer, 1980).

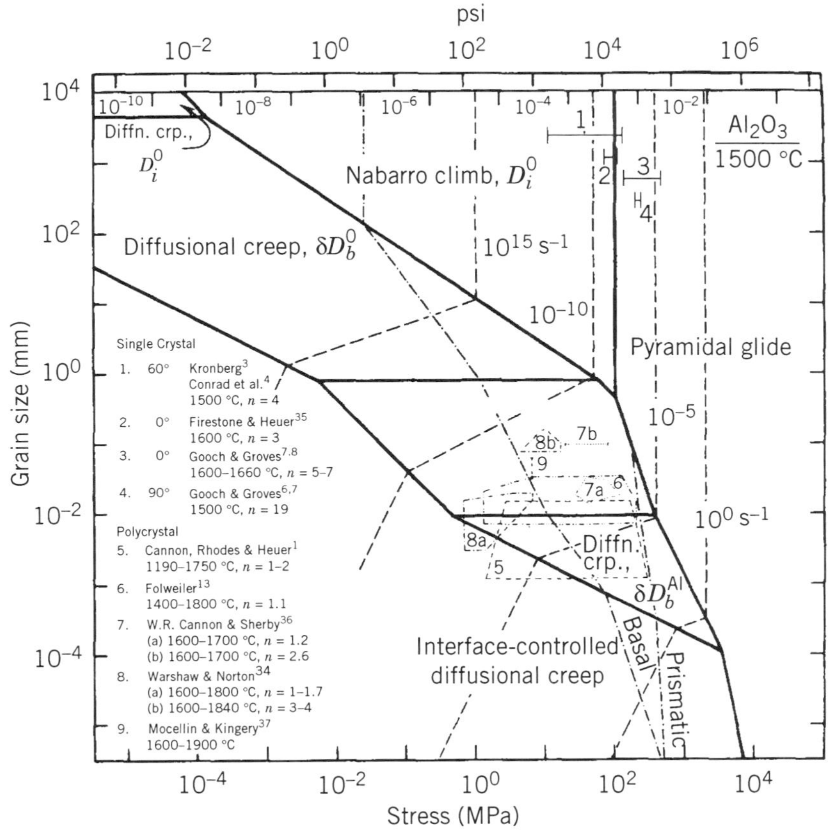

Figure 23.6 Deformation mechanism map for MgO-doped polycrystalline alumina at 1500 °C. (From A. H. Heuer, N. J. Tighe, and R. M. Cannon, "Plastic Deformation of Fine-Grained Alumina (Al_2O_3): II, Basal Slip and Nonaccommodated Grain-Boundary Sliding," *J. Am. Ceram. Soc.*, Vol. 63, 53–58, 1980. Reprinted by permission of the American Ceramic Society.)

Combining their observations with a thorough survey of previous work on high-temperature creep of polycrystalline alumina, Heuer, Tighe, and Cannon (1980) constructed a deformation map for MgO-doped alumina at 1500 °C, as shown in Figure 23.6. This map is expressed in terms of stress as a function of grain size with heavy lines indicating the locus of equal rates of creep from the adjacent mechanisms. Lines of constant strain rate are superimposed. For deformation-limited applications, this map is evidently a very useful starting point for a first estimate of allowable operating conditions, although the usual cautions apply concerning the sensitivity of mechanical behavior to other details of microstructure.

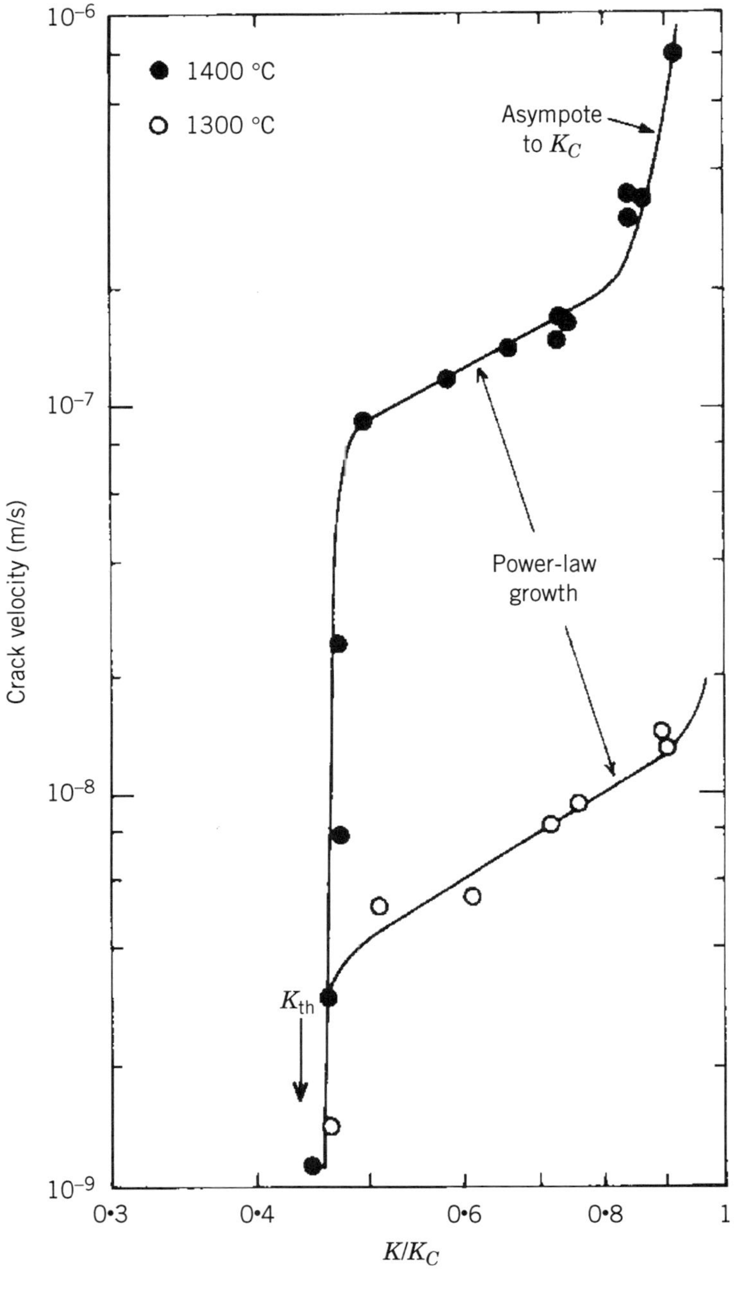

Figure 23.7 Crack velocity in MgO-doped polycrystalline alumina. (From W. Blumentahl and A. G. Evans, "High-Temperature Failure of Polycrystalline Alumina: II, Creep Crack Grwoth and Blunting," *J. Am. Ceram. Soc.*, Vol. 67, 751–759, 1984. Reprinted by permission of the American Ceramic Society.)

A study of damage and ultimate failure in MgO-doped, hot-pressed alumina undergoing creep at temperatures between 1300 and 1400 °C was made by Evans and his associates (Johnson, Dalgleish, and Evans, 1984). Crack nucleation was observed at several types of microstructural heterogeneities. The most important nucleation sites were chemical heterogeneities and large-grained zones. These heterogeneities produce local stress enhancement due to deviations in viscosity from that of the matrix material (Dalgleish, Johnson, and Evans, 1984). Indentation cracks were introduced into some specimens and their growth under creep conditions was studied, leading to the crack growth results shown in Figure 23.7 (Blumenthal and Evans, 1984). A sharp crack growth threshold near $K = 0.4\ K_{IC}$ is apparent followed by a power law crack growth region leading eventually to an asymptote to K_{IC}. The authors, however, express some caution concerning the use of K_{IC} as an appropriate loading factor for cracks developing under these conditions. Monkman–Grant behavior was observed as shown in Figure 23.8.

The creep and stress-rupture results discussed above were obtained on alumina doped with MgO to prevent grain growth. Xu and Solomon (1992) studied creep in tension and cavitation in pure alumina and measured the grain size as a function of time during the creep experiments. In some experiments

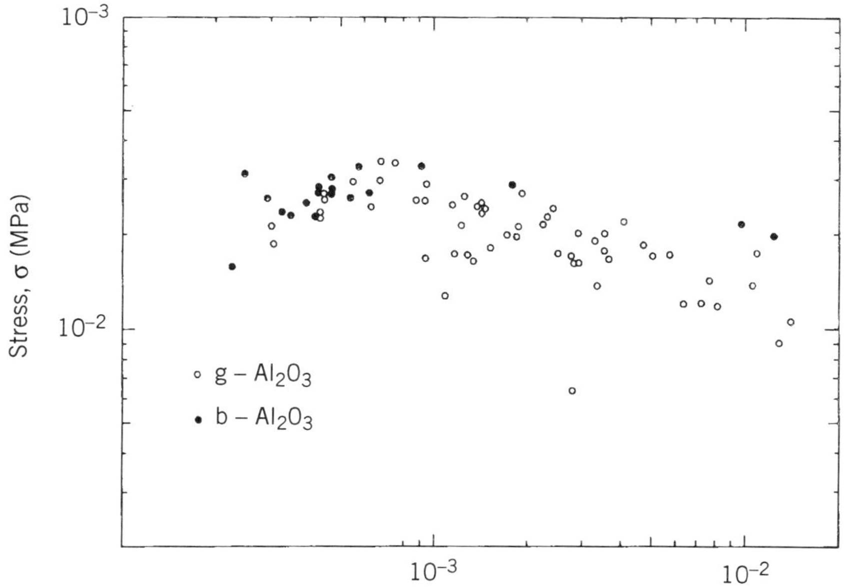

Figure 23.8 Monkman–Grant plot for MgO-doped alumina. (From S. M. Johnson, B. J. Dalgleish, and A. G. Evans, "High-Temperature Failure of Polycrystalline Alumina: III, Failure Times," *J. Am. Ceram. Soc.*, Vol. 67, 759–763, 1984. Reprinted by permission of the American Ceramic Society.)

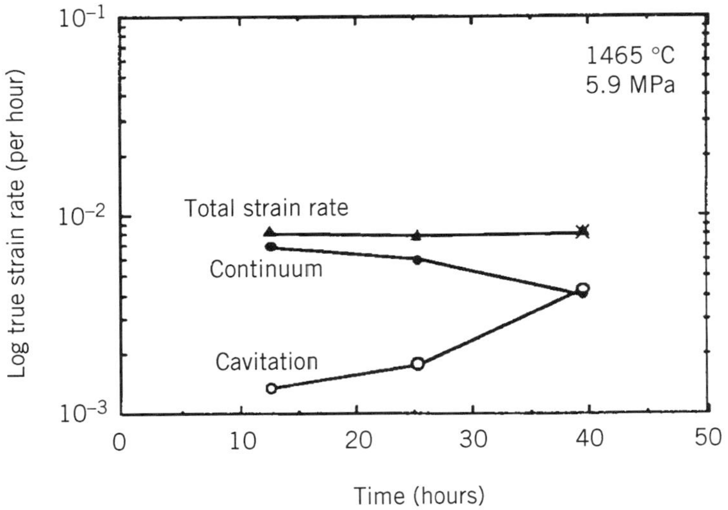

Figure 23.9 Separation of strain rate for pure alumina. (From A. Xu and A. A. Solomon, "Diffusional Creep and Cavitational Strains in High-Purity Alumina Under Tension and Subsequent Hydrostatic Compression," *J. Am. Ceram. Soc.*, Vol. 75, 985–995, 1992. Reprinted by permission of the American Ceramic Society.)

they also determined both the longitudinal strain (i.e., strain in the tensile testing direction), ϵ_{tot}, and the transverse strain, ϵ_{tr}. From these they calculated the volume-conserving strain in the longitudinal direction, ϵ_{con}, and the cavitational strain, ϵ_{cav}, (assuming that it contributed to the the longitudinal strain only) using the simplified method of separating strain proposed by Raj (1982):

$$\varepsilon_{con} = -2\varepsilon_{tr}$$

$$\varepsilon_{cav} = \varepsilon_{tot} - \varepsilon_{con}$$

The result for a specimen at 1465 °C under 5.9 MPa is shown in Figure 23.9. An approximately constant total (true) strain rate is seen to be the sum of two components. The diffusional component of strain rate decreases with time because of grain growth. The cavitation component of strain rate increases with time. Evidently grain growth introduces additional complexity, indicating again the need for caution in applying deformation mechanism maps. These authors successfully apply a theory of cavitation rates and note that exaggerated grain growth in high-density specimens can lead to early cavity coalescence and failure.

CHAPTER 24

MECHANICAL PROPERTIES OF SILICON NITRIDE AND RELATED MATERIALS

24.1 INTRODUCTION

This chapter presents selected experimental results on silicon nitride to typify the material. Silicon nitride is actually a variable material depending strongly on the sintering additives used and the type of microstructure developed during processing. Emphasis is here placed on commercial and research materials with the best mechanical properties so far developed. We attempt to provide typical data from recent papers on advanced forms of silicon nitride rather than present a historical account.

Silicon nitride is generally regarded as the leading candidate for structural use above 1000 °C although silicon carbide is also used. Silicon nitride has been the subject of much research and development; earlier work has been well summarized in a conference report (Riley, 1977) that includes a survey chapter on the mechanical properties of reaction-bonded silicon nitride (Davidge, 1977). Much work on silicon nitride has been reported in the symposium series on ceramics for heat engines (Bunk and Hausner, 1986; Tennery, 1989; Carlsson, Johansson, and Kahlman, 1992) and a survey of recent progress in self-reinforced silicon nitride was given by Pyzik and Carroll (1994).

24.2 SILICON NITRIDE AND RELATED SiAlONs

Silicon nitride (Si_3N_4) has two structural modifications, α and β, both of which are hexagonal. The α form is made up of alternate layers of the β form and its mirror image and has approximately twice the *c* axis length (a = 0.775 nm, c = 0.5168 nm for α and a = 0.7608, c = 0.2911 nm for β). The α form is thought to exist only when kinetic factors predominate. Powders of silicon nitride are generally in the α form unless specially treated. The β form is thought to be the stable form and solid specimens are predominantly in this form. Transformation from the α to β form takes place during sintering, gas pressure sintering, or hot pressing and is believed to result from solution and reprecipitation in the sintering aid, which is typically liquid at the processing temperature (Weiss, 1981).

Solid silicon nitride made from silicon nitride powder generally contains a few percent of sintering additives such as Al_2O_3 and Y_2O_3 and also contains some amount of SiO_2 either from some degree of oxidation or as a result of deliberate addition. These additives improve properties in at least two respects: allowing sintering to full or nearly full density and promoting growth of elongated grains that increase strength and toughness (Li and Yamanis, 1989). However, these additives increase the creep rate at high temperature and reduce the corresponding time to creep failure (Ferber, Jenkins, Nolan, and Yeckley, 1994). A compromise is thus required, which depends on the temperature and stress expected in the application under consideration.

A different route to produce solid silicon nitride is by forming a shape from pure silicon powder and then nitriding it at high temperature. The overall volume changes very little during nitriding. The resulting material has open porosity typically in the range of 15–20% by volume, which lowers its strength. It does have the advantage of being free of sintering additives so that its creep resistance is typically good. We shall, however, concentrate here on solid silicon nitride made from silicon nitride powder.

Monolithic silicon nitride (i.e., solid without toughening reinforcements) usually has a fracture toughness of 4–6 MPa · $m^{1/2}$. This is somewhat higher than most monolithic ceramics and is attributed to the tendency of β grains to grow with an elongated morphology. Attempts to toughen silicon nitride by reinforcement with SiC whiskers have produced toughness values around 10 MPa · $m^{1/2}$ (Shalek, Petrovic, Hurley, and Gac, 1986). These are interesting materials but do involve the use of whiskers with the attendant problems of cost, need for careful handling, and either hot-pressing or hot isostatic pressing to obtain full density. When sintering conditions in silicon nitride with no whisker additions are adjusted to promote development of this whisker-like morphology of some of the silicon nitride grains themselves, the material can be considered to be an "in situ-grown whisker-reinforced material" (Li and Yaumanis, 1989).

Silicon nitride forms solid solutions with (Al + O) substituting for (Si + N) and these are commonly termed β-SiAlON ceramics. The general formula for these is $Si_{3-x}Al_xO_xN_{4-x}$ with $0 \leq x \leq 2$ (Jack and Wilson, 1977; Weiss, 1981). These materials form a family of structural ceramics in themselves and have been reviewed recently by Ekstrom and Nygren (1992). It should be noted that the use of sintering additives to produce "silicon nitride" can produce some degree of this type of solid solution even when this is not deliberately sought.

24.3 HIGH-TOUGHNESS, IN SITU-REINFORCED SILICON NITRIDE

As an example of silicon nitride in an advanced state of development, we summarize some results on Allied-Signal's AS-700 material. This material is made by gas pressure sintering green billets formed by cold isostatic pressing. Table 24.1 compares some properties of this material with NGK's SN-84H silicon nitride.

Weibull modulus values of about 10 are said by the authors to be typical for comparable strong silicon nitrides without pronounced whisker-like grains so that the toughening by whisker-like grain growth has substantially improved the Weibull modulus.

R-curve behavior was found in AS-700 as shown in Figure 24.1 through 24.4. Figure 24.1 shows a pronounced R-curve behavior as measured by double cantilever beam experiments using specimens with 23-mm-long sawn notches. For such long cracks the critical stress intensity factor increases to about 10 MPa · $m^{1/2}$ for crack extension of about 1.5 mm. The rising part of the R-curve (out to the arrow in Figure 24.1) is well represented by

$$\Delta K = k(\Delta c)^{n_K} \tag{24.1}$$

as used in Section 9.4, where Δc is change in crack length. Krause (1988) has provided an analysis that allows calculation of k and n_K from data on strength of indented bars as a function of indentation load. The indented strength is predicted to be given by

$$\log \sigma = \log \alpha - \beta \log P \tag{24.2}$$

TABLE 24.1 Mechanical Properties of Silicon Nitrides

Ceramic	Density (g/cm^3)	Four-point Bend Strength at 23 °C (MPa)	Fracture Toughness (MPa · m$^{1/2}$)	Weibull Modulus
AS-700	3.31	550	10.6 ± 0.5	18.2
SN-84H	3.24	900	5.7 ± 0.1	N/A

Source: Li and Yamanis (1989).

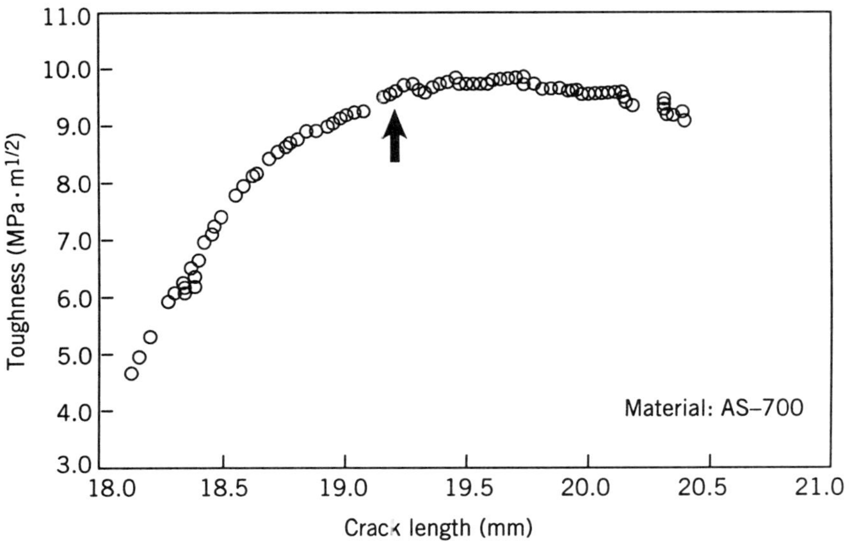

Figure 24.1 R-curve from double cantilever beam measurement for AS-700 silicon nitride. (From Li and Yamanis, 1988.)

where P is indentation load. Values of α and β can be calculated by fitting data to this equation and the R-curve parameters can then be calculated from

$$n_K = (1 - 3\beta)/(2 + 2\beta) \tag{24.3}$$

$$k = Y\alpha\,(\beta\,\gamma)^{-\beta}\,(1 + \beta)^{(1+\beta)} \tag{24.4}$$

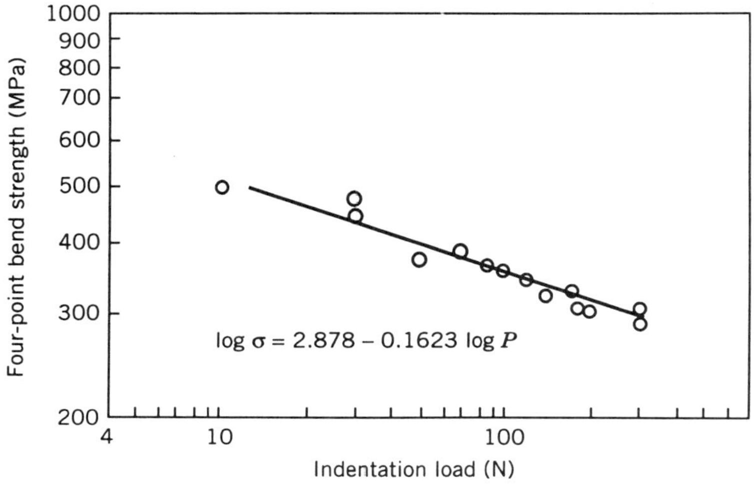

Figure 24.2 Indentation–strength plot for AS-700 silicon nitride. (From Li and Yamanis, 1988.)

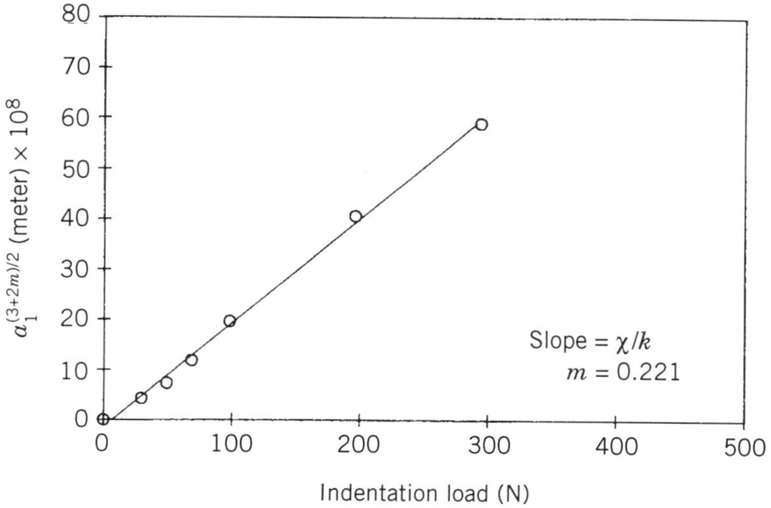

Figure 24.3 Initial crack length as a function of load for AS-700 silicon nitride. (From Li and Yamanis, 1988.)

where Y is the usual crack shape factor and

$$\gamma = P / c_i^{\frac{2}{1+\beta}} \tag{24.5}$$

where c_i is the initial crack length after indentation. The parameter n_K obtained from the slope of the straight line in Figure 24.2 and equation (24.3) is 0.221. A plot based on equation (24.5) for AS-700 is shown in Figure 24.3 and

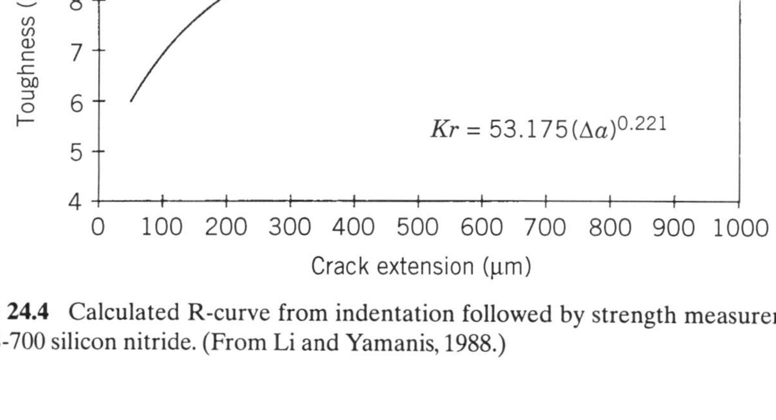

Figure 24.4 Calculated R-curve from indentation followed by strength measurement for AS-700 silicon nitride. (From Li and Yamanis, 1988.)

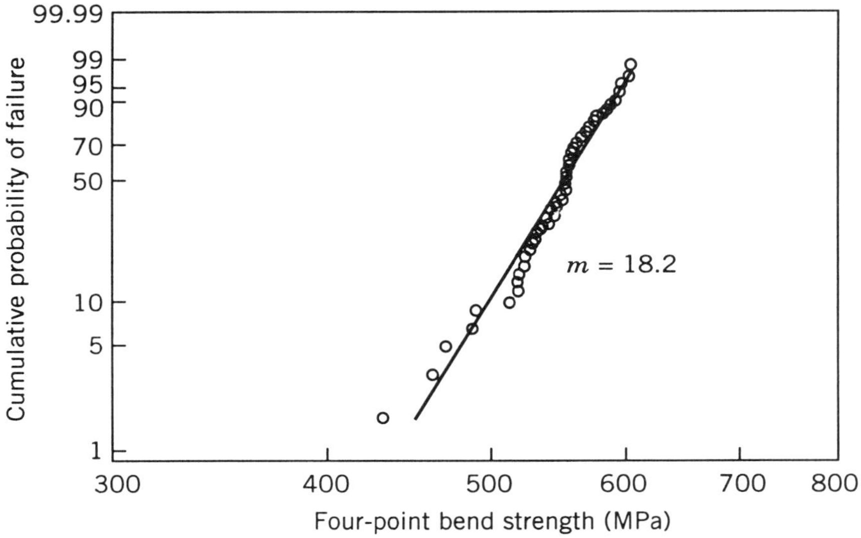

Figure 24.5 Weibull plot of strength distribution. (From Li and Yamanis, 1988.)

gives a straight line through the origin as predicted. From this slope and α from Figure 24.2 one can then calculate $k = 53.2$. With these values the R-curve for short cracks produced by the indenter can be calculated and is given in Figure 24.4. This shows that the predicted critical stress intensity value grows from 5.96 to 10.68 MPa $\cdot$ $m^{1/2}$ as the crack extends to the point of instability taken

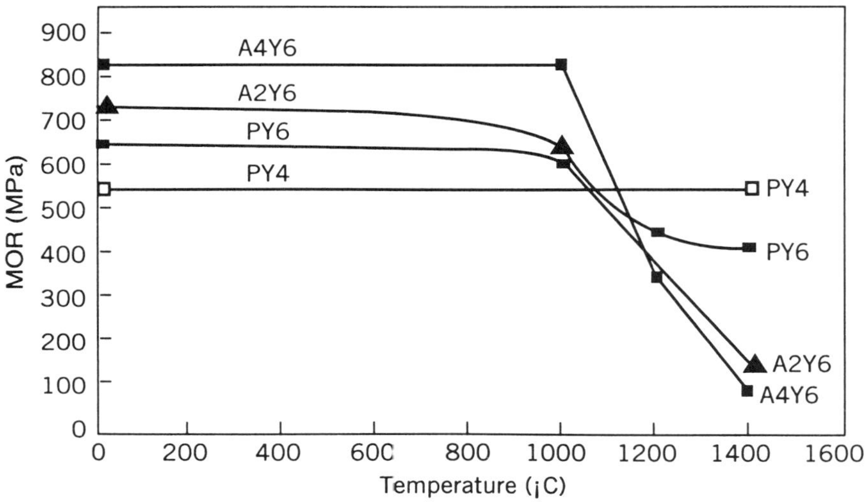

Figure 24.6 Modulus of rupture of silicon nitride A4Y6 indicates 4% alumina and 6% yttria. P indicates no alumina. (From Pasto, Van Schalkwyk, and Mahoney, 1989. Reprinted by permission of the American Ceramic Society.)

as three times the value of c_i = 250 μm. Figure 24.5 shows an m value of 18.5, considerably increased by R-curve behavior.

24.4 HIGH-TEMPERATURE FAILURE OF SILICON NITRIDE

We have already presented some data on high-temperature mechanical behavior of a particular silicon nitride (NC-130) and a fracture mechanism map in Chapter 21. Quinn (1982) reviewed the available data on stress rupture of silicon nitride. He concluded that at room temperature stress corrosion does take place but only for stresses very near the fast-fracture stress. Moisture-assisted attack causes surface flaws to extend along grain boundaries. At higher temperatures (typically above 800 °C) failure can occur by slow growth of preexisting flaws and at still higher temperature by creep rupture. The transition temperature between failure by slow crack growth and by creep rupture depends on the stress level, as shown in the fracture maps given in Chapter 20. In the creep-rupture range failure results from the formation of cracks by cavitation and their growth and/or consolidation. Grain boundary sliding and cavitation are the result of an oxide phase at the grain boundary. Considerable progress has been made in modifying the properties of grain boundaries to minimize creep.

Pasto, Van Schalkwyk, and Mahoney (1989) studied silicon nitride containing varying amounts of Y_2O_3 and Al_2O_3 as sintering aids. Typical compositions involved 4–6 wt. % of alumina while alumina content ranged from 0–4 wt. %. Their modulus of rupture (MOR) results are shown in Figure 24.6. Increasing the alumina content increases the room temperature strength but degrades the high-temperature strength. Increasing alumina content also greatly increased the creep rate at constant temperature, as shown in Figure 24.7. They selected

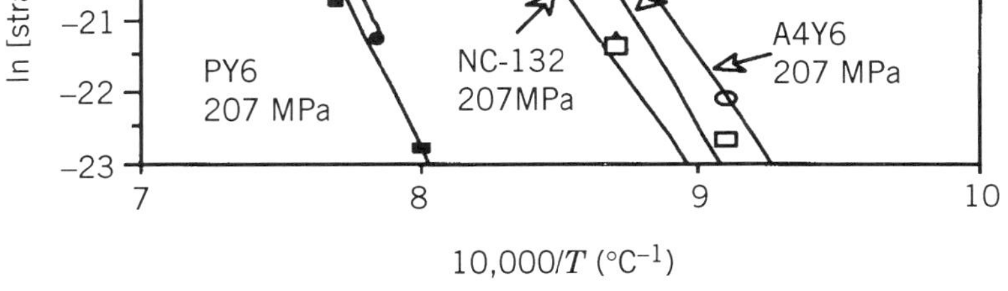

Figure 24.7 Creep rates as a function of temperature for various silicon nitrides. (From Pasto, Van Schalkwyk, and Mahoney, 1989. Reprinted by permission of the American Ceramic Society.)

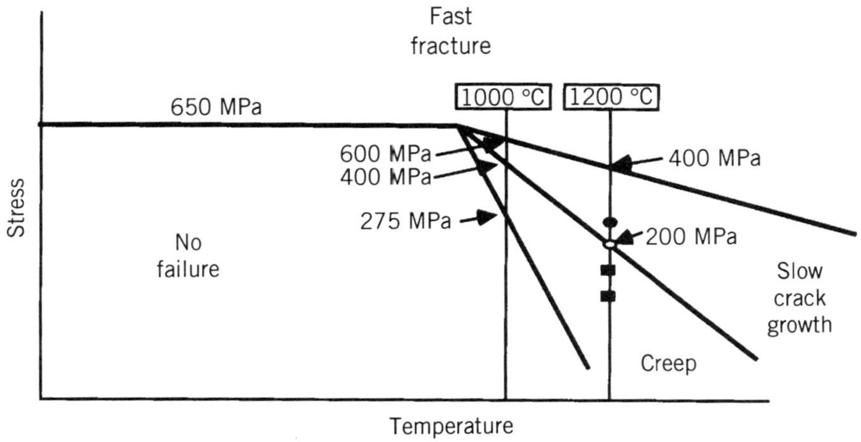

Figure 24.8 Schematic fracture mechanism map for silicon nitride with 1.5 wt. % alumina and 6 wt. % yttria. (From Pasto, Van Schalkwyk, and Mahoney, 1989. Reprinted by permission of the American Ceramic Society.)

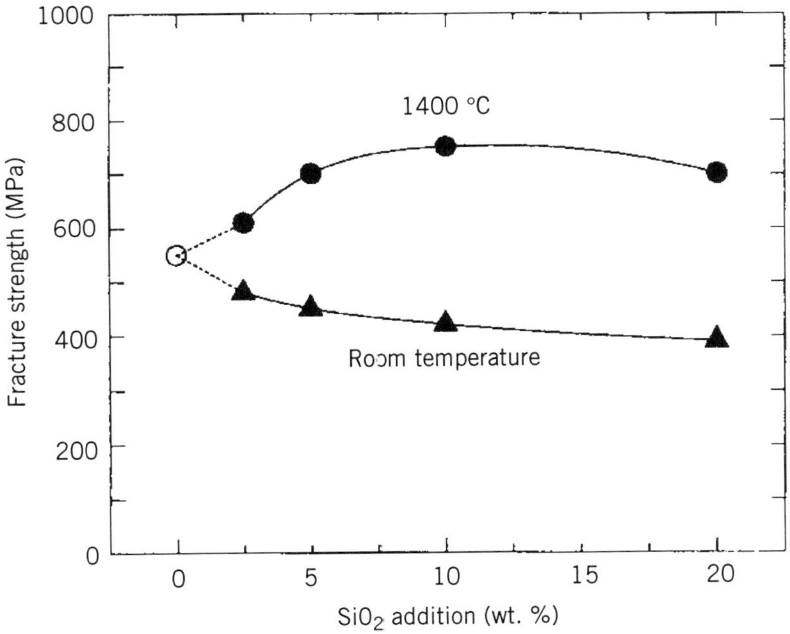

Figure 24.9 Strength of silicon nitride as a function of silica addition. (From J. Zeng, I. Tanaka, Y. Miyamoto, O. Yamada, and K. Niihara, "High-Temperature Strength and Cavitation Threshold of Silicon Nitride–Silica Ceramics," *J. Am. Ceram. Soc.*, Vol. 75, 195–200, 1992. Reprinted by permission of the American Ceramic Society.)

A1.5Y6 (i.e., silicon nitride with 1.5 wt. % alumina and and 6 wt. % yttria) as a good compromise and produced the schematic fracture mechanism map shown in Figure 24.8 for this material, not considering strain rate effects.

Zeng, Tanaka, Miyamoto, Yamada, and Niihara (1992) report that silicon nitride containing only silica as a sintering aid has a high cavitation threshold. They studied the effect on strength at high temperature of adding high-purity silica to high-purity silicon nitride. They found that the glassy silica was localized at intergranular pockets for amounts less than 10 wt. %. They found high short time strength at 1400 °C despite substantial amounts of silica and attributed this to a high cavitation threshold for cavitation of the high-purity silica phase. Figure 24.9 shows that the short time strength (measured with strain rate equal to 0.00015/s) is actually greater at 1400 °C than at room temperature. Figure 24.10

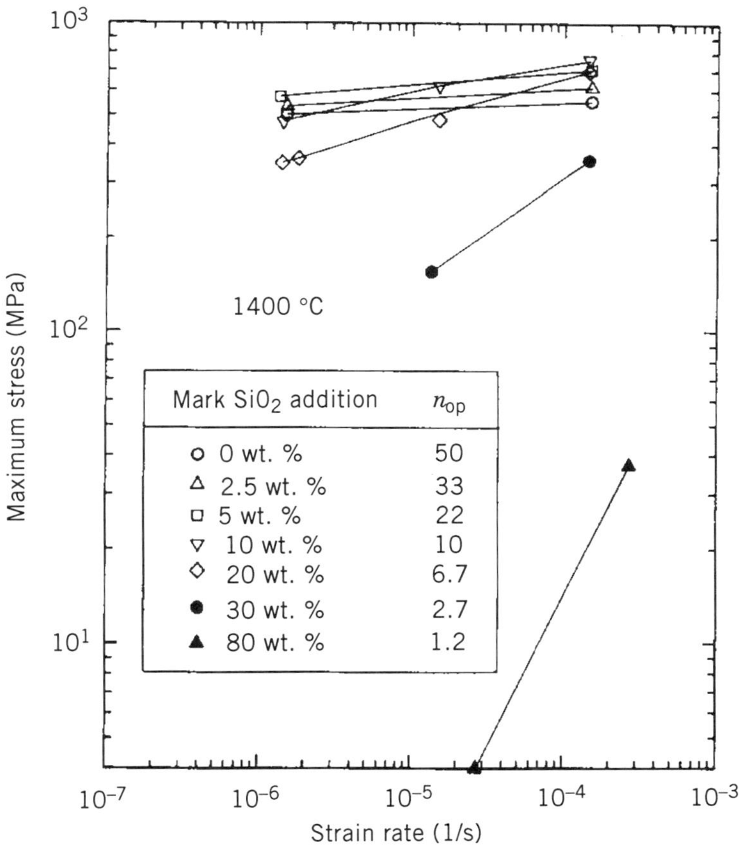

Figure 24.10 Stress in silicon nitride as a function of strain rate for various silica additions. (From J. Zeng, I. Tanaka, Y. Miyamoto, O. Yamada, and K. Niihara, "High-Temperature Strength and Cavitation Threshold of Silicon Nitride–Silica Ceramics," *J. Am. Ceram. Soc.*, Vol. 75, 195–200, 1992. Reprinted by permission of the American Ceramic Society.)

shows the maximum stress reached as a function of strain rate for various amounts of silica. The maximum stress varies relatively little with strain rate over two orders of magnitude and with silica addition up to 20 wt. %. Larger amounts of silica caused a large decrease in the maximum stress and much greater strain rate dependence. Microstructural examination showed that for up to 10 wt. % silica was localized primarily at grain triple points. Above 20 wt. % a thick intergranular glassy layer formed. A striking change in the shape of the stress–strain curve measured at a strain rate of 0.00015/s as the silica content is increased from 10 to 20 wt. % is evident in Figure 24.11. The investigators calculate a cavitation threshold that is somewhat dependent on strain rate and pocket size as shown in Figure 24.12 but is above 700 MPa at 1400 °C for all rates and sizes considered. The authors attribute the good strength retention to this high cavitation threshold. The authors attribute the loss of strength at silica contents above 20 wt. % to the fact that silica is then no longer constrained at intergranular pockets so that hydrostatic tension cannot develop, leading to strength degradation and macroscopic plastic deformation.

These results are interesting in terms of basic information and are relevant to practical situations requiring only the ability to sustain short time loads at high temperatures. However, a situation requiring sustained load-bearing capability for hundreds or thousands of hours is, of course, far removed from the experimental times used by these investigators. Confirmation of the extrapo-

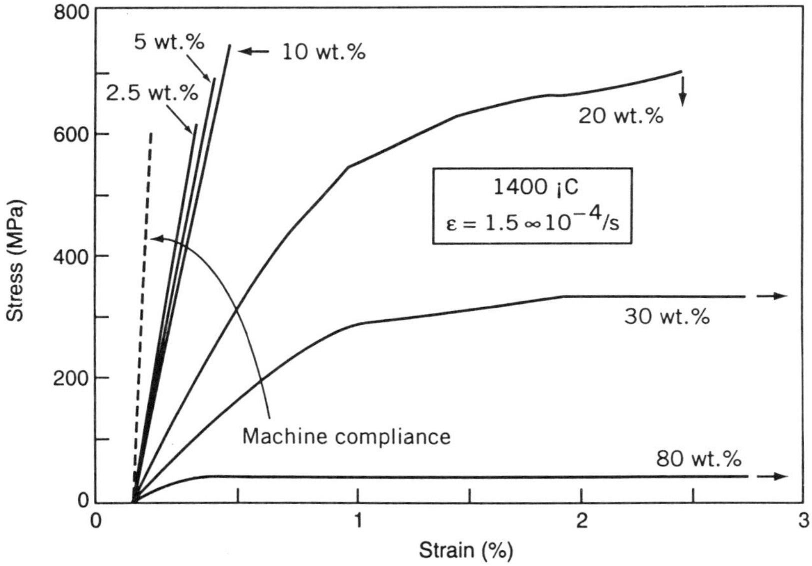

Figure 24.11 Stress–strain curves for silicon nitride at 1400 °C for various silica contents. (From J. Zeng, I. Tanaka, Y. Miyamoto, O. Yamada, and K. Niihara, "High-Temperature Strength and Cavitation Threshold of Silicon Nitride–Silica Ceramics," *J. Am. Ceram. Soc.*, Vol. 75, 195–200, 1992. Reprinted by permission of the American Ceramic Society.)

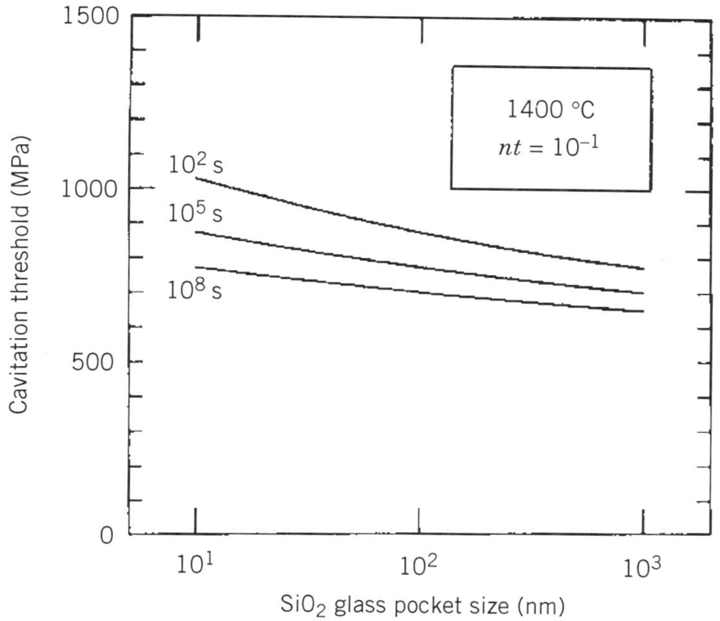

Figure 24.12 Calculated cavitation threshold for silica pockets in silicon nitride. (From J. Zeng, I. Tanaka, Y. Miyamoto, O. Yamada, and K. Niihara, "High-Temperature Strength and Cavitation Threshold of Silicon Nitride–Silica Ceramics," *J. Am. Ceram. Soc.*, Vol. 75, 195–200, 1992. Reprinted by permission of the American Ceramic Society.)

lation of the cavitation theory to much longer times and its implications for long time ability to carry loads remain to be studied.

We turn now to experiments on relatively long time load-bearing behavior of material representative of the most creep-resistant practical current silicon nitride. Ferber, Jenkins, Nolan, and Yeckley (1994) have studied creep and creep rupture of NT154 (a Saint-Gobain/Norton silicon nitride with about 4 wt. % yttria) and a newly developed version designated NT164. Both compositions were within the silicon nitride–silicon oxynitride–yttrium disilicate compatibility triangle and both were heat treated following hot pressing to crystallize a yttrium disilicate polymorph at the multigrain junctions. Creep curves were measured in tension at 1260 and 1370 °C and the results are shown in Figure 24.13. The creep rate of NT164 is less than that for NT154, as is evident, noting the change in time scale in the two plots. The results were analyzed in terms of the creep equations of the form discussed in earlier chapters,

$$\dot{\varepsilon}_s = A_0(\sigma/\sigma_0)^{n_{cr}} \exp(-Q_c/RT) \tag{24.6}$$

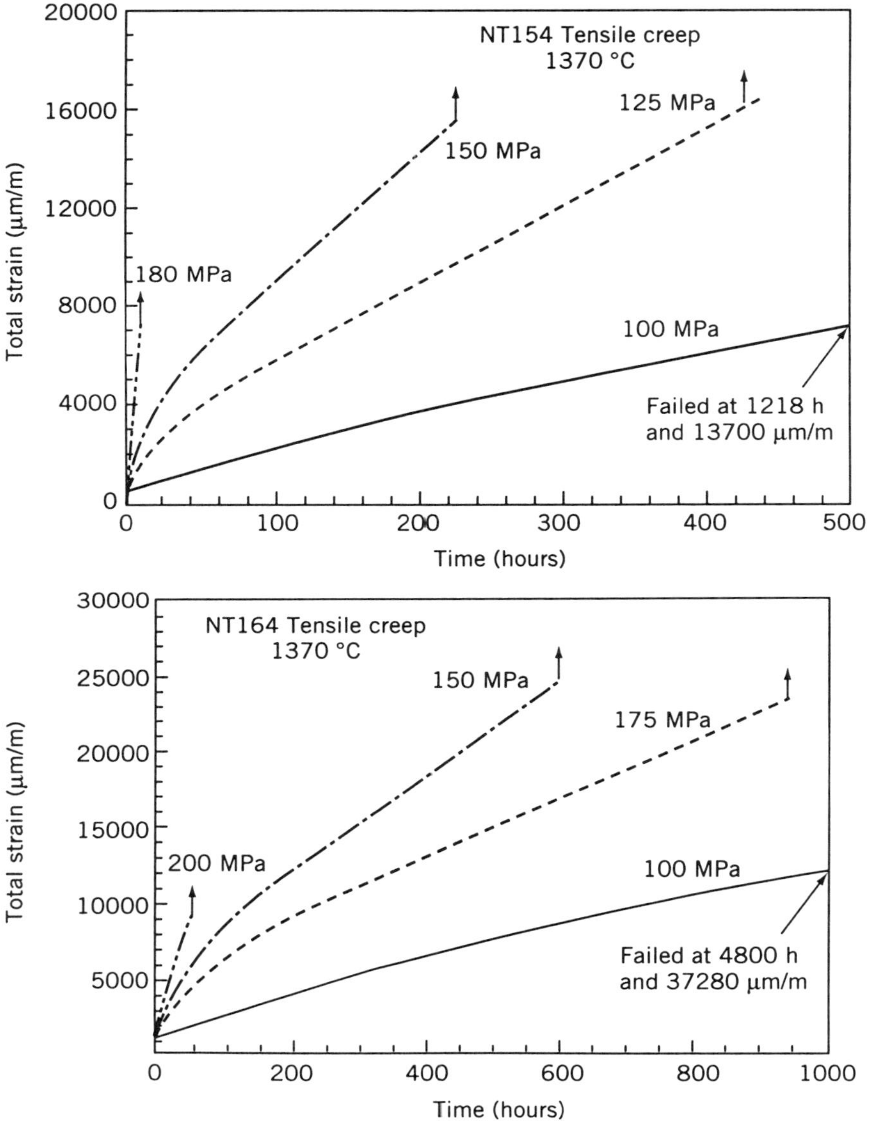

Figure 24.13 Tensile creep curves for NT154 (upper portion) and NT164 silicon nitride. (From M. K. Feber, M. G. Jenkins, T. A. Nolan, and R. L. Yeckley, "Comparison of the Creep and Creep Rupture Performance of Two HIPed Silicon Nitride Ceramics," *J. Am. Ceram. Soc.*, Vol. 77, 657–665, 1994. Reprinted by permission of the American Ceramic Society.)

where $\dot{\varepsilon}_s$ is the steady-state creep rate, A_0 is a parameter, $\sigma_0 = 1$ MPa, and n_{cr} is the creep exponent. The failure time is given in terms of similar parameters by

$$t_f = B_0(\sigma/\sigma_0)^{-N}\exp(Q_f/RT) \tag{24.7}$$

and the failure time is also expressed by the Monkman–Grant relationship:

$$t_f = C(\dot{\varepsilon}_s)^{-M} \tag{24.8}$$

When the time to failure is controlled by creep damage it is expected that $Q_c = Q_f$ and

$$N = n_{cr}M \tag{24.9}$$

The creep rate–stress equation is reasonably well obeyed except at the very highest stress values used, as shown in Figure 24.14. Tensile creep-rupture

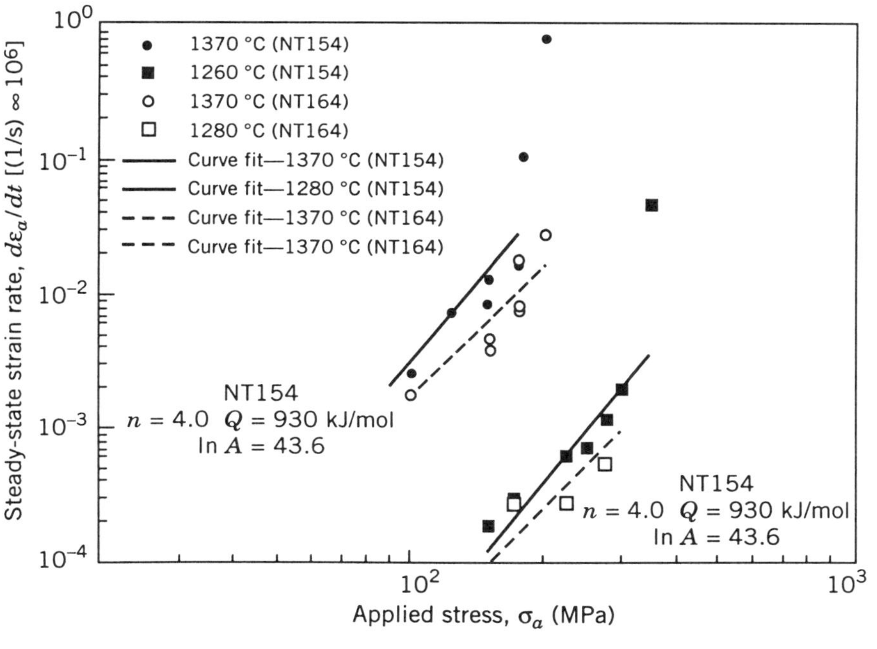

Figure 24.14 Stress and temperature dependence of creep rate for silicon nitride. (From M. K. Feber, M. G. Jenkins, T. A. Nolan, and R. L. Yeckley, "Comparison of the Creep and Creep Rupture Performance of Two HIPed Silicon Nitride Ceramics," *J. Am. Ceram. Soc.*, Vol. 77, 657–665, 1994. Reprinted by permission of the American Ceramic Society.)

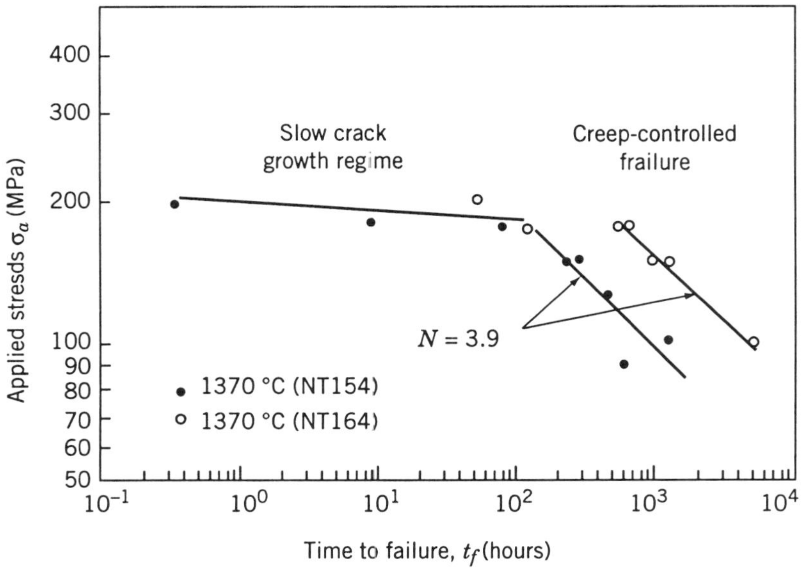

Figure 24.15 Creep life dependence on stress for silicon nitride. (From M. K. Feber, M. G. Jenkins, T. A. Nolan, and R. L. Yeckley, "Comparison of the Creep and Creep Rupture Performance of Two HIPed Silicon Nitride Ceramics," *J. Am. Ceram. Soc.*, Vol. 77, 657–665, 1994. Reprinted by permission of the American Ceramic Society.)

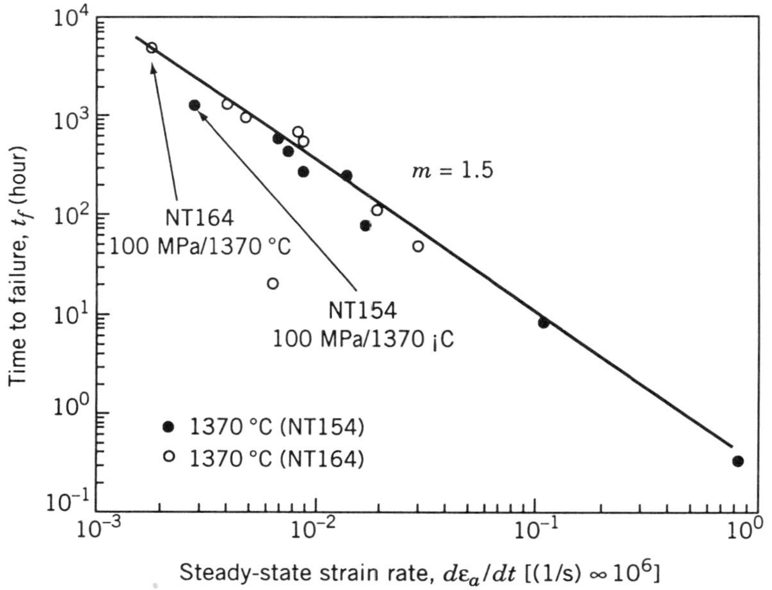

Figure 24.16 Monkman–Grant pilot for silicon nitride. (From M. K. Feber, M. G. Jenkins, T. A. Nolan, and R. L. Yeckley, "Comparison of the Creep and Creep Rupture Performance of Two HIPed Silicon Nitride Ceramics," *J. Am. Ceram. Soc.*, Vol. 77, 657–665, 1994. Reprinted by permission of the American Ceramic Society.)

behavior is shown in Figure 24.15 and is seen to fall into regions controlled by slow crack growth and by creep as expected. The Monkman–Grant relationship is obeyed, as shown in Figure 24.16. This silicon nitride is probably representative of the best commercial or near commercial material for creep and stress rupture available at the time of this writing.

CHAPTER 25

MECHANICAL PROPERTIES OF SILICON CARBIDE

25.1 INTRODUCTION

Although the use of silicon nitride is growing, silicon carbide is probably still the most widely used nonoxide ceramic. It has high hardness, moderate strength, and good strength retention to high temperatures. Silicon carbide exists in a cubic form, termed the beta form, and in a variety of hexagonal and rhombohedral modifications collectively termed the alpha form (Srinivasan, 1989). Carbothermal production at high temperatures generally produces the alpha form, which can have many polytypes corresponding to different packing sequences of parallel layers of tetrahedra. The hexagonal or rhombohedral axis in the alpha form corresponds to the [111] direction in the cubic form. The alpha forms are designated H or R for hexagonal or rhombohedral packing; for example, 6H designates hexagonal packing and 6 layers to repeat. Chemical conversion from polymer precursors usually produces the beta form, which converts to alpha at high temperatures, typically around 2000 °C.

Pure silicon carbide powder does not sinter (i.e., does not densify) without pressure (Richerson, 1992). Commercial silicon carbide products are made by three processes (Shaffer, 1991): second-phase bonding, reaction bonding, and sintering. In addition, silicon carbide parts can be made by chemical vapor deposition. Second-phase bonding is done by using a mixture of SiC powder with resin, glass, silicon nitride, clay, metal, or other material. Reaction bonding is done by mixing SiC powder with carbon or silicon metal powder. In reaction-bonded material, carbon can react with silicon carbide vapor to form a silicon carbide bond (termed self-bonded silicon carbide) or silicon can react with a nitrogen atmosphere to form a silicon nitride bond. Self-bonded material typically contains some free silicon in the pores. Sintering of fine SiC powders can be done by using boron and carbon together as sintering aids to produce fully dense material apparently by solid state sintering. Carbon apparently removes SiO_2 while boron goes into limited solid solution and enhances diffusion. Fully dense SiC can be made from either alpha or beta powder with the use of these sintering aids at a sintering temperature of about 2150 °C for 30 minutes. Reaction sintered silicon carbide has been made at 2000 °C by using alumina. This leads to the formation of a liquid and liquid-phase sintering followed by formation of Al_2OC, which forms a solid solution with SiC so that no glassy phase is left.

A material best considered as a type of composite is siliconized silicon carbide. In this material a substantial fraction of silicon (as high as 50 vol. %) is used to bond the SiC grains.

Typical advanced silicon carbide ceramics include: (1) Norton Company NC-430, a multiphase reaction-bonded (with 12 vol. % free silicon) α(6H)-SiC; (2) Carborundum Hexoloy, a pressureless-centered (with 0.4 wt. % B and 0.5 wt. % C) α(6H)-SiC; and (3) Synterials chemically vapor deposited 60% β/40% α(6H)-SiC (Nixon and Davis, 1992).

25.2 STRENGTH OF SILICON CARBIDE

The strength of siliconized silicon carbide tubes was measured by Darroudi, Tressler, and Kasprzyk (1993) using a C-ring technique with a slotted tube that measures the strength of the outer surface (Ferber, Tennery, Walters, and Ogle, 1980) and an O-ring technique that measures the strength at the inner surface (Ripperger and Davids, 1947). A tensile testing technique with flat, dogbone-shaped specimens was also used. This silicon carbide was made by a silicon melt infusion process using silicon carbide grains made by the Acheson process. The measured strength values are shown in Figure 25.1 and the tensile stress–strain curves in Figure 25.2. A short time strength of about 50 MPa in tension was found up to 1350 °C. The stress–strain curves were nearly linear at temperatures up to 1000 °C but showed evidence of substantial plastic deformation at higher temperatures.

A high-strength silicon carbide sintered from beta powder to form predominantly alpha-phase grains of average grain size of about 3.5 μm with occasional elongated grains has been reported by Woodilla, Buonomo, Bar-On, Katz, and Whalen (1993). They obtained a room-temperature flexural strength of 592 MPa with a Weibull modulus of 9.75. The material showed improved creep resistance, as shown in Figure 25.3.

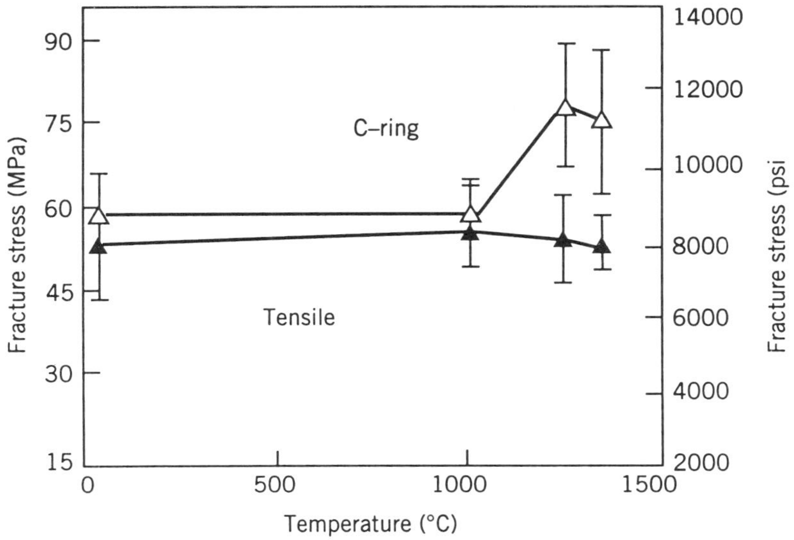

Figure 25.1 Strength as a function of temperature for siliconized silicon carbide. (From T. Darroudi, R. E. Tressler, and M. R. Kasprzyk, "Low-Cost Melt-Formed Siliconized Silicon Carbide Radiant Tube Materials," *J. Am. Ceram. Soc.*, Vol. 76, 173–179, 1993. Reprinted by permission of the American Ceramic Society.)

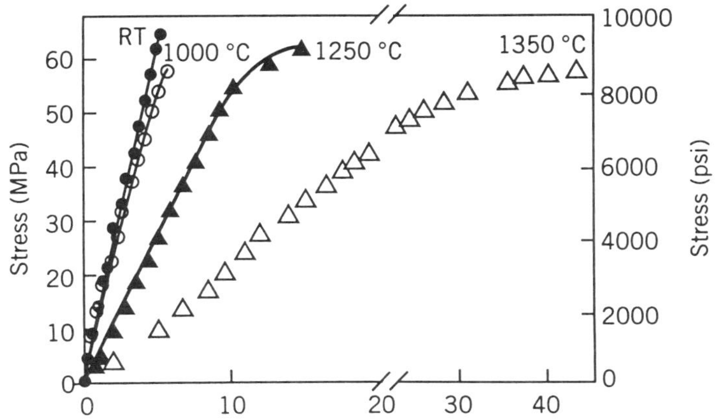

Figure 25.2 Stress–strain behavior of siliconized silicon carbide at various temperatures. (From T. Darroudi, R. E. Tressler, and M. R. Kasprzyk, "Low-Cost Melt-Formed Siliconized Silicon Carbide Radiant Tube Materials," *J. Am. Ceram. Soc.*, Vol. 76, 173–179, 1993. Reprinted by permission of the American Ceramic Society.)

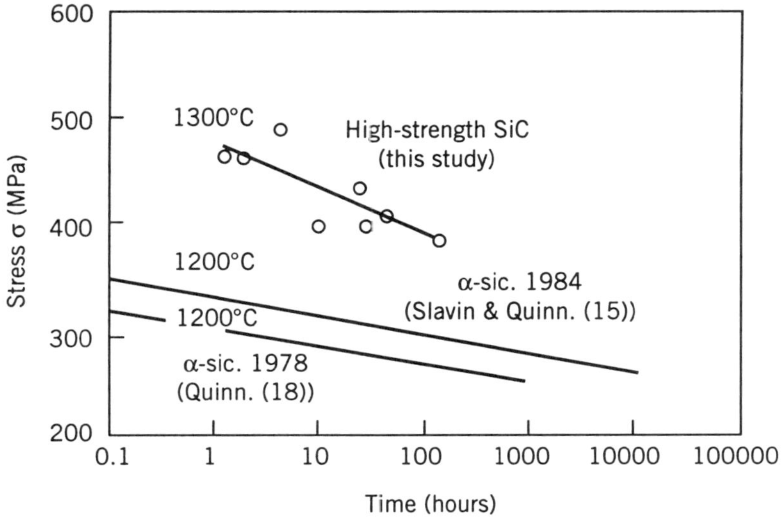

Figure 25.3 Stress–rupture behavior of various sintered SiC materials. (From D. Woodilla, M. Buonomo, I. Bar-On, R. N. Katz, and T. Whalen, "Elevated-Temperature Behavior of High-Strength Silicon Carbide," *J. Am. Ceram. Soc.*, Vol. 76, 249–252, 1993. Reprinted by permission of the American Ceramic Society.)

25.3 PLASTIC DEFORMATION OF SINGLE CRYSTAL SiC

Early experiments, reviewed by Corman (1992), indicated little or no plastic deformation in silicon carbide single crystals, but later work has shown the existence of slip on the (0001) plane above about 800 °C. Corman studied the compressive creep of α(6H)-SiC single crystals parallel to and at 45° to [0001] and found basal slip in both, indicating some off-axis stress in the parallel specimens. His steady-state creep rate results for 45° crystals are shown in Figure 25.4. His comparison with previously reported creep rates for various forms of polycrystalline silicon carbide are shown in Figure 25.5. The high creep rates found by him for single crystals in comparison for polycrystalline specimens are surprising. He considers the possibility that his high creep rates are due to greater purity and perfection of his single crystals. Higher creep resistance was found in highly oriented CVD SiC that deformed primarily by slip (Carter, Davis, and Bentley, 1984).

25.4 CREEP OF POLYCRYSTALLINE SiC

Wiederhorn, Roberts, Chuang, and Chuck (1988) and Hockey and Wiederhorn (1992) have studied the creep rates of siliconized silicon carbide in tension and compression at 1300 °C. Their material was fully dense and had about 67 vol.

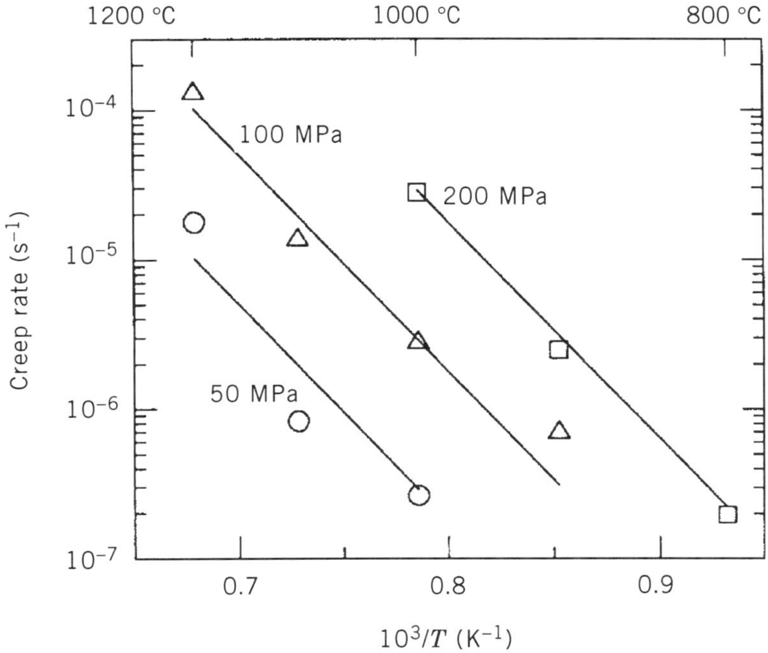

Figure 25.4 Steady-state creep rates for 45° single-crystal α-SiC. (From G. S. Corman, "Creep of 6H α-Silicon Carbide Single Crystals," *J. Am. Ceram. Soc.*, Vol. 75, 3421–3424, 1992. Reprinted by permission of the American Ceramic Society.)

% SiC grains 2–5 μm in size in a matrix of silicon. Their results for 1300 °C are shown in Figure 19.1. It is evident that a much larger creep rate occurs in tension than in compression. It is also evident that a change of mechanism is operating at low and high creep rates for both tension and compression. The creep asymmetry between tension and compression has been found in many two-phase ceramics as reviewed by Hockey and Wiederhorn for glass-ceramics, vitreous-bonded alumina, SiC whisker-reinforced silicon nitride, HIPed silicon nitride, SiAlON, reaction-bonded silicon nitride, and siliconized silicon carbide. In every case at least twice the stress was required in compression to give the same creep rate as in tension. When the second phase was not present, no creep asymmetry was found in studies on alumina and silicon carbide. The greater creep rate in tension has often been attributed to cavity formation. Hockey and Wiederhorn (1992) also found cavities in tension but not in compression. However, they note that cavity formation effectively ceases in their siliconized SiC at low tensile stresses so that cavity formation alone cannot provide a complete explanation for the creep asymmetry, at least in this material. They carried out creep experiments above the melting point of silicon and found that in compression the creep rate was consistent with an Arrhenius extrapolation of creep rate at temperatures below the melting point. In con-

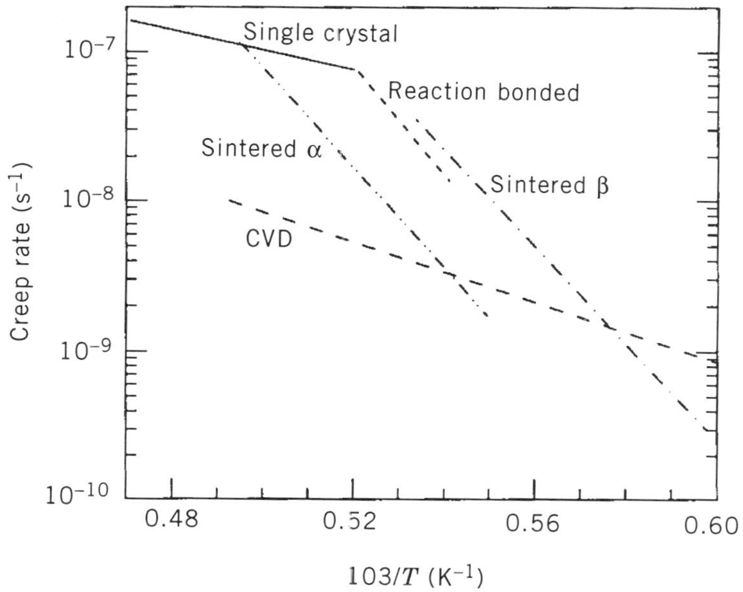

Figure 25.5 Creep rates for several forms of SiC. (From G. S. Corman, "Creep of 6H α-Silicon Carbide Single Crystals," *J. Am. Ceram. Soc.*, Vol. 75, 3421–3424, 1992. Reprinted by permission of the American Ceramic Society.)

trast, siliconized silicon carbide would not support a significant load in tension at temperatures above the melting point of silicon.

Hockey and Wiederhorn adapted from soil mechanics a model involving dilatancy to explain the tension–compression creep asymmetry in siliconized silicon carbide. They conclude that in this material creep is determined by interactions between the SiC grains. The dilational component of the applied stress seeks to separate grains under an applied tensile stress and tends to push them together under an applied compressive stress. Experiments on sand have shown that sand with 38% void content increased in volume by about 3% during the first 5% of axial strain; that is, it showed dilatancy. In contrast, sand with 45% void content did not show a volume increase. The siliconized silicon carbide used by Hockey and Wiederhorn had 33 vol. % silicon, which they treat as analogous to voids in sand. Their model considers that the silicon flows but that the creep rate is determined by the motion of the SiC particles and that this in turn is determined by the forces at contact points between particles. The process of sliding of the particles against each other is more difficult in compression than in tension.

Nixon and Davis (1992) have studied creep of alpha silicon carbide in compression in the range 1770–2020 K and 35–373 MPa. They used sintered α(6H)-SiC (Hexoloy) with three grain sizes: G1 = 3.5, G2 = 4.9, and G3 = 7.5 μm. Plots of their steady-state creep results are shown in Figure 25.6, which

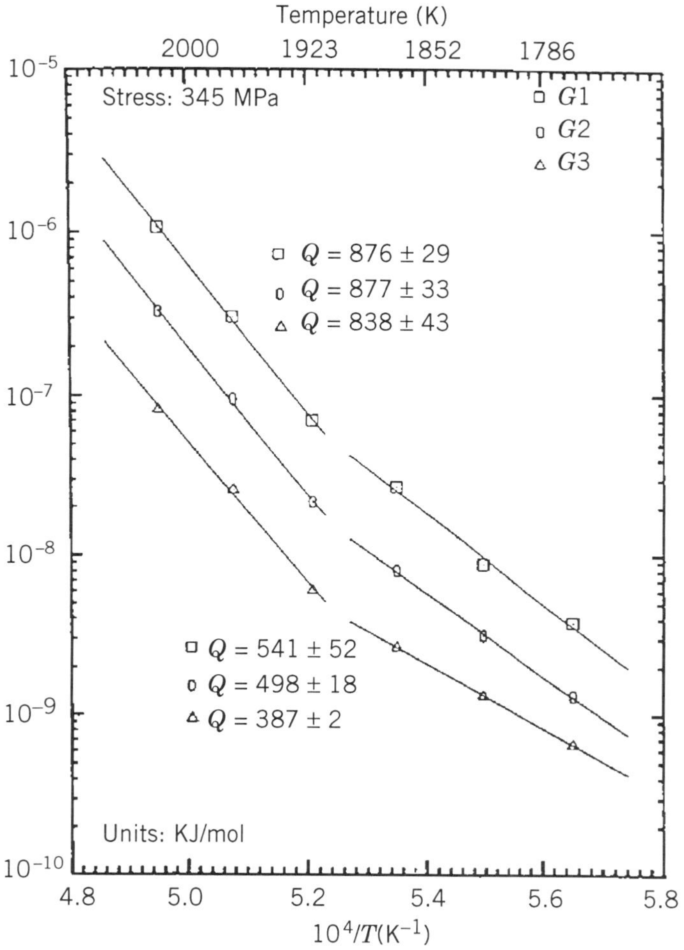

Figure 25.6 Steady-state creep rate of sintered α(6H)-SiC. (Nixon and Davis, 1992.)

demonstrates a change in behavior at about 1900 K. Activation energies above and below this temperature are also given. They concluded that two creep mechanisms were active: (1) grain boundary sliding accommodated by diffusion and (2) dislocation glide. The degree to which each is involved depends on the temperature and the microstructure.

CHAPTER 26

MECHANICAL PROPERTIES OF ZIRCONIA AND ZIRCONIA-TOUGHENED ALUMINA

26.1 INTRODUCTION

The transformation of tetragonal to monoclinic zirconia has already been discussed in Chapter 11. The basic theory of stress-induced transformation and the resulting toughening were given and the particle or grain-size dependence of the transformation was discussed. Here the experimental results on behavior of zirconia are briefly summarized in the light of the theories of fracture and toughening developed in earlier chapters. The Science and Technology of Zirconia series provides good basic references (Heuer and Hobbs, 1981; Claussen, Ruhle, and Heuer, 1983; Somiya, Yamamoto, and Yanagida, 1986; Badwal, Bannister, and Hannink, 1993). Additional excellent surveys include Green, Hannink, and Swain (1989) and Hannink and Swain (1994).

As mentioned in Section 11.2, polycrystalline zirconias are generally grouped into three categories: (1) partially stabilized zirconia, (2) tetragonal zirconia polycrystals, and (3) zirconia-toughened materials.

Partially stabilized zirconias are precipitation-toughened materials with enough of an additive in solid solution to partially stabilize the material in the cubic phase such as Mg–PSZ = magnesia partially stabilized zirconia and Ca–PSZ = calcia partially stabilized zirconia.

Tetragonal zirconia polycrystals are single-phase (or nearly single-phase) materials based on tetragonal solid solutions such as Y–TZP = yttria tetragonal zirconia polycrystals. These can be considered a type of partially stabilized zirconia but are singled out for special consideration.

Zirconia-toughened materials are materials that are toughened with a dispersion of zirconia. Examples include ZTA = zirconia toughened alumina and ZTM = zirconia toughened mullite.

A review of properties of single crystal and polycrystalline zirconias was published by Ingel, Lewis, Bender, and Rice (1983); Table 26.1 is taken from their paper.

Large values of toughness have been measured in various transformation-toughened zirconias and aluminas. Impressive and important as these values are, a note of caution is in order concerning their meaning. Not only the measurement method but the nature of the initiating crack can affect the value obtained. For example, a diamond-notched material, left unannealed, will yield a much higher initiating K_{IC} than a notched and annealed material due to the presence of an established shielding zone (R. H. J. Hannink, personal communication, 1995).

TABLE 26.1 Physical and Mechanical Properties of Zirconias

Single Crystals

Crystal	wt. % Additive	wt. % Precipitates	Density (kg/m^3)	Hardness (GPa)	Young's Modulus (GPa)	Flexural Strength (MPa)	Fracture Toughness (MPa · m$^{1/2}$)
Mg–PSZ	2.8	48	5790	14.4	200	685	4.82
Ca–PSZ	4.0	38	5850	17.1	210	661	3.97
Ca–CSZ	9	0	5680	17.2	210	241	2.54
Y–PSZ	5	52	6080	13.6	233	1384	6.92
Y–CSZ	20	0	5910	16.1	233	346	1.91

Polycrystalline Materials

Material	Young's Modulus (GPa)	Flexural Strength (MPa)	Fracture Toughness (MPa · m$^{1/2}$)
Mg–PSZ	200	430–700	4.7–15
Ca–PSZ	200–217	400–650	5.0–9.6
Y–PSZ	210–238	696–980	5.8–9.0

Source: Ingel, Lewis, Bender and Rice (1983) Reprinted by permission of the American Ceramic Society.

26.2 PARTIALLY STABILIZED ZIRCONIAS

Green, Hannink, and Swain (1989) consider coarse-grained (10–100 μm) PSZ as a major category of zirconia. Such materials are made typically from fine (less than 1 μm) powders of zirconia and stabilizer. The stabilizer is usually MgO, CaO, Y_2O_3, or a rare earth oxide. For material stabilized with MgO or CaO the heating curve for sintering ends inside the cubic phase field and the material is rapidly cooled to room temperature. Sintering in Mg–PSZ takes place primarily in the range 1150–1400 °C. Both stabilization in the cubic phase and grain growth occur above 1400 °C so that the need to achieve complete solid solution results in a coarse grain size. Subsequently, the material is aged in the mixed cubic/tetragonal field to cause precipitation of tetragonal precipitates within the cubic grains. Alternatively, a controlled cooling rate after sintering may be used to give sufficient time in the aging temperature range to develop optimum properties. The goals are to form a uniform distribution of tetragonal precipitates within the grains, to retain most of these in the tetragonal form at room temperature, and to avoid large tetragonal precipitates at the grain boundaries.

Commercial Ca–PSZ generally contains 3–4.5 wt. % CaO. After sintering the material is aged at 1300 °C to produce precipitates of tetragonal zirconia. The strength at room temperature depends strongly on the composition and aging time, as shown in Figure 26.1 (Garvie, Hannink, and Pascoe, 1978, as quoted by Green, Hannink, and Swain, 1989). In particular, the phenomenon of overaging is evident. The measured values of flexural strength with scatter for 3.7 wt. % Ca–PSZ as well as the fracture surface energy are shown in Figure 26.2 (Garvie, Hannink, and Urbani, 1980, as quoted by Green, Hannink, and Swain, 1989). As aging proceeds, the strength and fracture surface energy

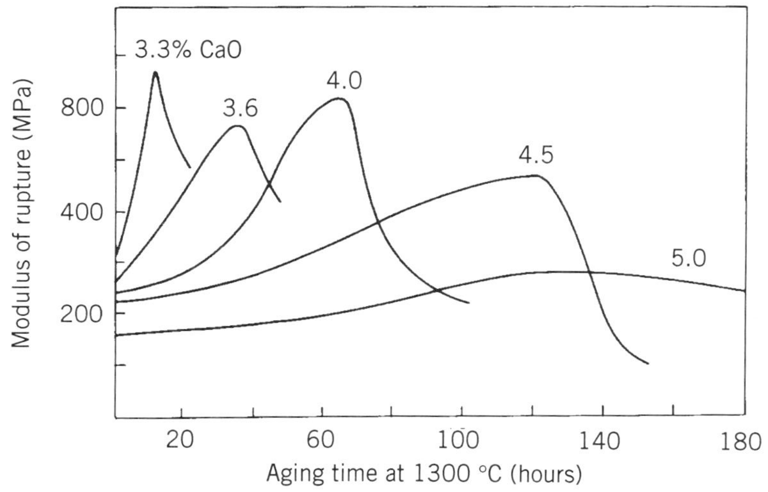

Figure 26.1 Flexural strength of Ca–PSZ as a function of aging time. (From Garvie, Hannink, and Pascoe, 1978.)

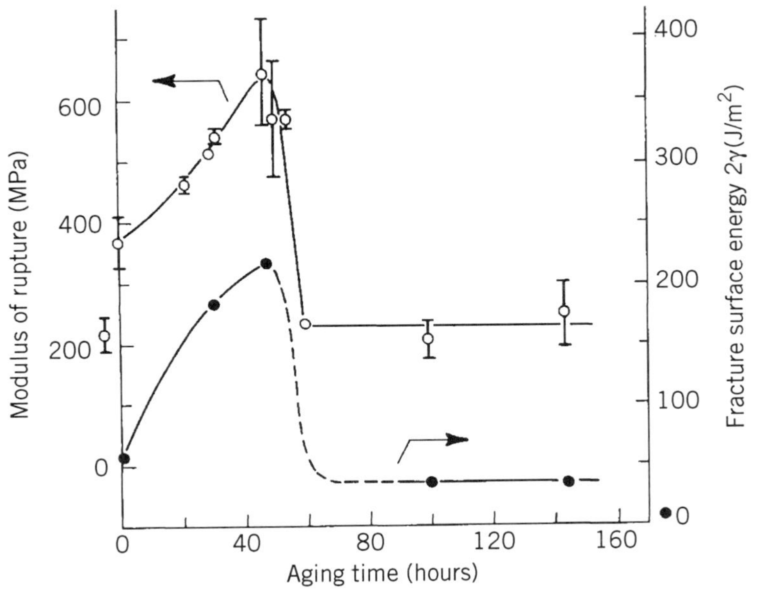

Figure 26.2 Flexural strength and fracture surface energy of 3.7 wt. % Ca–PSZ as a function of aging time. (Reprinted from *Ceramugica International*, Vol. 6, 19–24, 1980.)

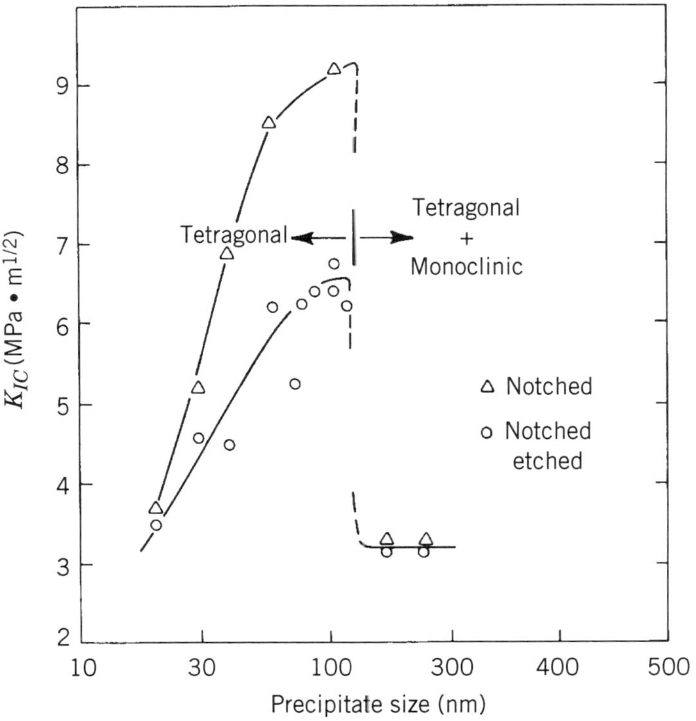

Figure 26.3 Critical stress intensity factor of 3.7 wt. % Ca–PSZ as a function of aging time. (From Swain, Hannink, and Garvie, 1983.)

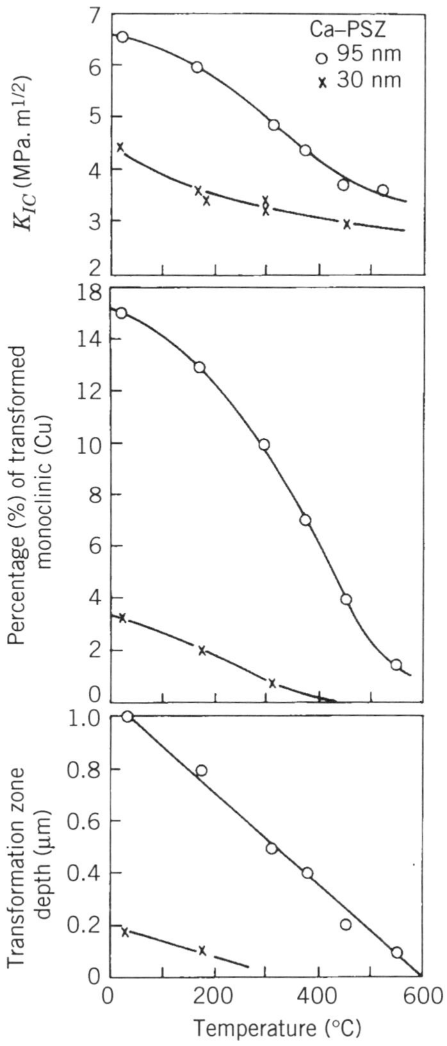

Figure 26.4 Temperature dependence for Ca–PSZ. (From Swain, Hannink, and Garvie, 1983.)

increase roughly in parallel and then drop abruptly. The microstructural feature responsible for this variation is precipitate size, as shown in Figure 26.3 (Swain, Hannink, and Garvie, 1980, as quoted by Green, Hannink, and Swain, 1989). Here the critical stress intensity factor is seen to rise initially with increasing precipitate size in accordance with the theory and experiments described in Section 11.5. It was found that the width of the transformation zone (the wake) increased with the size of the transforming particle. However, when a critical size is exceeded the tetragonal particle transforms spontane-

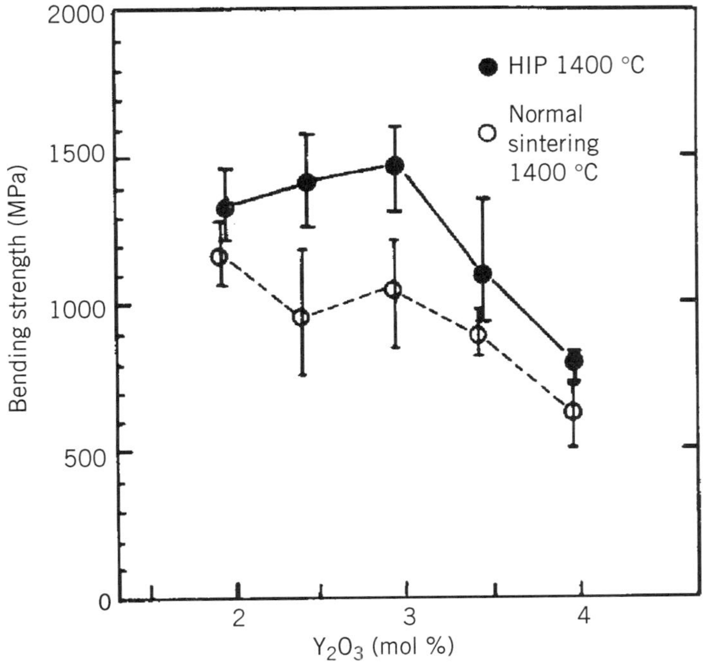

Figure 26.5 Flexural strength of Y–TZP as a function of yttria content. (From Tsukuma, Kubota, and Tsukidate, 1983. Reprinted by permission of the American Ceramic Society.)

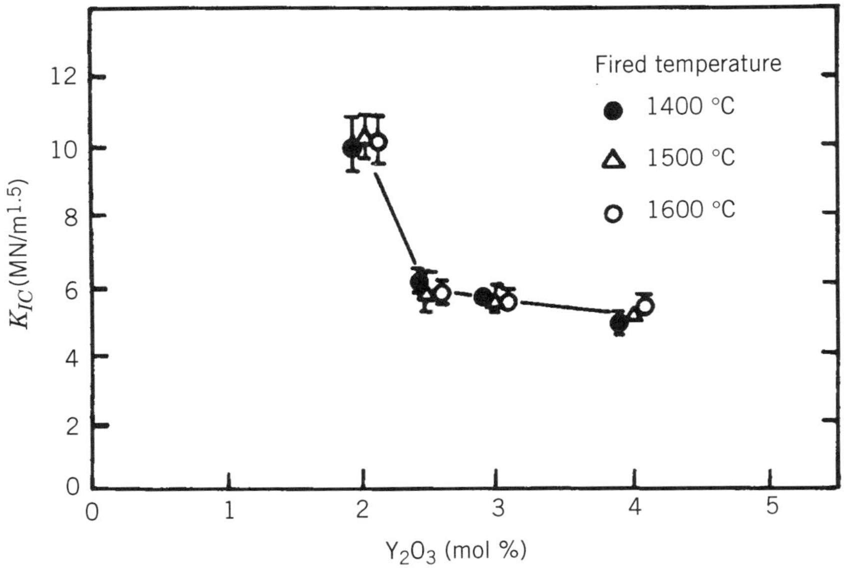

Figure 26.6 Fracture toughness of Y–TZP as a function of yttria content. (From Tsukuma, Kubota, and Tsukidate, 1983. Reprinted by permission of the American Ceramic Society.)

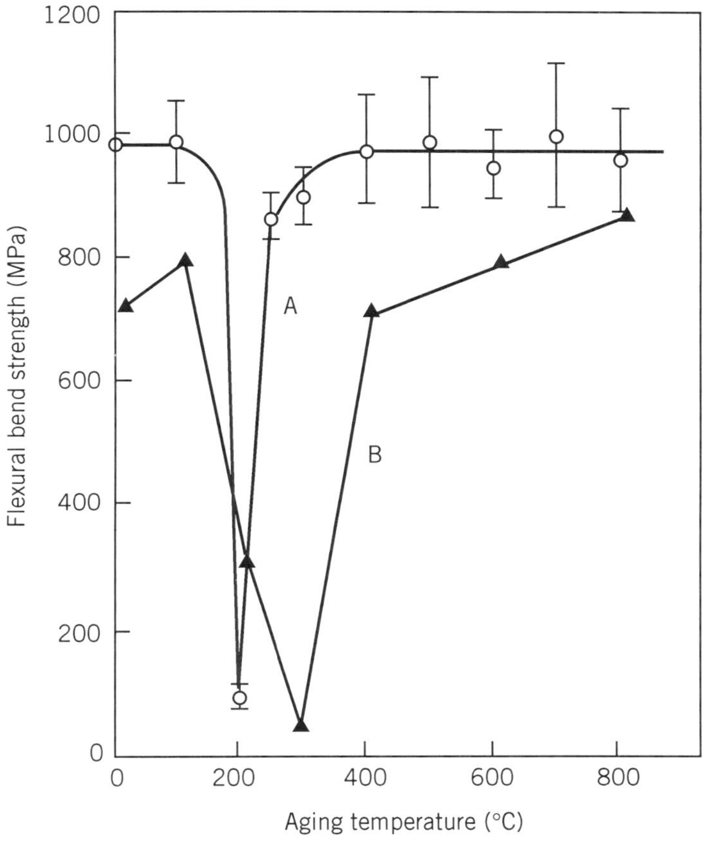

Figure 26.7 Effect of aging in water vapor on strength of Y–TZP of (A) 3 mol % for 2000 hr and (B) 4 mol % for 100 hr. (Reproduced with permission from the *Annual Review of Materials Science*, Vol. 24, © 1994, by Annual Reviews, Inc.)

ously without the application of stress so that the toughness falls abruptly. The maximum value of toughness occurs for a tetragonal particle size that maximizes the size of the transformation zone but that is not quite large enough to transform without the concentrated stress near a crack in a loaded sample.

PSZ that has been aged to produce maximum toughness at room temperature shows a rapid decrease of toughness with increasing temperature of toughness measurement. Figure 26.4 (Swain, Hannink, and Garvie, 1980, as quoted by Green, Hannink, and Swain, 1989) gives results on Ca–PSZ with two heat treatments to produce two tetragonal precipitate sizes. The larger size (95 nm) gives a toughness of 7.6 MPa $\cdot$ $m^{1/2}$ at room temperature in contrast to the smaller (30 nm) that gives only 4.4 MPa $\cdot$ $m^{1/2}$. With increasing temperature of measurement, the toughness decreases and is down to the value without transformation toughening by about 500 °C. This is explained by the decrease of the

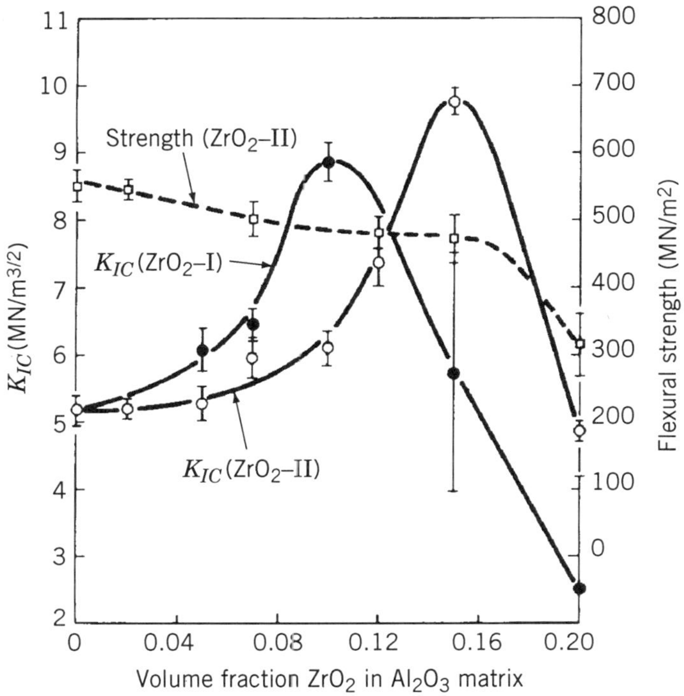

Figure 26.8 Flexure strength and toughness of alumina as a function of zirconia addition: I and II refer to 0.3- and 1.25-μm zirconia particles. (From N. Claussen, "Fracture Toughness of Al_2O_3 with an Unstabilized ZrO_2 Dispersed Phase," *J. Am. Ceram. Soc.*, Vol. 59, 49–51, 1976. Reprinted by permission of the American Ceramic Society.)

free energy of transformation with increasing temperature that results in the requirement of a higher local stress to cause the transformation, leading to a smaller transformation zone width and a smaller transformation-toughening contribution. Measured values of the percentage of zirconia transformed to monoclinic on the fracture surface and of the transformation zone depth are both shown in Figure 26.4 and support this explanation.

Mg–PSZ has more complex microstructure and resulting variation in mechanical properties; its behavior is well described by Green, Hannink, and Swain (1989). Commercial Mg–PSZ contain 8–10 mol % MgO. After sintering just inside the cubic field the material is either cooled rapidly (about 500 °C/hr for 9.7 mol % Mg–PSZ) and then heat treated (typically above 1400 °C) or a programmed cooling schedule is used to achieve heat treatment in a single firing. The precipitation behavior is complex and depends strongly on details of the heat treatment. Values at room temperature of flexural strength in excess of 700 MPa, critical stress intensity factor greater than 15 MPa · $m^{1/2}$, and Weibull modulus of about 30 can be achieved by a controlled cool to near

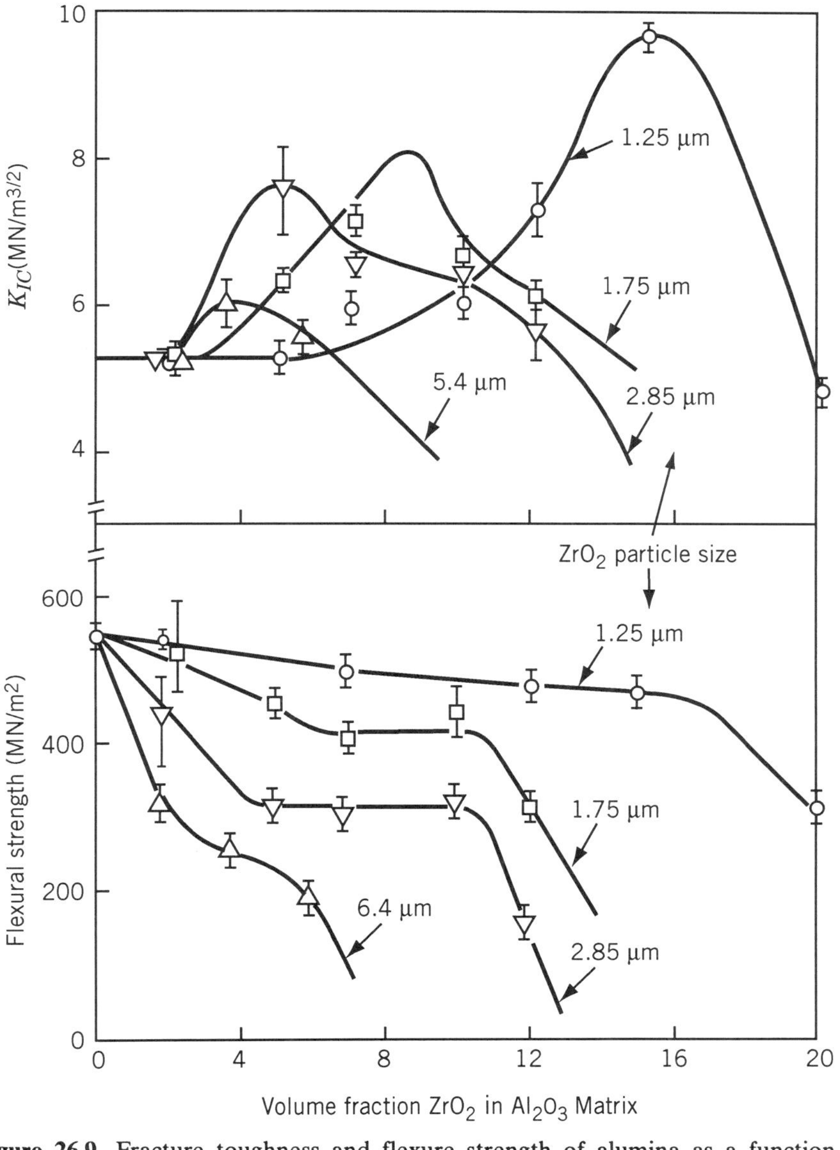

Figure 26.9 Fracture toughness and flexure strength of alumina as a function of zirconia content for various zirconia particle sizes. (From N. Claussen, X. Steeb, and X. Pabst, "Effect of Induced Microcracking on the Fracture Toughness of Ceramics," *Am. Ceram. Soc. Bull.*, Vol. 56, 559–562, 1977. Reprinted by permission of the American Ceramic Society.)

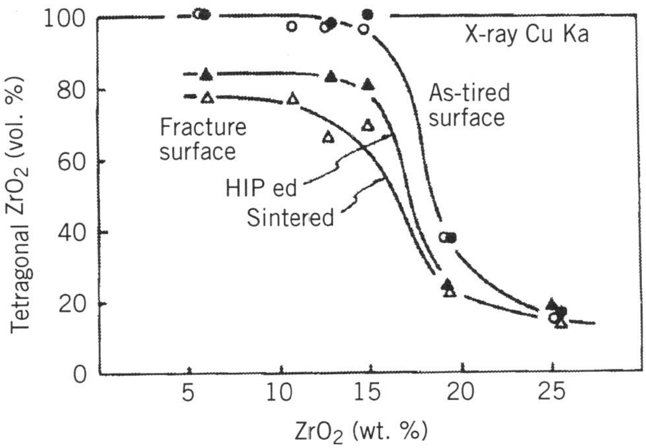

Figure 26.10 Tetragonal content as a function of total zirconia content. (From S. Hori, M. Yoshimura, and S. Somiya, "Strength–Toughness Relations in Sintered and Isostatically Hot-Pressed ZrO_2-Toughened Al_2O_3, *J. Am. Ceram. Soc.*, Vol. 69, 169–172, 1986. Reprinted by permission of the American Ceramic Society.)

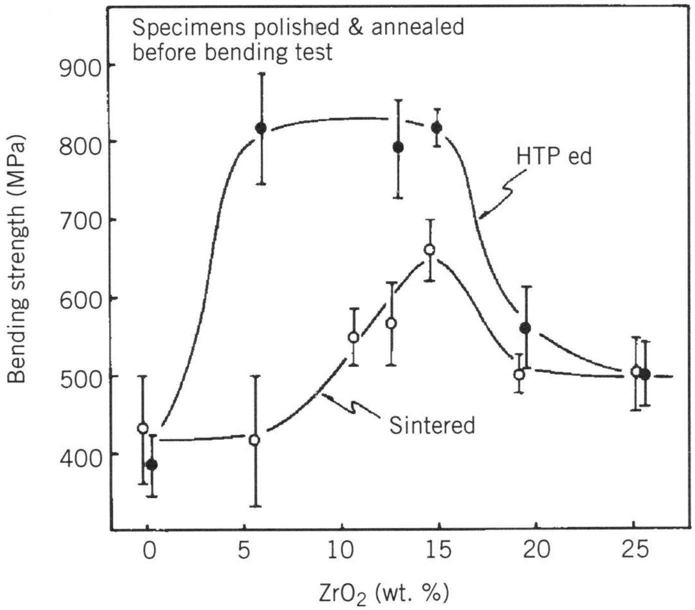

Figure 26.11 Flexure strength as a function of total zirconia content. (From S. Hori, M. Yoshimura, and S. Somiya, "Strength–Toughness Relations in Sintered and Isostatically Hot-Pressed ZrO_2-Toughened Al_2O_3, *J. Am. Ceram. Soc.*, Vol. 69, 169–172, 1986. Reprinted by permission of the American Ceramic Society.)

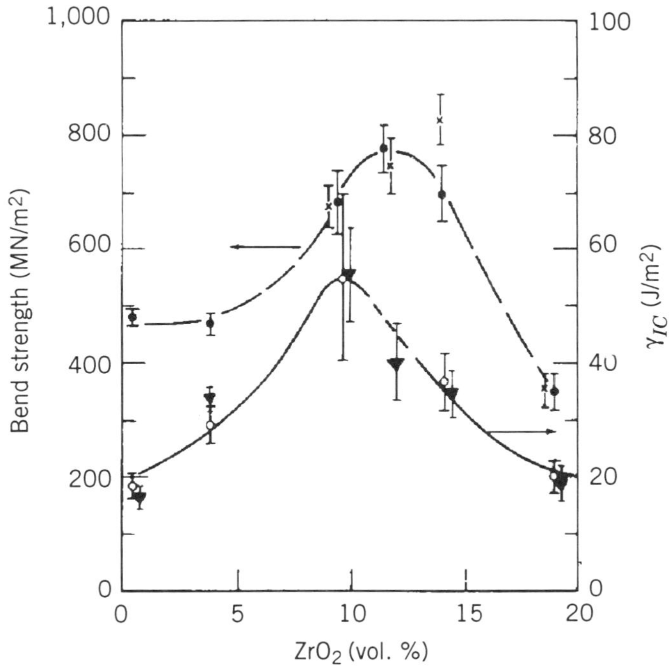

Figure 26.12 Strength and fracture surface energy of alumina as a function of vol. % zirconnia. (From P. R. Becher, "Transient Thermal Stress Behavior in ZrO_2-Toughened Al_2O_3, *J. Am. Ceram. Soc.*, Vol. 64, 37–39, 1981. Reprnted by permission of the American Ceramic Society.)

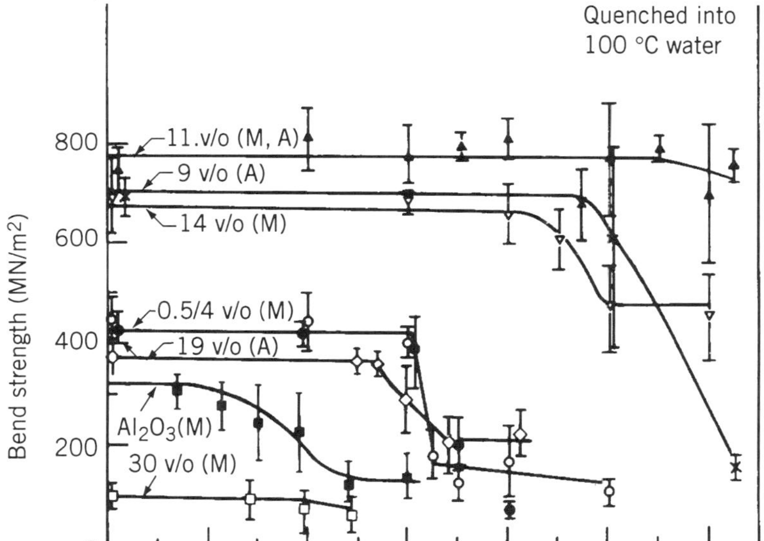

Figure 26.13 Bend strength as a function of thermal shock temperature for alumina with indicated vol. % of zirconia. (From P. R. Becher, "Transient Thermal Stress Behavior in ZrO_2-Toughened Al_2O_3, *J. Am. Ceram. Soc.*, Vol. 64, 37–39, 1981. Reprnted by permission of the American Ceramic Society.)

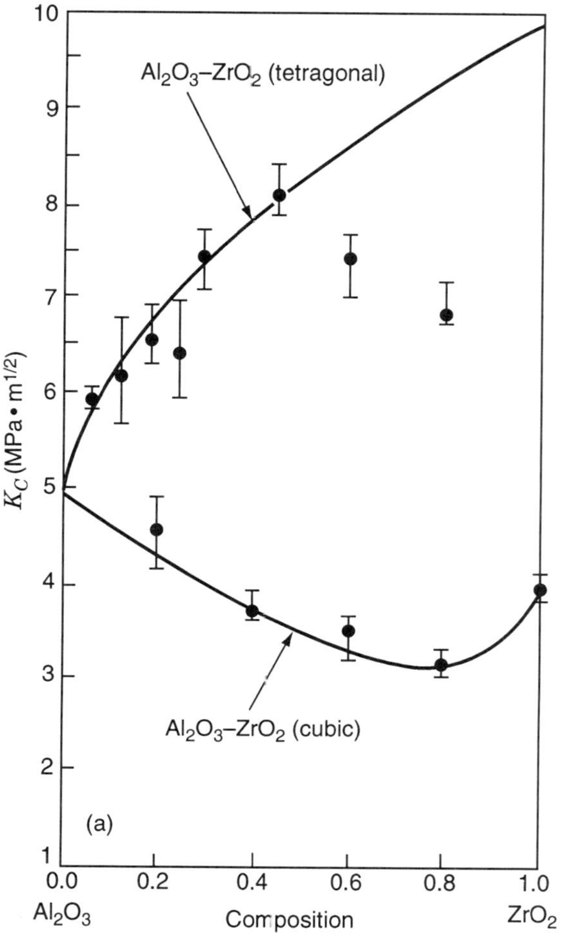

Figure 26.14 Critical stress intensity factor as a function of composition. (From Lange, 1982a.)

room temperature from the solution treatment temperature followed by aging at 1100 °C (R. H. J. Hannink, personal communication, 1995).

Both Mg–PSZ and Ca–PSZ show R-curve behavior, but especially strong R-curve effects are found in Mg–PSZ. The R-curve is strongly dependent on microstructure as influenced by composition and heat treatment. Swain and Hannink (1983) state that R-curve behavior in Mg–PSZ may occur in three ways: (1) crack extension into material with highly metastable T-ZrO_2 precipitates, (2) crack branching and microcracking in subeutectoid aged material, and (3) crack deflection and crack branching resulting from large precipitates in overaged material. The strongest R-curve behavior occurs in material with highly metastable precipitates. Increasing critical stress intensity factor values

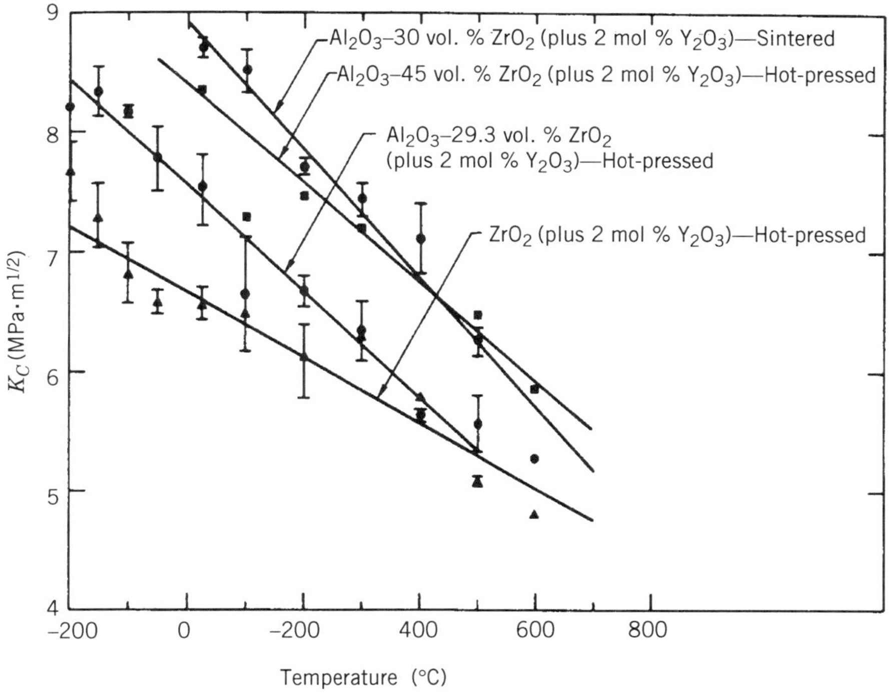

Figure 26.15 Critical stress intensity factor as a function of temperature. (From Lange, 1982b.)

from 4.5 to 12 MPa · m$^{1/2}$ have been found as cracks extend with most of the increase in the first 50 μm of crack extension.

The high strength and high toughness values achievable in PSZ make it of great interest for structural applications. However, consideration must be given to effects that can limit its usefulness: temperature dependence and fatigue. The temperature dependence of toughness in Ca–PSZ was illustrated in Figure 26.4. Cyclic fatigue of ceramics was discussed in Section 14.3, including examples of behavior of Mg–PSZ.

26.3 YTTRIA TETRAGONAL ZIRCONIA POLYCRYSTALS

Ideally, TZP materials are single-phase tetragonal, but in practice some amount of a second crystalline phase (generally cubic) may be present and some glassy grain boundary phase usually also exists. Green, Hannink, and Swain (1989) consider Y–TZP materials to be fine-grained polycrystalline zirconia with 1.75–3.5 mol % yttria. Strength and toughness values of sintered and HIPed Y–TZP are shown in Figures 26.5 and 26.6 (Tsukuma, Kubota, and Tsukidate, 1983). A combination of strength around 1400 MPa and toughness

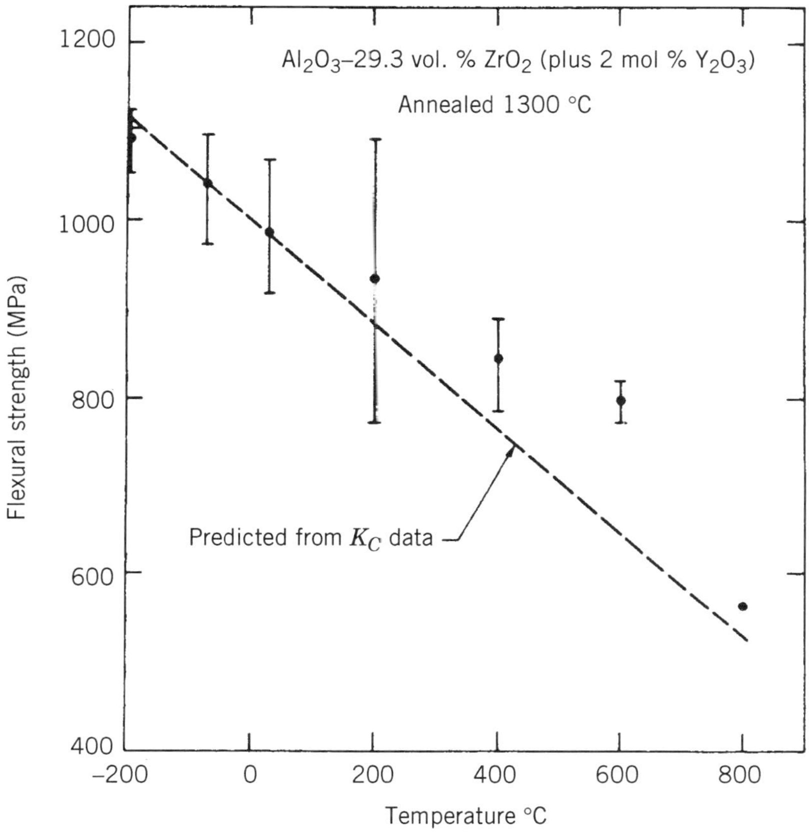

Figure 26.16 Strength as a function of temperature for alumina with 29.3 vol. % zirconia containing 2 mol % yttria. (From Lange, 1982b.)

above 10 MPa · $m^{1/2}$ is possible. Virkar and Matsumoto (1986) have suggested stress-induced switching of tetragonal axes (ferroelastic domains) as a toughening mechanism. However, Y–TZP is subject to serious strength degradation by water vapor in the 200–300 °C temperature range, as shown in Figure 26.7.

26.4 ZIRCONIA-TOUGHENED CERAMICS

A variety of zirconia-dispersed ceramics (ZDC) have been made but we concentrate on zirconia-toughened alumina (ZTA) as a particularly successful and useful material. Claussen (1976) showed that fracture toughness could be increased by dispersing fine particles of pure zirconia in polycrystalline alumina (Figures 26.8 and 26.9). Strength and toughness follow quite different curves with the peak toughness being dependent on zirconia particle size.

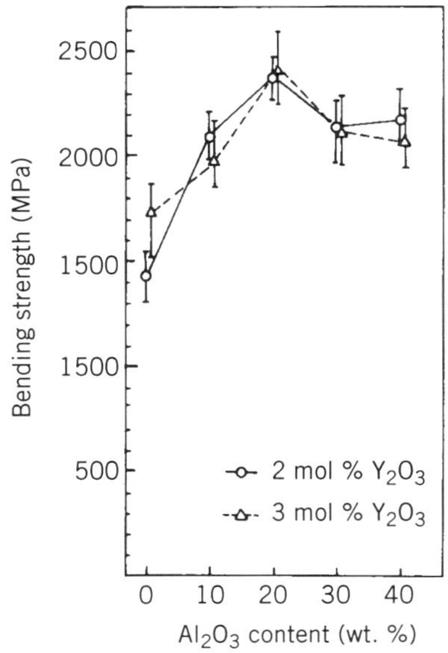

Figure 26.17 Bending strenth as a function of alumina content in zirconia alloyed with yttria. (From K. Tsukuma and K. Ueda, "Strength and Fracture Toughness of Isostatically Hot-Pressed Composites of Al_2O_3 and Y_2O_3–Partially-Stabilized ZrO_2," *J. Am. Ceram. Soc.*, Vol. 68, C-4–C-5, 1985. Reprinted by permission of the American Ceramic Society.)

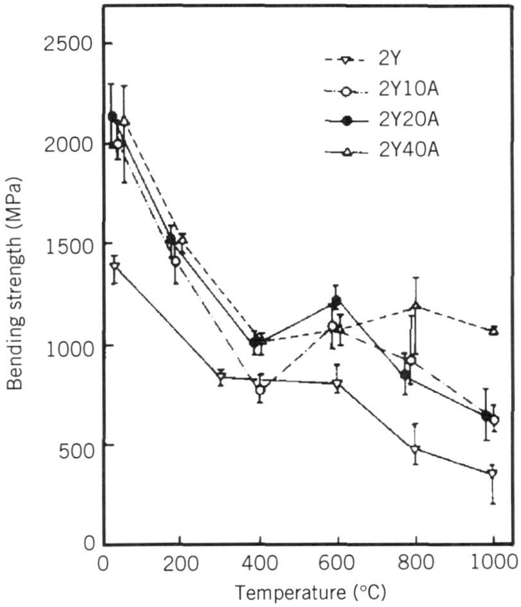

Figure 26.18 Bending strength as a function of temperature for isostatically hot-pressed zirconia alloyed with 2 mol % yttria and with a 20 wt. % alumina dispesion. (From K. Tsukuma, K. Ueda, K. Matsushita, and M. Shimada, "High-Temperature Strength and Fracture Toughness of Y_2O_3–Partially Stabilized ZrO_2/Al_2O_3 Composites," *J. Am. Ceram. Soc.*, Vol. 68, C-56, 1985. Reprinted by permission of the American Ceramic Society.)

Claussen, Steeb, and Pabst (1977) showed that when the zirconia inclusions are above a critical particle size the fracture toughness is inversely proportional to the inclusion size and proportional to the volume fraction of the inclusions up to a maximum, above which the toughness falls with further increase in inclusion volume. Green (1982) showed that the maximum zirconia inclusion size at which tetragonal zirconia was retained at room temperature (i.e., did not transform to monoclinic) decreased with increasing volume fraction of zirconia. For zirconia inclusions with a size distribution of 0.2–0.6 μm, the transition from retained tetragonal to monoclinic is essentially complete in going from 15 wt. % zirconia to 20 wt. %, as shown in Figure 26.10. The strength of hot-pressed alumina with 5–15 wt. % zirconia is doubled, as shown in Figure 26.11, but falls abruptly when more zirconia is used. Becher (1981) has reported similar results for strength and fracture surface energy of alumina as a function of volume percent zirconia, as shown in Figure 26.12. The diameter of his zirconia grains varied from 0.5 μm for 0.5 vol. % to 1–2 μm for 30 vol. % zirconia. Becher also showed that for zirconia contents that maximized strength the thermal shock resistance was much improved. For alumina with 11.5 vol. % zirconia there was little or no decrease in strength as a result of quenching from 1200 °C, as shown in Figure 26.13.

The limitation of about 15 wt. % maximum of zirconia in alumina can be removed by using a dispersion of zirconia particles alloyed with yttria rather than pure zirconia. The addition of yttria lowers the driving force for the tetragonal to monoclinic phase transformation and allows much higher vol-

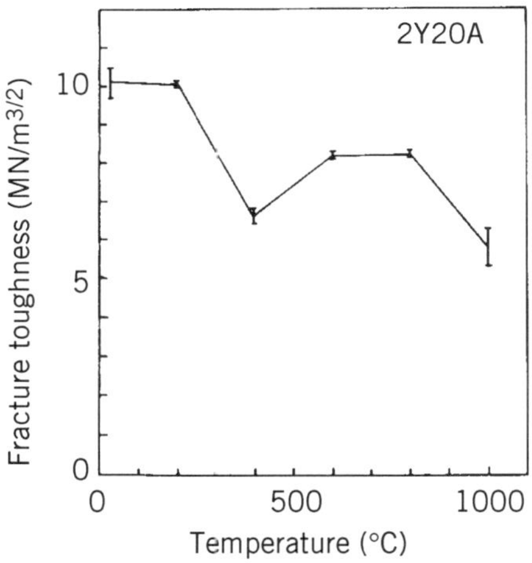

Figure 26.19 Fracture toughness as a function of temperature for isostatically hot-pressed zirconia alloyed with 2 mol % yttria and with a 20 wt. % alumina dispersion. (From K. Tsukuma, K. Ueda, K. Matsushita, and M. Shimada, "High-Temperature Strength and Fracture Toughness of Y_2O_3–Partially Stabilized ZrO_2/Al_2O_3 Composites," *J. Am. Ceram. Soc.*, Vol. 68, C-56, 1985. Reprinted by permission of the American Ceramic Society.)

ume fractions of zirconia to be used. Lange (1982a) was able to fabricate solids using any mixture of alumina and zirconia with only 2 mol % yttria in the zirconia. The stress intensity factor is shown as a function of composition in Figure 26.14 and as a function of temperature in Figure 26.15. The temperature dependence of the strength is shown in Figure 26.16.

Very strong ceramics have been made by dispersing alumina in zirconia alloyed with yttria (Lange, 1982; Tsukuma and Ueda, 1985; Tsukuma, Ueda, Matsushita, and Shimada, 1985). The bending strength as a function of alumina addition is shown in Figure 26.17. The average bending strength of zirconia with 20 wt. % alumina grains was 2400 MPa and the fracture toughness measured by the chevron notch technique was 17 MPa · $m^{1/2}$. The temperature dependence of the bending strength for zirconia alloyed with 2 mol % yttria is shown in Figure 26.18 for various amounts of alumina grains up to 40 wt. %. The strength of the very high strength material drops to about 40% of the room temperature value by 400 °C but the impressively high strength of 1000 MPa is retained at 1000 °C in the material with 10% alumina. The temperature dependence of the toughness is shown in Figure 26.19 and is still around 6 MPa · $m^{1/2}$ at 1000 °C for the 20% alumina material.

CHAPTER 27

MECHANICAL PROPERTIES OF POROUS CERAMICS

27.1 INTRODUCTION

Many ceramics have some degree of porosity. In some cases (such as ceramics used as thermal insulators, filters, or catalyst supports) a high degree of porosity is essential for the function that they perform. Mechanical properties including elastic moduli, strength, and toughness decrease with increasing porosity. Treatments often consider isolated pores in a solid at one extreme and consider a solid, such as a foam, made up of thin-walled cells at the other extreme (Gibson and Ashby, 1988).

27.2 EFFECT OF POROSITY ON ELASTIC CONSTANTS

It is intuitively evident that porosity will reduce elastic moduli and strength. This effect is usually discussed in terms of the relationship between the volume fraction porosity, P, and the relative elastic modulus values, the ratios of the elastic moduli at porosity P to their values at porosity $P = 0$. Considerable research has been done in fitting empirical curves and developing theoretical models. The problem is difficult because pores differ in shape depending on the processing. It is therefore not to be expected that a simple model based, for example, on spherical pores will accurately represent the behavior of real ceramics. However, the various relationships do provide a useful approximate description. Such relationships have been reviewed by Wachtman (1969) and R. W. Rice (1977). Some of the most widely used relationships are discussed here.

Mackenzie (1950) derived the following expressions for the bulk and shear moduli at small amounts of porosity in terms of their values, B_0 and G_0, at zero porosity:

$$\frac{1}{B} = \frac{1}{B_0(1-P)} + \frac{3P}{4\mu_0(1-P)} + \text{ terms of order } P^3 \tag{27.1}$$

$$\mu = \mu_0\left[1 - \frac{5P(3B_0 + 4\mu_0)}{9B_0 + 8\mu_0}\right] + \text{terms of order } P^2 \tag{27.2}$$

Spriggs (1961) and Spriggs and Brissette (1962) studied the effect of porosity up to fairly large values and proposed the following empirical equations for Young's modulus and the shear modulus:

$$E = E_0 \exp(-bP) \tag{27.3}$$

$$\mu = \mu_0 \exp(-bP) \tag{27.4}$$

Here b is a parameter to be fitted empirically; it is usually fitted independently to Young's modulus and shear modulus data and so usually has somewhat different values for the two moduli. The b parameter is typically about 4 for oxides studied in the porosity range 0–40%, as shown in Table 27.1.

Hasselman (1962) suggested for the bulk modulus and the shear modulus the equations

$$B = B_0\left[1 - \frac{AP}{1-(A+1)P}\right] \tag{27.5}$$

$$\mu = \mu_0\left[1 - \frac{AP}{1-(A+1)P}\right] \tag{27.6}$$

where A is a parameter to be fitted empirically.

For small values of b any of these equations reduces to a linear decrease of relative elastic moduli with volume fraction porosity. That is, for Young's modulus the equations take the form

$$\frac{E}{E_0} = 1 - bP \tag{27.7}$$

TABLE 27.1 Parameters in Spriggs' Equation for Porosity Dependence of Elastic Moduli

Young's Modulus			
Material	E_0	b for Young's Modulus	Reference
Aluminum oxide	4102 kilobars	3.92	Knudsen, 1961
Magnesium oxide	46.08 million psi	4.74	Spriggs, Brissette, and Vasilos, 1962
Shear Modulus			
Material	μ_0	b for Shear Modulus	Reference
Magnesium oxide	20.24 million psi	4.9	Spriggs Brissette, and Vasolos, 1962

and similarly for the other moduli. The value of b for Young's modulus is typically around 4 (Spriggs, 1961; Knudson, 1962; Spriggs and Brissette, 1962).

Nielson (1982, 1984) derived the following equation for Young's modulus of a solid with pore volume fraction P:

$$E = E_0 \frac{(1-P)^2}{1 + (\frac{1}{\rho} - 1)P} \tag{27.8}$$

Here ρ is the Nielson shape factor. Nielson does not give an equation for ρ but states that a value of 0.4 or less is appropriate for materials with porosity that is characteristic of joined particles; a shape factor for porosity that resembles ribbons should have a shape factor in the range 0.3–0.7, and a shape factor of 0.6–1.0 is characteristic of pores enveloped by a solid phase.

The slopes of the Spriggs and Nielson equations for P near zero are

$$\left(\frac{d(E/E_0)}{dP}\right)_{\text{Spriggs for } P\to 0} = -b \tag{27.9}$$

$$\left(\frac{d(E/E_0}{dP}\right)_{\text{Nielson for } P\to 0} = \frac{1}{\rho} - 1 \tag{27.10}$$

Plots of the Spriggs and Nielson equations with b and ρ chosen to give equal initial slopes give very closely the same curve up to about 40% porosity. Either equation can be used within experimental accuracy in this porosity range. A more rigorous test of the equations is how well they represent the data at porosity values above 40%. On a linear plot the differences between the two equations may not appear great. However, a better test is to plot experimental data as log (E/E_0) as a function of P. On such a plot the Spriggs equation gives a straight line. Actual data at high porosity typically show that the experimental values fall progressively below this line as the porosity increases above 40%. Work by Ashkin, Haber, and Wachtman (1990) on silica with porosity ranging from 0.2 to 1 showed that Young's modulus was represented over this entire range fairly well by Nielson's equation with a single shape factor value of 0.4. Closer inspection indicates that the shape factor may be nearer to 0.6 for porosity in the range 0.2–0.5 for some specimens. As noted earlier, it is reasonable that pore shape should depend on the method of preparation. Furthermore, it should change as sintering proceeds from a very porous body to one with low porosity.

Additional theoretical equations for the porosity dependence of elastic moduli have been reviewed by Ramakrishnan and Arunachalam (1993). R. W. Rice (1993) has extensively considered the dependence of properties of ceramics, including elastic properties, on porosity and has recently presented a new approach based on mapping the pore and solid shapes.

If one thinks of a porous solid as being analogous to a building, it is plausible that the mechanical behavior of the solid will depend strongly on the details of arrangement of the solid components just as the stiffness and strength of a building or bridge depend on the arrangement of the structural members. Accurate treatment evidently demands detailed knowledge. An important class of porous solids not properly regarded as solids with distributed porosity is the class of cellular solids. Their properties have been reviewed by Gibson and Ashby (1988), who give the following definition: "a cellular solid is one made up of an interconnected network of solid struts or plates which form the edges and faces of cells." The reader is referred to their work for further information.

REFERENCES

Abramowitz, M., and Stegun, I. A., *Handbook of Mathematical Functions with Formulas, Graphs, and Mathematical Tables*, National Bureau of Standards Applied Mathematics Series 55, June 1964.

Amazigo, J. C., and Budiansky, B., "Interaction of Particulate and Transformation Toughening," *J. Mech. Phys. Solids*, Vol. 36, 581–595, 1988.

Anderson, O. L., "A Proposed Law of Corresponding States for Oxide Compounds," *J. Geophys. Res.*, Vol. 70, 4963–4971, 1966a.

Anderson, O. L., "Derivation of Wachtman's Equation for the Temperature Dependence of Elastic Moduli of Oxide Compounds," *Phys. Rev.*, Vol. 144, 553–557, 1966b.

Anderson, O. L., *Equations of State of Solids for Geophysics and Ceramic Science*, Oxford University Press, New York, 1995.

Andreasen, J. G., "Statistics of Brittle Failure in Multiaxial Stress States," *J. Am. Ceram. Soc.*, Vol. 76, 2933–2935, 1993.

Anstis, G. R., Chantikul, P., Lawn, B. R., and Marshall, D. B., "A Critical Evaluation of Indentation Techniques for Measuring Fracture Toughness: I, Direct Crack Measurements," *J. Am. Ceram. Soc.*, Vol. 64, 533–538, 1981.

Argon, A. S., "Inelastic Deformation and Fracture in Oxide, Metallic, and Polymeric Glasses," pp. 79–132 in Uhlmann, D. R., and Kreidl, N. J., editors, *Glass Science and Technology, Volume 5 Elasticity and Strength in Glasses*, Academic Press, New York, 1980.

Ashby, M. F., "First Report on Deformation Maps," *Acta Metall.*, Vol. 20, 877–897, 1972.

Ashby, M. F., Gandhi, C. and Taplin, D. M. R., "Fracture Mechanism Maps and Their Construction for F.C.C. Metals and Alloys," *Acta Metall.*, Vol. 27, 699–720, 1979.

Ashby, M. F., and Verral, R. A., Diffusion Accomodated Flow and Superplasticity, *Acta. Metall.*, Vol. 21, 193–163, 1973.

Ashby, M. F., and Verrall, R. A., "Micromechanisms of Flow and Fracture and Their Relevance to the Rheology of the Upper Mantle," *Philos. Trans. R. Soc. London*, Vol. 288A, 59–95, 1978.

Ashkin, D., Haber, R. A., and Wachtman, J. B., "Elastic Properties of Porous Silica Derived from Colloidal Gels," *J. Am. Ceram. Soc.*, Vol. 73, 3376–3381, 1990.

ASTM, "Biaxial Flexure Strength (Modulus of Rupture) of Ceramic Substrates," pp. 783–789 in *1978 Annual Book of ASTM Standards, Part 43, Electronics*, ANSI/ASTM F394-78, American Society for Testing and Materials, Philadelphia, 1978.

ASTM, *Standard Test Method for Dynamic Young's Modulus, Shear Modulus, and Poisson's Ratio for Advanced Ceramics by Sonic Resonance*, ASTM C 1198 - 91, American Society for Testing and Materials, Philadelphia, 1991.

Aveston, J., Cooper, G. A., and Kelly, A., "Single and Multiple Fracture," pp. 15–26 in *Proceedings of the Conference on the Properties of Fibre Composites: Conference Proceedings*, National Physical Laboratory, IPC Science & Technology Press, Guildford, UK, 1971.

Badwal, S. P. S., Bannister, M. J., and Hannink, R. H. J., *Science and Technology of Zirconia V*, Technomic Publishing Co., Lancaster, PA, 1993.

Baratta, F. I., "Requirements for Flexure Testing of Brittle Materials," pp. 194–222 in Freiman, S., and Hudson, C., editors, *Methods for Assessing the Structural Reliability of Brittle Materials*, ASTM STP 844, American Society for Testing and Materials, Philadelphia, 1982.

Bartholomew, R. F., and Garfinkel, H. M., "Chemical Strengthening of Glass," pp. 217–270 in Uhlmann, D. R., and Kreidl, N. J., editors, *Glass Science and Technology, Volume 5 Elasticity and Strength in Glasses*, Academic Press, New York, 1980.

Batdorf, S. B., and Crose, J. G., "A Statistical Theory for the Fracture of Brittle Structures Subjected to Nonuniform Polyaxial Stresses," *J. Appl. Mech.*, June, 459–464, 1974.

Batdorf, S. B., and Chang, D. J., "On the Relation Between the Fracture Statistics of Volume Distributed and Surface-Distributed Cracks," *Int. J. Fract.*, Vol. 15, 191–199, 1979.

Batdorf, S. B., and Heinisch, H. L. Jr., "Weakest-Link Theory Reformulated for Arbitrary Fracture Criterion," *J. Am. Ceram. Soc.*, Vol. 61, 355–358, 1978.

Batdorf, S. B., and Sines, G., "Combining Data for Improved Weibull Parameter Estimation," *J. Am. Ceram. Soc.*, Vol. 63, 214–218, 1980.

Battelle Memorial Institute, *Engineering Property Data on Selected Ceramics: Volume I, Nitrides; Volume 2, Carbides; Volume 3, Single Oxides*, Metals and Ceramics Information Center, Battelle Columbus Division, report MCIC-HB-07, August 1979. Reprinted in July 1987 and distributed by the American Ceramic Society.

Beall, G. H., "Design and Properties of Glass-Ceramics," *Annu. Rev. Mater. Sci.*, Vol. 22, 91–119, 1992.

Becher, P. F., "Transient Thermal Stress Behavior in ZrO_2-Toughened Al_2O_3," *J. Am. Ceram. Soc.*, Vol. 64, 37–39, 1981.

Becher, P. F., "Microstructural Design of Toughened Ceramics," *J. Am. Ceram. Soc.*, Vol. 73, 255–269, 1991.

Becher, P. F., Fuller, E. R. Jr., and Angelini, P., "Matrix-Grain-Bridging Contributions to the Toughenss of Whisker-Reinforced Ceramics," *J. Am. Ceram. Soc.*, Vol. 74, 2131–2135, 1991.

Becher, P. F., Hsueh, C.-H., Angelini, P, and Tiegs, T. N., "Toughening Behavior of Whisker-Reinforced Ceramic Matrix Composites," *J. Am. Ceram. Soc.*, Vol. 71, 1050–1061, 1988.

Becher, P. F., Hsueh, C. H., Angelini, P., and Tiegs, T. N., "Theoretical and Experimental Analysis of the Toughening Behavior of Whisker Reinforcement in Ceramic Matrix Composites," *Mater. Sci. Eng.*, Vol. A107, 257–259, 1989.

Becher, P. F., and Swain, M. V., "Grain-Size Dependent Transformation Behavior in Polycrystalline Tetragonal Zirconia," *J. Am. Ceram. Soc.*, Vol. 75, 493–502, 1992.

Becher, P. F., Swain, M. V., and Somiya, S., "Advanced Structural Ceramics," *Mater. Res. Soc. Symp. Proc.*, Vol. 78, 1987.

Becher, P. F., and Tiegs, T. N., "Toughening Behavior Involving Multiple Mechanisms: Whisker Reinforcement and Zirconia Toughening," *J. Am. Ceram. Soc.*, Vol. 70, 651–654, 1987

Bennison, S. J., and Lawn, B. R., "Role of Interfacial Grain-Bridging Sliding Friction in the Crack-Resistance and Strength Properties of Nontransforming Ceramics," *Acta Metall.*, Vol. 37, 2659–2671, 1989.

Beyerle, D. S., Spearing, S. M., and Evans, A. G., "Damage Mechanisms and the Mechanical Properties of a Laminated 0/90 Ceramic/Matrix Composite," *J. Am. Ceram. Soc.*, Vol. 75, 3321–3330, 1992.

Bilde-Sørenson, J. B., Lawler, B. F., Geipel, T., Pirouz, P., Heuer, A. H., and Lagerlöf, K. P. D., On Basal Slip and Twinning in Sapphire (α-Al_2O_3: I. Basal Slip Revisited, *Acta Metall.*, in press, 1995.

Birch, J. M., Wilshire, B., and Godfrey, D. J., "Deformation and Fracture Processes During Creep of Reaction Bonded and Hot Pressed Silicon Nitride," *Proc. Br. Ceram. Soc.*, Vol. 26, 141–154, 1978.

Bleise, D., and Steinbrech, R. W., "Flat R-Curve from Stable Propagation of Indentation Cracks in Coarse-Grained Alumina," *J. Am. Ceram. Soc.*, Vol. 77, 315–322, 1994.

Blumenthal, W., and Evans, A. G., "High-Temperature Failure of Polycrystalline Alumina: II, Creep Crack Growth and Blunting," *J. Am. Ceram. Soc.*, Vol. 67, 751–759, 1984.

Boley, B. A., and Weiner, J. H., *Theory of Thermal Stresses*, Wiley, New York, 1960.

Born, M., and Huang, K., *Dynamical Theory of Crystal Lattices*, Oxford University Press, New York, 1954.

Bradt, R. C., Evans, A. G., Hasselman, D. P. H., and Lange, F. F., *Fracture Mechanics of Ceramics, Volume 5 Concepts, Surface Flaws, Statistics, and Microcracking*, Plenum Press, New York, 1983a.

Bradt, R. C., Evans, A. G., Hasselman, D. P. H., and Lange, F. F., *Fracture Mechanics of Ceramics, Volume 6 Concepts, Measurements, Transformations, and High-Temperature Fracture*, Plenum Press, New York, 1983b.

Bradt, R. C., Hasselman, D. P. H., and Lange, F. F., *Fracture Mechanics of Ceramics, Volume 1 Concepts, Flaws and Fractography*, Plenum Press, New York, 1973a.

Bradt, R. C., Hasselman, D. P. H., and Lange, F. F., *Fracture Mechanics of Ceramics, Volume 2 Microstructure, Materials, and Applications*, Plenum Press, New York, 1973b.

Bradt, R. C., Hasselman, D. P. H., and Lange, F. F., *Fracture Mechanics of Ceramics, Volume 3 Flaws and Testing*, Plenum Press, New York, 1978a.

Bradt, R. C., Hasselman, D. P. H., and Lange, F. F., *Fracture Mechanics of Ceramics, Volume 4 Crack Growth and Microstructure*, Plenum Press, New York, 1978b.

Bradt, R. C., Hasselman, D. P. H., Munz, D., Sakai, M., and Shevchenko, V. Y., editors, *Fracture Mechanics of Ceramics, Volume 10 Fracture Fundamentals, High-Temperature Deformation, Damage, and Design*, Plenum Press, New York, 1992.

Bradt, R. C., and Tressler, R. E., *Fractography of Glass*, Plenum Press, New York, 1994.

Brennan, J. J., and Prewo, K. M., "Silicon Carbide Fiber Reinforced Glass-Ceramic Matrix Composites Exhibiting High Strength and Toughness," *J. Mater. Sci.*, Vol. 17, 2371–2383, 1982.

Brinkman, C. R., and Duffy, S. F., editors, *Life Prediction Methodologies and Data for Ceramic Materials*, ASTM Publication Code Number 04–12010-09, American Society for Testing and Materials, Philadelphia, 1994.

Broek, D., *Elementary Engineering Fracture Mechanics*, 4th ed., Martinus Nijhoff Publishers, Dordrecht, 1986.

Brown, W. F. Jr., and Srawley, J. E., pp. 12–14 in *Plane Strain Crack Toughness Testing of High Strength Metallic Materials*, ASTM Special Technical Publication 410, American Society for Testing and Materials, Philadelphia, 1966.

Brühner-Foit, A., and Munz, D., "Statistical Analysis of Flexure Strength Data," International Energy Agency Annex II, Subtask 4, June 1989.

Budiansky, B., "On the Theoretical Toughness and Strength of Ceramic Composites," MECH-216 Report from Harvard University; also in *Proceedings of IUTAM Symposium of Fracture of Brittle Disordered Materials: Concrete, Rock, and Ceramics*, Sept. 20–24, 1993, Brisbane.

Budiansky, B., and Amazigo, J. C., "Toughening by Aligned, Frictionally Constrained Fibers," *J. Mech. Phys. Solids*, Vol. 37, 93–109, 1989.

Budiansky, B., Amazigo, J. C., and Evans, A. G., "Small-Scale Crack Bridging and the Fracture Toughness of Particulate-Reinforced Ceramics," *J. Mech. Phys. Solids*, Vol. 36, 167–187, 1988.

Budiansky, B., Hutchinson, J. W., and Evans, A. G., "Matrix Fracture in Fiber-Reinforced Ceramics," *J. Mech. Phys. Solids*, Vol. 34, 167–189, 1986.

Budiansky, B., Hutchinson, J. W., and Lambropoulos, J. C., "Continuum Theory of Dilatant Transformation Toughening in Ceramics," *Int. J. Solids Structures*, Vol. 19, 337–355, 1983.

Budiansky, B., and Stump, D. M., "Theory of Ceramic/Zirconia Composites: Toughening, Strengthening and Synergism," pp. 317–325 in P. Vincenzini, editor, *Advanced Structural Inorganic Composites*, Elsevier Science Publishers B. V., Amsterdam, 1991.

Budiansky, B., and Truskinovsky, L., "On the Mechanics of Stress-Induced Phase Transformation in Zirconia," *J. Mech. Phys. Solids*, Vol. 41, 615–630, 1993.

Bunk, W., and Hausner, H., *Ceramic Materials and Components for Engines, Proceedings of the Second International Symposium*, Verlag Deutsche Keramische Gesellschaft, Bad Honnef, 1986.

Butler, E. P., and Fuller, E. R. Jr., "Ceramic-Matrix," pp. 77–108 in *Composite Materials*, Vol. 7 of *Kirk–Othmer Encyclopedia of Chemical Technology*, 4th ed., Wiley, New York, 1993.

Cahn, R. W., "Twinned Crystals," *Adv. Phys.*, Vol. 3, 363–445, 1954.

Cai, H., Kalceff, M. A. S., and Lawn, B. R., "Deformation and Fracture of Mica-Containing Glass-Ceramics in Hertzian Contacts," *J. Mater. Res.*, Vol. 9, 121–129, 1994.

Campbell, G. H., Ruhle, M., Dalgleish, B. J., and Evans, A. G., "Whisker Toughening: A Comparison Between Aluminum Oxide and Silicon Nitride Toughened with Silicon Carbide," *J. Am. Ceram. Soc.*, Vol. 73, 521–530, 1990.

Cannon, R. M., Rhodes, W. H., and Heuer, A. H., "Plastic Deformation of Fine-Grained Alumina (Al_2O_3): I, Interface-Controlled Diffusional Creep," *J. Am. Ceram. Soc.*, Vol. 63, 46–53, 1980.

Cannon, W. R., "Transformation Toughened Ceramics for Structural Aplications," pp. 195–288 in Wachtman, J. B. Jr., editor, *Structural Ceramics*, Vol. 29 of *Treatise on Materials Science and Technology*, edited by H. Herman, Academic Press, Orlando, 1989.

Cannon, W. R., and Langdon, T. G., "Review: Creep of Ceramics: Part 1 Mechanical Characteristics," *J. Mater. Sci.*, Vol. 18, 1–50, 1983.

Cannon, W. R., and Langdon, T. G., "Review: Creep of Ceramics: Part 2 An Examination of Flow Mechanisms," *J. Mater. Sci.*, Vol. 23, 1–20, 1988.

Cao, H. C., Bischoff, E., Sbaizero, O., Ruhle, M., and Evans, A. G., Marshall, D. B., and Brennan, J. J., "Effect of Interfaces on the Properties of Fiber-Reinforced Ceramics," *J. Am. Ceram. Soc.*, Vol. 73, 1691–1699, 1990.

Cao, H., and Thouless, M. D., "Tensile Tests of Ceramic-Matrix Composites: Theory and Experiment," *J. Am. Ceram. Soc.*, Vol. 73, 2091–2094, 1990.

Carlsson, R., Johansson, T., and Kahlman, L., *4th International Symposium on Ceramic Materials and Components for Engines*, Elsevier Applied Science, London, 1992.

Carroll, D. F., and Tressler, R. E., "Accumulation of Creep Damage in a Siliconized Silicon Carbide," *J. Am. Ceram. Soc.*, Vol. 71, 472–477, 1988.

Carroll, D. F., and Tressler, R. E., "Effect of Creep Damage on the Tensile Creep Behavior of a Siliconized Silicon Carbide," *J. Am. Ceram. Soc.*, Vol. 72, 49–53, 1989.

Carroll, D. F., Wiederhorn, S. M., and Roberts, D. E., "Technique for Tensile Testing of Ceramics," *J. Am. Ceram. Soc.*, Vol. 72, 1610–1614, 1989.

Carslaw, H. S., and Jaeger, J. C., *Conduction of Heat in Solids*, 2nd ed., Oxford University Press, Oxford, UK, 1959.

Carter, C. H. Jr., Davis, R. F., and Bentley, J., "Kinetics and Mechanisms of High-Temperature Creep in Silicon Carbide: II, Chemically Vapor Deposited," *J. Am. Ceram. Soc.*, Vol. 67, 732–740, 1984.

Chan, K. S., and Page, R. A., "Creep Damage Development in Structural Ceramics," *J. Am. Ceram. Soc.*, Vol. 76, 803–826, 1993.

Chantikul, P., Anstis, G. R., Lawn, B. R., and Marshall, D. B., "A Critical Evaluation of Indentation Techniques for Measuring Fracture Toughness: II, Strength Method," *J. Am. Ceram. Soc.*, Vol. 64, 539–543, 1981.

Chantikul, P., Bennison, S. J., and Lawn, B. R., "Role of Grain Size in the Strength and R-Curve Properties of Alumina," *J. Am. Ceram. Soc.*, Vol. 72, 2419–2427, 1990.

Chao, L.-Y., and Shetty, D. K., "Equivalence of Physically Based Statistical Fracture Theories for Reliability Analysis of Structural Ceramics in Multiaxial Loading," *J. Am. Ceram. Soc.*, Vol. 73, 1917–1921, 1990.

Chao, L.-Y., and Shetty, D. K., "Reliability Analysis of Structural Ceramics Subjected to Biaxial Flexure," *J. Am. Ceram. Soc.*, Vol. 74, 333–344, 1991.

Charalambides, P. G., "Steady-State Mechanics of Delamination Cracking in Laminated Ceramic-Matrix Composites," *J. Am. Ceram. Soc.*, Vol. 74, 3066–3080, 1991.

Charalambides, P. G., and Evans, A. G., "Debonding Properties of Residually Stressed Brittle-Matrix Composites," *J. Am. Ceram. Soc.*, Vol. 42, 746–753, 1989.

Charalambides, P. G., and Evans, A. G., "Debonding Properties of Residually Stressed Brittle-Matrix Composites," *J. Am. Ceram. Soc.*, Vol. 72, 746–753, 1989.

Charles, R. J., and Hillig, W. S., pp. 511–527 in *Symposium on Mechanical Strength of Glass and Ways of Improving It*, Florence, Italy, Sept. 25–29, 1962. Union Scientifique Continentale du Verre, Charleroi, Belgium, 1962.

Chawla, K. K., *Ceramic Matrix Composites*, Chapman & Hall, London, 1993.

Chen, C.-F., and Chuang, T.-J., "Improved Analysis for Flexural Creep with Application to Sialon Ceramics," *J. Am. Ceram. Soc.*, Vol. 73, 2366–2373, 1990.

Chen, C.-F., Wiederhorn, S. M., and Chuang, T.-J., "Cavitation Damage During Flexural Creep of SiAlON–YAG Ceramics," *J. Am. Ceram. Soc.*, Vol. 74, 1658–1662, 1991.

Chen, I. W., "Model of Transformation Toughening in Brittle Materials," *J. Am. Ceram. Soc.*, Vol. 74, 2564–2572, 1991

Chen, I-W., and Chiao, Y. H., "Martensitic Nucleation in ZrO_2," *Acta Metall.*, Vol. 31, 1627–1638, 1983.

Chen, Z., and Mecholsky, J. J. Jr., "Toughening by Metallic Lamina in Nickel/Alumina Composites," *J. Am. Ceram. Soc.*, Vol. 76, 1258–1264, 1993.

Cho, C., Holmes, J. W., and Barber, J. R., "Distribution of Matrix Cracks in a Uniaxial Ceramic Composite," *J. Am. Ceram. Soc.*, Vol. 75, 316–324, 1992.

Chokshi, A. H., and Langdon, T. G., "Characteristics of Creep Deformation in Ceramics," *Mater. Sci. Technol.*, Vol. 7, 577–584, 1991.

Chou, T.-W., *Microstructural Design of Fiber Composites*, Cambridge University Press, New York, 1992.

Chou, Y.-S., and Green, D. J., "Silicon Carbide Platelet/Alumina Composites: III, Toughening Mechanisms," *J. Am. Ceram. Soc.*, Vol. 76, 1985–1992, 1993.

Christensen, R. M., and Waal, F. W., "Effective Stiffness of Randomly Oriented Fiber Composites," *J. Comp. Mater.*, Vol. 6, 518–532, 1972.

Chuang, T.-J., "A Diffusive Crack Growth Model for Creep Fracture," *J. Am. Ceram. Soc.*, Vol. 65, 93–103, 1982. "Correction," *J. Am. Ceram. Soc.*, Vol. 65, 226, 1982.

Chuang, T.-J., "Estimation of Power-Law Creep Parameters from Bend Test Data," *J. Mater. Sci.*, Vol. 21, 165–175, 1986.

Chuang, T.-J., Tressler, R. E., and Minford, E. J., "On the Static Fatigue Limit at Elevated Temperatures," *Mater. Sci. Eng.*, Vol. 82, 187–195, 1986.

Chuang, T.-J., and Wiederhorn, S. M., "Damage-Enhanced Creep in a Siliconized Silicon Carbide: Mechanics of Deformation," *J. Am. Ceram. Soc.*, Vol. 71, 595–601, 1988.

CIAC Newsletter, Ceramics Information Analysis Newsletter, Vol. 4, No. 3, CINDAS Purdue University, June, 1994

Clarke, D. R., "Microfracture in Brittle Solids Resulting from Anisotropic Shape Changes," *Acta Metall.*, Vol. 28, 913–924, 1980.

Clarke, F. J. P., "Residual Strain and the Fracture Stress–Grain Size Relationship in Brittle Solids," *Acta Metall.*, Vol. 12, 139–143, 1964.

Claussen, N., "Fracture Toughness of Al_2O_3 with an Unstabilized ZrO_2 Dispersed Phase," *J. Am. Ceram. Soc.*, Vol. 59, 49–51, 1976.

Claussen, N., "Microstructural Design of Zirconia-Toughened Ceramics," pp. 325–351 in Claussen, N., Ruhle, M., and Heuer, A., editors, *Science and Technology of Zirconia II*, Advances in Ceramics Vol. 12, American Ceramic Society, Westerville, OH, 1983.

Claussen, N., Ruhle, M., and Heuer, A., editors, *Science and Technology of Zirconia II*, Advances in Ceramics Vol. 12, American Ceramic Society, Westerville, OH, 1983.

Claussen, N., Steeb, J., and Pabst, R. F., "Effect of Induced Microcracking on the Fracture Toughness of Ceramics," *Am. Ceram. Soc. Bull.*, Vol. 56, 559–562, 1977.

Coble, R. L., "A Model for Boundary Diffusion Controlled Creep in Polycrystalline Materials," *J. Appl. Phys.*, Vol. 34, 1679–1682, 1963.

Cook, R. F., and Clarke, D. R., "Fracture Stability, R-Curves and Strength Variability," *Acta Metall.*, Vol. 36, 555–562, 1988.

Cook, R. F., and Lawn, B. R., "A Modified Indentation Toughness Technique," *J. Am. Ceram. Soc.*, Vol. 66, C200–C201, 1983.

Cook, R. F., Lawn, B. R., and Fairbanks, C. J., "Microstructure–Strength Properties in Ceramics: I, Effect of Crack Size on Toughness," *J. Am. Ceram. Soc.*, Vol. 68, 604–615, 1985a.

Cook, R. F., Lawn, B. R., and Fairbanks, C. J., "Microstructure–Strength Properties in Ceramics: II, Fatigue Relations," *J. Am. Ceram. Soc.*, Vol. 68, 616–623, 1985b.

Cook, R. F., Liniger, E. G., Steinbrech, R. W., and Deuerler, F., "Sigmoidal Indentation–Strength Characteristics of Polycrystalline Alumina," *J. Am. Ceram. Soc.*, Vol. 77, 303–314, 1994.

Cook, R. F., and Pharr, G. M., "Direct Observation and Analysis of Indentation Cracking in Glasses and Ceramics," *J. Am. Ceram. Soc.*, Vol. 73, 787–817, 1990.

Corman, G. S., "Creep of 6H α-Silicon Carbide Single Crystals," *J. Am. Ceram. Soc.*, Vol. 75, 3421–3424, 1992.

Cottrell, A. H., *Dislocations and Plastic Flow in Crystals*, Clarendon Press, Oxford, 1953.

Courtney, T. H., *Mechanical Behavior of Materials*, McGraw-Hill, New York, 1990.

Cox, B. N., "Extrinsic Factors in the Mechanics of Bridged Cracks," *Acta Metall. Mater.*, Vol. 39, 1189–1201, 1991.

Cox, B. N., and Lo, C. S., "Load Ratio, Notch, and Scale Effects for Bridged Cracks in Fibrous Composites," *Acta Metall. Mater.*, Vol. 40, 69–80, 1992.

Cox, B. N., Marshall, D. B., and Thouless, M. D., "Influence of Statistical Fiber Strength Distribution on Matrix Cracking in Fiber Composites," *Acta Metall.*, Vol. 37, 1933–1943, 1989.

Creyke, W. E. C., Sainsbury, I. E. J., and Morrell, R., *Design with Non-ductile Materials*, Applied Science Publishers, London, 1982.

Cui, Y. L., "Interaction of Fiber Toughening and Transformation Toughening," *J. Mech. Phys. Solids*, Vol. 40, 1837–1850, 1992.

Cui, Y. L., and Budiansky, B., "Steady-State Matrix Cracking of Ceramics Reinforced by Aligned Fibers and Transforming Particles," *J. Mech. Phys. Solids*, Vol. 41, 615–630, 1993.

Dalgleish, B. J., Johnson, S. M., and Evans, A. G., "High-Temperature Failure of Polycrystalline Alumina: I, Crack Nucleation," *J. Am. Ceram. Soc.*, Vol. 67, 741–750, 1984.

Darroudi, T., Tressler, R. E., and Kasprzyk, M. R., "Low-Cost Melt-Formed Siliconized Silicon Carbide Radiant Tube Materials," *J. Am. Ceram. Soc.*, Vol. 76, 173–179, 1993.

Dauskardt, R. H., James, M. R., Porter, J. R., and Ritchie, R. O., "Cyclic Fatigue-Crack Growth in a SiC-Whisker-Reinforced Alumina Ceramic Composite: Long- and Short-Crack Behavior," *J. Am. Ceram. Soc.*, Vol. 75, 759–771, 1992.

Dauskardt, R. H., Marshall, D. B., and Ritchie, R. O., "Cyclic Fatigue-Crack Propagation in Magnesia-Partially-Stabilized Zirconia Ceramics," *J. Am. Ceram. Soc.*, Vol. 73, 893–903, 1990.

Davidge, R. W., "Mechanical Properties of Reaction Bonded Silicon Nitride," pp. 541–560 in Riley, F. L., editor, *Nitrogen Ceramics*, Nordhoff International Publishing, Leyden, 1977.

Davidge, R. W., *Mechanical Behavior of Ceramics*, Cambridge University Press, New York, 1979.

Davidge, R. W., McLaren, J. R., and Tappin, B., "Strength–Probability–Time (SPT) Relationships in Ceramics," *J. Mater. Sci.*, Vol. 8, 1699–1705, 1973.

Davidge, R. W., and Tappin, G., "Thermal Shock and Fracture in Ceramics," *Trans. Br. Ceram. Soc.*, Vol. 68, 405–422, 1967.

Davidge, R. W., and Tappin, G., "Internal Strain Energy and the Strength of Brittle Materials," *J. Mater. Sci.*, Vol. 3, 297–301, 1968.

de With, G., and Wagemans, H. H. M., "Ball-on-Ring Test Revisited," *J. Am. Ceram. Soc.*, Vol. 72, 1538–1541, 1989.

Dickson, R. W., and Wachtman, J. B., "An Alumina Standard Reference Material for Resonance Frequency and Dynamic Elastic Moduli Measurement I. For Use at 25 °C," *J. Res. Natl. Bur. Stand.*, Vol. 75A, 155–162, 1971.

Dieter, G. E., *Mechanical Metallurgy*, 3rd ed., McGraw-Hill, New York, 1986.

Doremus, R. H., *Glass Science*, Wiley, New York, 1973.

Doremus, R. H., "Fracture and Fatigue of Glass," pp. 169–239 in Tomozawa, M., and Doremus, R. H., editors, *Treatise on Materials Science and Technology, Vol. 22 Glass III*, Academic Press, New York, 1982.

Dörre, E., and Hubner, H., *Alumina*, Springer-Verlag, Berlin, 1984.

Ekstrom, T., and Nygren, M., "SiAlON Ceramics," *J. Am. Ceram. Soc.*, Vol. 75, 259–276, 1992.

Evans, A. G., "The Strength of Brittle Materials Containing Second Phase Dispersions," *Philos. Mag.*, Vol. 26, 1327–1344, 1972.

Evans, A. G., "A Simple Method for Evaluating Slow Crack Growth in Brittle Materials," *Intl. J. Fract.*, Vol. 9, 267–275, 1973.

Evans, A. G., "A General Approach for the Statistical Analysis of Multiaxial Fracture," *J. Am. Ceram. Soc.*, Vol. 61, 302–308, 1978a.

Evans, A. G., "Microfracture from Thermal Expansion Anisotropy—I. Single Phase Systems," *Acta. Metall.*, Vol. 26, 1845–1853, 1978b.

Evans, A. G., "A Dimensional Analysis of the Grain-Size Dependence of Strength," *J. Am. Ceram. Soc.*, Vol. 63, 115–116, 1980.

Evans, A. G., "Perspective on the Development of High-Toughness Ceramics," *J. Am. Ceram. Soc.*, Vol. 73, 187–206, 1990.

Evans, A. G., and Blumenthal, W., "High Temperature Failure in Ceramics," pp. 423–448 in Bradt, R. C., Evans, A. G., Hasselman, D. P. H., and Lange, F. F., editors, *Fracture Mechanics of Ceramics*, Vol. 6, Plenum Press, New York, 1983.

Evans, A. G., and Cannon, R. M., "Toughening of Brittle Solids by Martensitic Transformations," *Acta Metall.*, Vol. 34, 761–800, 1986.

Evans, A. G., and Charles, E. A., "Fracture Toughness Determinations by Indentation," *J. Am. Ceram. Soc.*, Vol. 59, 371–372, 1976.

Evans, A. G., and Dalgleish, B. J., "Some Aspects of the High Temperature Performance of Ceramics and Ceramic Composites," pp. 929–955 in Wilshire, B., and Evans, R. W., editors, *Creep and Fracture of Engineering Materials and Structures*, The Institute of Metals, London, 1987.

Evans, A. G., Domergue, J.-M., and Vagaggini, E., "Methodology for Relating the Tensile Constitutive Behavior of Ceramic-Matrix Composites to Constituent Properties," *J. Am. Ceram. Soc.*, Vol. 77, 1425–1435, 1994.

Evans, A. G., and Faber, K. T., "Crack-Growth Resistance of Microcracking Brittle Materials," *J. Am. Ceram. Soc.*, Vol. 64, 255–260, 1984.

Evans, A. G., and Fuller, E. R., "Crack Propagation in Ceramic Materials Under Cyclic Loading Conditions," *Metall. Trans.*, Vol. 5, 27–33, 1974.

Evans, A. G., and Heuer, A. H., "Review—Transformation Toughening in Ceramics: Martensitic Transformations in Crack-Tip Stress Fields," *J. Am. Ceram. Soc.*, Vol. 68, 241–253, 1980.

Evans, A. G., and Langdon, T. G., "Structural Ceramics," *Prog. Mater. Sci.*, Vol. 21, 171–441, 1976.

Evans, A. G., and Marshall, D. B., " The Mechanical Behavior of Ceramic Matrix Composites," *Acta Metall.*, Vol. 37, 2567–2583, 1989.

Evans, A. G., and McMeeking, R. M., "On Toughening of Ceramics by Strong Reinforcements," *Acta Metall.*, Vol. 34, 2435–2441, 1986.

Evans, A. G., and Wiederhorn, S. M., "Proof Testing of Ceramic Materials—An Analytical Basis for Failure Prediction," *Int. J. Fract.*, Vol. 10, 379–392, 1974.

Evans, R. W., and Wilshire, B., *Introduction to Creep*, The Institute of Metals, London, 1993.

Faber, K. T., and Evans, A. G., "Crack Deflection Processes—I. Theory," *Acta Metall.*, Vol. 31, 565–576, 1983a.

Faber, K. T., and Evans, A. G., "Crack Deflection Processes—II. Experiment," *Acta Metall.*, Vol. 31, 577–584, 1983b.

Ferber, M. K., Jenkins, M. G., Nolan, T. A., and Yeckley, R. L., "Comparison of the Creep and Creep Rupture Performance of Two HIPed Silicon Nitride Ceramics," *J. Am. Ceram. Soc.*, Vol. 77, 657–665, 1994.

Ferber, M. K., Tennery, V., Waters, S., and Ogle, J., "Fracture Strength Characterization of Tubular Ceramic Materials Using a Simple C-ring Geometry," *J. Mater. Sci.*, Vol. 21, 2628–2632, 1980.

Ferguson, W. J., and Rice, R. W., "Effect of Microstructure on the Ballistic Performance of Alumina," pp. 261–270 in Kriegel, W. W., editor, *Materials Science Research, Vol. 5, Ceramics in Severe Environments*, Plenum Press, New York, 1971.

Fett, T., and Munz, D. "Subcritical Crack Growth of Macrocracks in Alumina with R-Curve Behavior," *J. Am. Ceram. Soc.*, Vol. 75, 958–963, 1992.

Fett, T., and Munz, D., "Lifetime Prediction for Ceramic Materials Under Constant and Cyclic Load," pp. 161–174 in Brinkman, C. R., and Duffy, S. F., editors, *Life Prediction Methodologies and Data for Ceramic Materials*, ASTM STP 1201, American Society for Testing and Materials, Philadelphia, 1994.

Firestone, R. F., and Heuer, A. H., "Creep Deformation of 0° Sapphire," *J. Am. Ceram. Soc.*, Vol. 59, 24–29, 1976.

Flinn, B. D., Lo, C. S., Zok, F. W., and Evans, A. G., "Fracture Resistance Characteristics of a Metal-Toughened Ceramic," *J. Am. Ceram. Soc.*, Vol. 76, 369–375, 1993.

Frechette, V. D., *Failure Analysis of Brittle Materials*, American Ceramic Society, Westerville, OH, 1990.

Frenkel, J., "Zur Theorie der Elastizitatsgrenze und der Festigkeit kristallinischerKorper," *Z. Phys.*, Vol. 37, 572–609, 1926.

Frost, H. J., and Ashby, M. F., *Deformation-Mechanism Maps—The Plasticity and Creep of Metals and Ceramics*, Pergamon Press, New York, 1982.

Fulcher, G. S., "Analysis of Recent Measurements of the Viscosity of Glasses," *J. Am. Ceram. Soc.*, Vol. 8, 339–355, 1925.

Fuller, E. R. Jr., Freiman, S. W., Quinn, J. B., Quinn, G. D., and Carter, W. C., "Fracture Mechanics Approach to the Design of Glass Aircraft Windows: A Case Study," *Proc. Soc. Photo-Opt. Instrum. Eng.*, Vol. 2286, 419–430, 1994.

Fuller, E. R., Lawn, B. R., and Cook, R. F., "Theory of Fatigue for Brittle Flaws Originating from Residual Stress Concentrations," *J. Am. Ceram. Soc.*, Vol. 66, 314–321, 1993.

Fuller, E. R. Jr., Wiederhorn, S. M., Ritter, J. E. Jr., and Oates, P. B., "Proof Testing of Ceramics, Part 2 Theory," *J. Mater. Sci.*, Vol. 15, 2282–2295, 1980.

Fullman, R. L., "Measurement of Particle Sizes in Bodies," *Trans. AIME*, Vol. 197, 447–452, 1953.

Gandhi, C., and Ashby, M. F., "Fracture Mechanism Maps for Materials Which Cleave: F.C.C., B.C.C. and H.C.P. Metals and Ceramics," *Acta Metall.*, Vol. 27, 1565–1602, 1979.

Gardon, R., "Thermal Tempering of Glass," pp. 145–216 in Uhlmann, D. R., and Kreidl, N. J., editors, *Glass Science and Technology Volume 5 Elasticity and Strength in Glasses*, Academic Press, New York, 1980.

Garvie, R. C., Hannink, R. H., and Pascoe, R. T., "Ceramic Steel?," *Nature (London)*, Vol. 285(5537), 703–704, 1975.

Garvie, R. C., Hannink, R. H. J., and Pascoe, R. T., U.S. Patent 4,067,745, 1978.

Garvie, R. C., Hannink, R. H. J., and Urbani, C., *Ceramurgia Int.*, Vol. 6, 19–24, 1980.

Geipel, T., Bilde-Sørensen, J. B., Lawler, B. F., Pirouz, P., Lagerlöf, K. D. P., and Heuer, A. H., "On Basal Slip and Basal Twinning in Sapphire (α-Al_2O_3): III. HRTEM of the Twin/Matrix Interface," *Acta Metall.*, in press, 1995.

Geller, R. F., and Yavorsky, P. J., "Effects of Some Oxide Additions on the Thermal Length Changes of Zirconia," *J. Res. Natl. Bur. Stand.*, Vol. 35, 87–110, 1945.

Giannakopoulos, A. E., and Breder, K., "Synergism of Toughening Mechanisms in Whisker-Reinforced Ceramic-Matrix Composites," *J. Am. Ceram. Soc.*, Vol. 74, 194–202, 1991.

Gibson, L. J., and Ashby, M. F., *Cellular Solids—Structure and Properties*, Pergamon Press, New York, 1988.

Gilman, J. J., "The Strength of Ceramic Crystals," p. 240 in Klingsberg, C., editor, *The Physics and Chemistry of Ceramics*, Gordon & Breach, New York, 1963.

Gilman, J. J., and Johnston, W. G., "Behavior of Individual Dislocations in Strain-Hardened LiF Crystals," *J. Appl. Phys.*, Vol. 31, 687–692, 1960.

Gordon, R. S., "Mass Transport in the Diffusional Creep of Ionic Solids," *J. Am. Ceram. Soc.*, Vol. 56, 147–152, 1973.

Gordon, R. S., "Ambipolar Diffusion and Its Application to Diffusional Creep," pp. 445–464 in Cooper, A. R., and Heuer, A. H., editors, *Mass Transport Phenomena in Ceramics*, Plenum, New York, 1975.

Grain, C. F., "Phase Relations in the ZrO_2–MgO System," *J. Am. Ceram. Soc.*, Vol. 50, 288–290, 1967.

Grathwohl, G., "Creep and Fracture of Hot-Pressed Silicon Nitride with Natural and Artificial Flaws," pp. 565–577 in Wilshire, B., and Owen, R., editors, *Creep and Fracture of Engineering Materials and Structures*, Pineridge, Swansen, U.K., 1984.

Green, D. J., "Critical Microstructures for Microcracking in Al_2O_3–ZrO_2 Composites," *J. Am. Ceram. Soc.*, Vol. 65, 610–614, 1982.

Green, D. J., Hannink, R. H. J., and Swain, M. V., *Transformation Toughening of Ceramics*, CRC Press, Boca Raton, FL, 1989.

Green, D. J., Nicholson, P. S., and Embury, G. H., "Fracture of a Brittle Particulate Composite—Part 2 Theoretical Aspects," *J. Mater. Sci.*, Vol. 14, 1657–1661, 1979.

Griffith, A. A., "The Phenomena of Rupture and Flow in Solids," *Philos. Trans. R. Soc. Lond.*, Vol. A221, 163–198, 1920.

Griffith, A. A., "The Theory of Rupture," p. 55 in Biezeno, C. B., and Burgers, J. M., editors, *Proceedings of the First International Congress on Applied Mechanics*, 1924.

Groves, G. W., and Kelly, A., "Independent Slip Systems in Crystals," *Philos. Mag.*, Vol. 8, 877–887, 1963.

Gui, F., Li, M., and Reece, M. J., "Role of Crack-Bridging Ligaments in the Cyclic Fatigue Behavior of Alumina," *J. Am. Ceram. Soc.*, Vol. 75, 2976–2984, 1992.

Guiberteau, F., Padture, N. P., Cai, H., and Lawn, B. R., "Indentation Fatigue: A Simple Cyclic Hertzian Test for Measuring Damage Accumulation in Polycrystalline Ceramics," *Philos. Mag.*, Vol A68, 1003–1016, 1993.

Guiberteau, F., Padture, N. P., and Lawn, B. R., "Effect of Grain Size on Hertzian Contact Damage in Alumina," *J. Am. Ceram. Soc.*, Vol. 77, 1825–1831, 1994.

Hall, E. O., "The Deformation and Ageing of Mild Steel: III Discussion of Results," *Proc. Phys. Soc.*, Vol. B64, 747–753, 1951.

Hamilton, G. M., and Goodman, L. E., "The Stress Field Created by a Circular Sliding Contact," *J. Appl. Mech.*, Vol. 6, 371–376, 1966.

Hannink, R. H. J., "Microstructural Development of Sub-eutectoid Aged MgO–ZrO_2 Alloys," *J. Mater. Sci.*, Vol. 18, 457–470, 1983.

Hannink, R. H. J., and Swain, M. V., "Magnesia–Partially Stabilized Zirconia: The Influence of Heat Treatment on Thermomechanical Properties," *J. Aust. Ceram. Soc.*, Vol. 18, 53–62, 1983.

Hannink, R. H. J., and Swain, M. V., "Progress in Transformation Toughening of Ceramics," pp. 359–408 in Flemings, M. C., Wachtman, J. B. Jr., Kaufmann, E. N., and Giordmaine, J. A., editors, *Annual Review of Materials Science, Keynote Topic: Structural Materials*, Vol. 24, Annual Reviews, Palo Alto, CA, 1994.

Hashida, T., and Li, V. C., "New Development of the J-Based Fracture Testing Technique for Ceramic-Matrix Composites," *J. Am. Ceram. Soc.*, Vol. 77, 1553–1561, 1994.

Hashin, Z., and Rosen, B. W., "The Elastic Moduli of Fiber-Reinforced Materials," *J. Appl. Mech.*, Vol. 33, 223, 1964.

Hasselman, D. P. H., "On the Porosity Dependence of the Elastic Moduli of Polycrystalline Refractory Materials," *J. Am. Ceram. Soc.*, Vol. 45, 452, 1962.

Hasselman, D. P. H., "Unified Theory of Thermal Shock Fracture Initiation and Crack Propagation in Brittle Ceramics," *J. Am. Ceram. Soc.*, Vol. 52, 600–604, 1969.

Hasselman, D. P. H., "Thermal Stress Crack Stability and Propagation in Severe Thermal Environments," pp. 89–103 in Kriegel, W. W., and Palmour, Hayne III, editors, *Ceramics in Severe Environments*, Plenum Press, New York, 1971.

Hasselman, D. P. H., "Thermal Stress Resistance of Engineering Ceramics," *Mater. Sci. Eng.*, Vol. 71, 251–264, 1985.

Hasselman, D. P. H., and Fulrath, R. M., "Proposed Fracture Theory of Dispersion-Strengthened Glass Matrix," *J. Am. Ceram. Soc.*, Vol. 49, 68–72, 1966.

Hasselman, D. P. H., and Venkateswaran, A., "Effect of Cracks on Mechanisms and Kinetics of Creep Deformation of Brittle Ceramics," pp. 525–546 in Tressler, R. E., and Bradt, R. C., editors, *Deformation of Ceramic Materials, II*, Plenum Press, New York, 1984.

Hayhurst, D., Leckie, F. A., and Evans, A. G., "Component Design-Based Model for Deformation and Rupture of Tough Fiber-Reinforced Metal and Ceramic-Matrix Composites," *Proc. R. Soc. London*, Vol. A434, 369–381, 1991.

Hazzledine, "Work Hardening in Easy Glide," *Can. J. Phys.*, Vol. 45, 765–775, 1967.

He, M.-Y., and Hutchinson, J. W., "Kinking of a Crack Out of an Interface," *J. Appl. Mech.*, (*Trans. ASME*), Vol. 56, 270–278, 1989.

Hearmon, R. F. S., *An Introduction to Applied Anisotropic Elasticity*, Oxford University Press, New York, 1961.

Hearmon, R. F. S., *The Elastic Constants of Nonpiezo Crystals*, Landolt–Bornstein Tables, New Series, Group III, Vol. I, 1966.

Herring, C., "Diffusional Viscosity of a Polycrystalline Solid," *J. Appl. Phys.*, Vol. 21, 437–445, 1950.

Hertzberg, R. W., *Deformation and Fracture Mechanics of Engineering Materials*, 3rd ed., Wiley, New York, 1989.

Heuer, A. H., "Deformation Twinning in Corundum," *Philos. Mag.*, Vol. 13, 379–393, 1966.

Heuer, A. H., "Transformation Toughening in ZrO_2-Containing Ceramics," *J. Am. Ceram. Soc.*, Vol. 70, 689–698, 1987.

Heuer, A. H., Claussen, N., Kriven, W. M., and Ruhle, W. M., "Stability of Tetragonal ZrO_2 Particles in Ceramic Matrices," *J. Am. Ceram. Soc.*, Vol. 65, 642–650, 1982.

Heuer, A. H., and Hobbs, L. W., *Science and Technology of Zirconia*, Advances in Ceramics, Vol. 3, American Ceramic Society, Westerville, OH, 1981.

Heuer, A. H., Lange, F. F., Swain, M. V., and Evans, A. G., "Transformation Toughening: An Overview," *J. Am. Ceram. Soc.*, Vol. 69, i–iv, 1986.

Heuer, A. H., Lanteri, V., and Dominguez-Rodriguez, A., "High Temperature Precipitation Hardening of Y_2O_3 Partially-Stabilized ZrO_2 (Y-PSZ) Crystals, *Acta Metall.*, Vol. 37, 559–567, 1989.

Heuer, A. H., Tighe, N. J., and Cannon, R. M., "Plastic Deformation of Fine-Grained Alumina (Al_2O_3): II, Basal Slip and Nonaccommodated Grain-Boundary Sliding," *J. Am. Ceram. Soc.*, Vol. 63, 53–58, 1980.

Hillig, W. S. "A Methodology for Estimating the Mechanical Properties of Oxides at High Temperatures," *J. Am. Ceram. Soc.*, Vol. 76, 129–138, 1993.

Hillig, W. S., and Charles, R. J., pp. 682–705 in Zackay, V. F., editor, *High Strength Materials*, Wiley, New York, 1965.

Hirth, J. P., and Lothe, J., *Theory of Dislocations*, 2nd ed., Wiley, New York, 1982.

Hockey, B. J., "Plastic Deformation of Aluminum Oxide by Indentation and Abrasion," *J. Am. Ceram. Soc.*, Vol. 54, 223–231, 1971.

Hockey, B. J., and Wiederhorn, S. M., "Effect of Microstructure on the Creep of Siliconized Silicon Carbide," *J. Am. Ceram. Soc.*, Vol. 75, 1822–1830, 1992.

Hollenberg, G. W., Terwilliger, G. R., and Gordon, R. S., "Calculation of Stresses and Strains in Four-Point Bending Creep Tests," *J. Am. Ceram. Soc.*, Vol. 54, 196–199, 1971.

Holloway, D. G., "A Look at the History of Glass Strength," pp. 1–18 in Kurkjian, C. R., editor, *Strength of Inorganic Glass*, NATO Conference Series, VI, Materials Science, Vol. 11, Plenum Press, New York, 1983.

Hori, S., Yoshimura, M., and Somiya, S., "Strength–Toughness Relations in Sintered and Isostatically Hot-Pressed ZrO_2-Toughened Al_2O_3," *J. Am. Ceram. Soc.*, Vol. 69, 169–172, 1986.

Hu, X.-Z., and Mai, Y.-W., "Crack-Bridging Analysis for Alumina Ceramics Under Monotonic and Cyclic Loading," *J. Am. Ceram. Soc.*, Vol. 75, 848–893, 1992.

Hu, X.-Z., Mai, Y.-W., and Cotterell, B., "A Statistical Theory of Time-Dependent Fracture for Brittle Materials," *Philos. Mag. A*, Vol. 58, 299–324, 1988.

Hüebner, H., and Jillek, W., "Sub-critical Crack Extension and Crack Resistance in Polycrystalline Alumina," *J. Mater. Sci.*, Vol. 12, 117–125, 1977.

Hull, D., *Introduction to Dislocations*, Pergamon Press, New York, 1965.

Hull, D., and Bacon, D. J., *Introduction to Dislocations*, 3rd ed., Pergamon Press, New York, 1982.

Hutchinson, J. W., "Crack Tip Shielding by Micro-cracking in Brittle Solids," *Acta Metall.*, Vol. 35, 1605–1619, 1987.

Ingel, R. P., Lewis, D., Bender, B. A., and Rice, R. W., "Physical, Microstructural, and Thermomechanical Properties of ZrO_2 Single Crystals," pp. 408–414 in Claussen, N., Ruhle, M., and Heuer, A., editors, *Science and Technology of Zirconia II*, Advances in Ceramics Vol. 12, American Ceramic Society, Westerville, OH, 1983.

Inglis, C. E., "Stresses in a Plate Due to the Presence of Cracks and Sharp Corners," *Trans. Inst. Nav. Archit.*, Vol. 55, 219–230, 1913.

Irwin, G. R., "Analysis of Stresses and Strains Near the End of a Crack Traversing a Plate," *J. Appl. Mech.*, Vol. 24, 361–364, 1957.

ISO Doc. ISO/TC 206, Strategic Policy Statement, April 4, 1994.

Jack, K. H., and Wilson, W. I., "Ceramics Based on the Si-Al-O-N and Related Systems," *Nature (London)*, Vol. 238, 28–29, 1977.

Jackman, E. A., and Roberts, J. P., "Strength of Single-Crystal and Polycrystalline Corundum," *Philos. Mag.*, Vol. 46, 809–811, 1955.

Jacobs, D. S., and Chen, I.-W., "Mechanical and Environmental Factors in the Cyclic and Static Fatigue of Silicon Nitride," *J. Am. Ceram. Soc.*, Vol. 77, 1153–1161, 1994.

Jakus, K., "Proof Testing—A Tool for Quality Assurance," pp. 195–204 in Boyd, D. C., and MacDowell, J. F., editors, *Commercial Glasses*, Advances in Ceramics Vol. 18, American Ceramic Society, Westerville, OH, 1986.

Jakus, K., Coyne, D. C., and Ritter, J. E. Jr., "Analysis of Fatigue Data for Lifetime Predicitons for Ceramic Materials," *J. Mater. Sci.*, Vol. 13, 2071–2080, 1978.

Jakus, K., and Ritter, J. E. Jr., "Lifetime Prediction for Ceramics Under Random Loads," *Res. Mechanica*, Vol. 2, 39–52. 1981.

Jakus, K., Ritter, J. E. Jr., Service, T., and Sonderman, D., "Evaluation of Bimodal Concurrent Flaw Distributions," *J. Am. Ceram. Soc.*, Vol. 64, C-174–C-175, 1981.

Jakus, K., Ritter, J. E. Jr., and Sullivan, J. M., "Dependency of Fatigue Predictions on the Form of the Crack Velocity Equation," *J. Am. Ceram. Soc.*, Vol. 64, 372–374, 1981.

Jakus, K., and Wiederhorn, S. M., "Creep Deformation of Ceramics in Four-Point Bending," *J. Am. Ceram. Soc.*, Vol. 71, 832–836, 1988.

Japan Fine Ceramics Association, *FC Annual Report for Overseas Readers*, 1987.

Jayatilaka, Ayal de S., *Fracture of Engineering Brittle Materals*, Applied Science Publishers, London, 1979.

Johnson, C. A., "Fracture Statistics of Multiple Flaw Distributions," pp. 365–386 in Bradt, R. C., Evans, A. G., Hasselman, D. P. H., and Lange, F. F., editors, *Fracture Mechanics of Ceramics, Volume 5 Surface Flaws, Statistics, and Microcracking*, Plenum Press, New York, 1983.

Johnson, C. A., and Tucker, W. T., "Weibull Estimators for Pooled Fracture Data," pp. 250–264 in Brinkman, C. R., and Duffy, S. F., editors, *Life Prediction Methodologies and Data for Ceramic Materials*, ASTM STP 1201, American Society for Testing and Materials, Philadelphia, 1993.

Johnson, S. M., Dalgleish, B. J., and Evans, A. G., "High-Temperature Failure of Polycrystalline Alumina: III, Failure Times," *J. Am. Ceram. Soc.*, Vol. 67, 759–763, 1984.

Johnston, W. G., and Gilman, J. J., "Dislocation Velocities, Dislocation Densities, and Plastic Flow in Lithium Fluoride Crystals," *J. Appl. Phys.*, Vol. 30, 129–144, 1959.

Kajihara, K., Yoshizawa, Y., and Sakuma, T., "Superplastic Behavior of a Fine-Grained Yttria-Stabilized, Tetragonal Zirconia Polycrystal," *Acta Metall. Mater.*, Vol. 38, 1121, 1990.

Karandikar, P. G., Chou, T.-W., and Parvizi-Majidi, A., "Mechanical Properties," in *Handbook on Continuous Fiber Reinforced Ceramic Matrix Composites*, Lehman, R. L., technical editor, El-Rahaiby, S. K., managing editor, and Wachtman, J. B., American Ceramic Society editor, jointly published by Purdue University and the American Ceramic Society, 1996.

Kelly, A., and MacMillan, N, H., *Strong Solids*, 3rd edition, Oxford University Press, Oxford, 1986.

Kendall, K., Alford, N. M., and Birchall, J. D., "Weibull Modulus of Toughened Ceramics," *Mater. Res. Soc. Symp. Proc.*, Vol. 78, 181–187, 1987.

Keyes, B. L. P., "Ceramic Technology Project Database: September 1993 Summary Report," Oak Ridge National Laboratory Report ORNL/M-3155, January 1994. Also personal communication with the author, August 1994.

King, A. G., and Yavorsky, P. J., "Stress Relief Mechanisms in Magnesia- and Yttria-Stabilized Zirconia," *J. Am. Ceram. Soc.*, Vol. 51, 38–42, 1968.

Kirchner, H. P., *Strengthening of Ceramics—Treatments, Tests and Design Applications*, Marcel Dekker, New York, 1979.

Kirchner, H. P., and Conway, J. C. Jr., "Criteria for Crack Branching in Cylindrical Rods: I, Tension," *J. Am. Ceram. Soc.*, Vol. 70, 413–418, 1987a.

Kirchner, H. P., and Conway, J. C. Jr., "Criteria for Crack Branching in Cylindrical Rods: II, Flexure," *J. Am. Ceram. Soc.*, Vol. 70, 419–425, 1987b.

Kirchner, H. P., and Kirchner, J. W., "Fracture Mechanics of Fracture Mirrors," *J. Am. Ceram. Soc.*, Vol. 62, 198–202, 1979.

Kittel, C., *Introduction to Solid State Physics*, Wiley, New York, 1953.

Knehens, R., and Steinbrech, R., "Memory Effect of Crack Resistance During Slow Crack Growth in Notched Al_2O_3 Bend Specimens," *J. Mater. Sci. Lett.*, Vol. 1, 327–329, 1982.

Knowles, K. M., and Yang, X. F., "Mathematical Modeling of the Strength and Toughness of Unidirectional Fiber-Reinforced Ceramics," *Ceram. Eng. Sci. Proc.*, Vol. 12, 1387–1388, 1991.

Knudsen, F. P., "Effect of Porosity on Young's Modulus of Alumina," *J. Am. Ceram. Soc.*, Vol. 45, 94–95, 1962.

Kovar, D., and Readey, M. J., "Role of Grain Size in Strength Variability of Alumina," *J. Am. Ceram. Soc.*, Vol. 77, 1928–1938, 1994.

Krause, J. T., Testardi, L. R., and Thurston, R. N., "Deviations from Linearity in the Dependence of Elongation upon Force for Fibres of Simple Glass Formers and of Glass Optical Lightguides," *Phys. Chem. Glasses*, Vol. 20, 135–139, 1979.

Krause, R. F. Jr., "Rising Fracture Toughness from the Bending Strength of Indented Alumina Beams," *J. Am. Ceram. Soc.*, Vol. 71, 338–343, 1988.

Kriven, W. M., "Possible Alternative Transformation Tougheners to Zirconia: Crystallographic Aspects," *J. Am. Ceram. Soc.*, Vol. 71, 1021–1030, 1988.

Krohn, D. A., and Hasselman, D. P. H., "Relation of Flaw Size to Mirror in the Fracture of Glass," *J. Am. Ceram. Soc.*, Vol. 54, 411, 1971.

Kronberg, M. L., "Plastic Deformation of Single Crystals of Sapphire: Basal Slip and Twinning," *Acta Metall.*, Vol. 5, 507–524, 1957.

Kronberg, M. L., "Dynamical Flow Properties of Single Crystals of Sapphire: I," *J. Am. Ceram. Soc.*, Vol. 45, 274–279, 1962.

Kurkjian, C. R., *Strength of Inorganic Glass*, NATO Conference Series, VI, Materials Science, Vol. 11, Plenum Press, New York, 1983.

Kurkjian, C. R., "Mechanical Stability of Oxide Glasses," *J. Non-Crystall. Solids*, Vol. 102, 71–81, 1988.

Kuszyk, J. A., and Bradt, R. C., "Influence of Grain Size on Effects of Thermal Expansion Anisotropy on $MgTi_2$," *J. Am. Ceram. Soc.*, Vol. 56, 420–423, 1973.

Lagerlof, K. P. D., Heuer, A. H., Castaing, J., Riviere, J. P. and Mitchell, T. E., "Slip and Twinning in Sapphire (α-Al_2O_3)," *J. Am. Ceram. Soc.*, Vol. 77, 385–397, 1994.

Lagerlof, K. P. D., Lawlor, B. F., Geipel, T., Pirouz, P., Heuer, A. H., and Bilde-Sorensen, J. B., "Basal Twinning in Sapphire (α-Al_2O_3), *Solid State Phenomena*, Vol. 35–36, 607–612, 1994.

Lamon, J., "Statistical Approaches to Failure for Ceramic Reliability Assessment," *J. Am. Ceram. Soc.*, Vol. 71, 106–112, 1988.

Langdon, T. G., "Deformation Mechanism Maps for Applications at High Temperatures," *Ceramurgia Int.*, Vol. 6, 11–88, 1980.

Langdon, T. G., "The Mechanical Properties of Superplastic Materials," *Metall. Trans. A*, Vol. 13A, 689–701, 1982.

Langdon, T. G., "Superplastic Ceramics," *J. Mater.*, Vol. 42, 8–13, 1990.

Langdon, T. G., "The Physics of Superplastic Deformation," *Mater. Sci. Eng.*, Vol. A137, 1–11, 1991a.

Langdon, T. G., "Superplasticity: An Historical Perspective," pp. 3–12 in Hori, S., Tokizane, M., and Furushiro, N., editors, *Superplasticity in Advanced Materials*, the Japan Society for Research on Superplasticity, Tokyo, 1991b.

Langdon, T. G., "The Role of Grain Boundaries in High Temperature Deformation," *Mater. Sci. Eng.*, Vol. A166, 67–79, 1993a.

Langdon, T. G., "The Mechanical Properties of Superplastic Materials," pp. 265–281 in Vincenzini, P., editor, *Innovative Materials, Prospects and Problems in a Competitive Industrial Context*, Techna, Faenza, 1993b.

Langdon, T. G., "An Evaluation of the Strain Contributed by Grain Boundary Sliding in Superplasticity," *Mater. Sci. Eng.*, Vol. A174, 225–230, 1994a.

Langdon, T. G., "A Unified Approach to Grain Boundary Sliding in Creep and Superplasticity," *Acta Metall. Mater.*, Vol. 42, 2437–2443, 1994b.

Langdon, T. G., and Mohamed, F. A., "Deformation Mechanism Maps for Ceramics," *J. Mater. Sci.*, Vol. 11, 317–327, 1976.

Langdon, T. G., and Mohamed, F. A., "Deformation Mechanism Maps for Ceramics," *J. Mater. Sci.*, Vol. 13, 473–482, 1978.

Lange, F. F., "The Interaction of Crack Front with Second-Phase Dispersion," *Philos. Mag.*, 22, 983–992, 1970.

Lange, F. F., "Fracture Energy and Strength Behavior of a Sodium Borosilicate Glass–Al_2O_3 Composite System," *J. Am. Ceram. Soc.*, Vol. 54, 614–620, 1971.

Lange, F. F., "Transformation Toughening, Part 4 Fabrication, Fracture Toughness and Strength of Al_2O_3–ZrO_2 Composites," *J. Mater. Sci.*, Vol. 17, 247–254, 1982a.

Lange, F. F., "Transformation Toughening, Part 5 Effect of Temperature and Alloy on Fracture Toughness," *J. Mater. Sci.*, Vol. 17, 255–263, 1982b.

Lange, F. F. "Powder Processing Science and Technology for Increased Reliability," *J. Am. Ceram. Soc.*, Vol. 71, 3–15, 1989.

Larson, D. C., and Adams, J. W., "Property Screening and Evaluation of Ceramic Turbine Materials," Air Force Materials Laboratory Report AFWAL-TR-83-4141, April 1984.

Larson, D. R., Coppola, J. A., Hasselman, D. P. H., and Bradt, R. C., "Fracture Toughness and Spalling Behavior of High-Al_2O_3 Refractories," *J. Am. Ceram. Soc.*, Vol. 57, 417–421, 1974.

Lathabai, S., Rödel, J, and Lawn, B. R., "Cyclic Fatigue from Frictional Degradation at Bridging Grains in Alumina," *J. Am. Ceram. Soc.*, Vol. 74, 1340–1348, 1991.

Lawn, B. R., *Fracture of Brittle Solids*, Cambridge Press, Cambridge, 1993.

Lawn, B. R., Evans, A. G., and Marshall, D. B., "Elastic/Plastic Indentation Damage in Ceramics: The Median/Radial Crack System," *J. Am. Ceram. Soc.*, Vol. 63, 574–581, 1980.

Lawn, B. R., Freiman, S. W., Baker, T. L., Cobb, D. D., and Gonzalez, A. C., "Study of Microstructural Effects in the Strength of Alumina Using Controlled Flaws," *J. Am. Ceram. Soc.*, Vol. 67, 667–669, 1984.

Lawn, B. R., Padture, N. P., Cai, H., and Guiberteau, F., "Making Ceramics Ductile," *Science*, Vol. 263, 1114–1116, 1994.

Lawn, B. R., Padture, F., Guiberteau, F., and Cai, H. "A Model for Microcrack Initiation and Propagation Beneath Hertzian Contacts in Polycrystalline Ceramics," *Acta Metall. Mater.*, Vol. 42, 1683–1693, 1994.

Lawn, B. R., and Wilshaw, T. R., "Indentation Fracture; Principles and Applications," *J. Mater. Sci.*, Vol. 10, 1049–1081, 1975.

Lehman, R. L., technical editor, El-Rahaiby, S. K., managing editor, and Wachtman, J. B., American Ceramic Society editor, *Handbook on Continuous Fiber Reinforced Ceramic Matrix Composites*, jointly published by Purdue University and the American Ceramic Society, 1996.

Lewis, D. III, "Continuous Fiber Reinforced Ceramic Matrix Composites: A Historical Review," in *Handbook on Continuous Fiber Reinforced Ceramic Matrix Composites*, Lehman, R. L., technical editor, El-Rahaiby, S. K., managing editor, and Wachtman, J. B., American Ceramic Society editor, jointly published by Purdue University and the American Ceramic Society, 1996.

Lewis, D., and Rice, R. W., "Thermal Shock Fatigue of Monolithic Ceramics and Ceramic–Ceramic Particulate Composites," *Ceram. Eng. Sci. Proc.*, Vol. 2, 712–718, 1981.

Li, C.-W., and Yamanis, J., "Super-tough Silicon Nitride with R-Curve Behavior," *Ceram. Eng. Sci. Proc.*, Vol. 10, 632–645, 1989.

Lutz, E. H., Claussen, N., and Swain, M. V., "K^R-Curve Behavior of Duplex Ceramics," *J. Am. Ceram. Soc.*, Vol. 74, 11–18, 1991.

Lutz, E. H., Swain, M. V., and Claussen, N., "Thermal Shock Behavior of Duplex Ceramics," *J. Am. Ceram. Soc.*, Vol. 74, 19–24, 1991.

Ma, Y., and Langdon, T. G., "An Examination of the Implications of Void Growth in Submicrometer and Nanocrystalline Structures," *Mater. Sci. Eng.*, Vol. A168, 225–230, 1993.

Mackenzie, J. K., "The Elastic Constants of a Solid Containing Spherical Holes," *Proc. Phys. Soc.*, Vol. B63, 2, 1950.

Maehara, Y., and Langdon, T. G., "Review Superplasticity in Ceramics," *J. Mater. Sci.*, Vol. 25, 2275–2286, 1990.

Mai, Y.-W., and Lawn, B. R., "Crack-Interface Grain Bridging as a Fracture Resistance Mechanism in Ceramics: II, Theoretical Fracture Mechanics Model," *J. Am. Ceram. Soc.*, Vol. 70, 289–294, 1987.

Mandelbrot, B. B., *The Fractal Geometry of Nature*, W. H. Freeman, New York, 1982.

Marsh, D. M., *Proc. R. Soc.*, "Plastic Flow in Glass," Vol. A279, 420–435, "Plastic Flow and Fracture of Glass," Vol. A282, 33–43, 1964.

Marshall, D. B., "An Improved Biaxial Flexure Test for Ceramics," *Am. Ceram. Soc. Bull.*, Vol. 59, 551–553, 1980.

Marshall, D. B., "Strength Characteristics of Transformation-Toughened Zirconia," *J. Am. Ceram. Soc.*, Vol. 69, 173–180, 1986.

Marshall, D. B., Cox, B. N., and Evans, A. G., "The Mechanics of Matrix Cracking in Brittle-Matrix Fiber Composites," *Acta Metall.*, Vol. 33, 2013–2021, 1985.

Marshall, D. B., and Evans, A. G., "Failure Mechanisms in Ceramic-Fiber/Ceramic-Matrix Composites," *J. Am. Ceram. Soc.*, Vol. 68, 225–231, 1985.

Marshall, D. B. and Evans, A. G., "The Tensile Strength of Uniaxially Reinforced Ceramic Fiber Composites," pp. 1–15 in Bradt, R. C., Evans, A. G. Lange, F. F., and Hasselman, D. P. H., editors, *Fracture Mechanics of Ceramics, Volume 7*, Plenum Press, New York, 1986.

Marshall, D. B., and Evans, A. G., "The Influence of Residual Stress on the Toughness of Reinforced Brittle Materials," *Mater. Forum*, Vol. 11, 304–312, 1988.

Marshall, D. B., Evans, A. G., and Drory, M. D., "Transformation Toughening in Ceramics," pp. 289–307 in Bradt, R. C., Evans, A. G., Hasselman, D. P. H., and Lange, F. F., editors, *Fracture Mechanics of Ceramics, Volume 6, Measurements, Transformations, and High-Temperature Fracture*, Plenum Press, New York, 1983.

Marshall, D. B., Lawn, B. R., and Chantikul, P., "Residual Stress Effects in Sharp-Contact Cracking: I. Indentation Fracture Mechanics, II. Strength Degradation," *J. Mater. Sci.*, Vol. 14, 2110–1012 and 2225–2235, 1979.

Martinez-Fernandez, J., Jimenez-Melendo, M., Dominguez-Rodriguez, A., Cordier, P., Lagerlof, K. P. D., and Heuer, A. H., "High-Temperature Precipitation Hardening in Y_2O_3 Partially-Stabilized ZrO_2 (Y-PSZ) Single Crystals—III. Effect of Solute Composition and Orientation on the Hardening," *Acta Metall. Mater.*, Vol. X, in press, 1995.

Martinez-Fernandez, J., Jimenez-Melendo, M., Dominguez-Rodriguez, A., Lagerlof, K. P. D., and Heuer, A. H., "High-Temperature Precipitation Hardening in Y_2O_3 Partially-Stabilized ZrO_2 (Y-PSZ) Single Crystals—II. A Quantitative Model for the Hardening," *Acta Metall. Mater.*, Vol. 41, 3171–3180, 1993.

Matsui, M., Ishida, Y., Soma, T., and Oda, I., "Ceramic Turbocharger Rotor Design Considering Long Term Durability," pp. 1043–1062 in Bunk, W., and Hausner, H., editors, *Ceramic Materials and Components for Engines—Proceedings of the Second International Symposium*, Verlag Deutsche Keramische Gesellschaft, Bad Honnef, 1986.

Matsuo, Y., and Sasaki, H., "Effect of Grain Size on Microcracking in Lead Titanate Ceramics," *J. Am. Ceram. Soc.*, Vol. 49, 229–230, 1966.

Matthewson, M. J., "Design Practices for Glass and Glass Fibers," pp. 741–745 in Schneider, S. J., volume chairman, *Engineered Materials Handbook, Vol. 4, Ceramics and Glasses*, ASM International, Materials Park, OH, 1991.

Matthewson, M. J., *Optical Fiber Mechanical Testing Techniques*, Critical Reviews of Optical Science and Technology, Vol. CR50, Society of Photo-Optical Instrumentation Engineers, Bellingham, WA, 1994.

McCartney, L. N., "Mechanics of Matrix Cracking in Brittle-Matrix Fiber-Reinforced Composites," *Proc. R. Soc. London*, Vol. A409, 329–350, 1987.

McLean, A. F., and Hartsock, D. L., "Design with Structural Ceramics," pp. 27–98 in Wachtman, J. B. Jr., editor, *Structural Ceramics*, Academic Press, New York, 1989.

McLean, A. F., and Hartsock, D. L., "An Overview of the Ceramic Design Process," pp. 676–689 in Schneider, S. J., volume chairman, *Engineered Materials Handbook, Vol. 4, Ceramics and Glasses*, ASM International, Materials Park, OH, 1991.

McMeeking, R. M., and Evans, A. G., "Mechanics of Transformation-Toughening in Brittle Materials," *J. Am. Ceram. Soc.*, Vol. 65, 242–246, 1982.

McMillan, P. W., *Glass Ceramics*, 2nd ed., Academic Press, New York, 1979.

Mecholsky, J. J., "Fracture Surface Analysis of Optical Fibers," pp. 663–668 in Schneider, S. J., volume chairman, *Engineered Materials Handbook, Vol. 4, Ceramics and Glass*, ASM International, Materials Park, OH, 1991.

Mecholsky, J. J., private communication with the author, 1995.

Mecholsky, J. J. Jr., and Freiman, S. W., "Relationship Between Fractal Geometry and Fractography," *J. Am. Ceram. Soc.*, Vol. 74, 3136–3138, 1991.

Mecholsky, J. J., Freiman, S. W., and Morey, S. M., "Fractographic Analysis of Optical Fibers," *Am. Ceram. Soc. Bull.*, Vol. 56, 1016–1017, 1977.

Mecholsky, J. J., Gonzalez, A. C., and Freiman, S. W., "Fractographic Analysis of Delayed Failure in Soda-Lime Glass," *J. Am. Ceram. Soc.*, Vol. 62, 577–580, 1979.

Mecholsky, J. J., and Powell, S. R. Jr., *Fractography of Ceramic and Metal Failures*, ASTM STP 827, American Society for Testing and Materials, Philadelphia, 1984.

Mecholsky, J. J., Rice, R. W., and Freiman, S. W., "Prediction of Fracture Energy and Flaw Size in Glasses from Measurements of Mirror Size," *J. Am. Ceram. Soc.*, Vol. 57, 440–443, 1974.

Michalske, T. A., "Quantitative Fracture Surface Analysis," pp. 652–662 in Schneider, S. J., volume chairman, *Engineered Materials Handbook, Vol. 4, Ceramics and Glass*, ASM International, Materials Park, OH, 1991.

Michalske, T. A., and Freiman, S. W., "A Molecular Mechanism for Stress Corrosion in Vitreous Silica," *J. Am. Ceram. Soc.*, Vol. 66, 284–288, 1983.

Monkman, F. C., and Grant, N. J., "An Empirical Relationship between Rupture Life and Minimum Creep Rate in Creep-Rupture Tests," *Proc. ASTM*, Vol. 56, 593, 1956.

Morrell, R., *Handbook of Properties of Technical & Engineering Ceramics Part 1 An Introduction for the Engineer and Designer*, Her Majesty's Stationery Office, London, 1985.

Morrell, R., *Handbook of Properties of Technical & Engineering Ceramics Part 2 Data Reviews Section 1 High-Alumina Ceramics*, Her Majesty's Stationery Office, London, 1987.

Morrell, R., and Ashbee, K. H. G., "High Temperature Creep of Lithium Zinc Silicate Glass-Ceramics, Part I, General Behavior and Creep Mechanisms," *J. Mater. Sci.*, Vol. 8, 1253–1270, 1973.

Morscher, G. N., and DiCarlo, J. A., "A Simple Test for Thermomechanical Evaluation of Ceramic Fibers," *J. Am. Ceram. Soc.*, Vol. 75, 136–140, 1992.

Munro, R. G., "Ceramic Properties Database Systems," pp. 690–693 in Schneider, S. J., volume chairman, *Engineered Materials Handbook, Vol. 4, Ceramics and Glasses*, ASM International, Materials Park, OH, 1991.

Munro, R. G., and Hubbard, C. R., "Property Database for Gas-Fired Applications of Ceramics," *Ceram. Bull.*, Vol. 68, 2084–2090, 1989.

Munro, R. G., Hwang, F. Y., and Hubbard, C. R., "The Structural Ceramics Database: Technical Foundations," *J. Res. Natl. Inst. Stand. Tech.*, Vol. 94, 37–47, 1989.

Munz, D., Bubsey, R. T., and Shannon, J. L. Jr., "Fracture Toughness Determination of Al_2O_3 Using Four-Point-Bend Specimens with Straight-Through and Chevron Notches," *J. Am. Ceram. Soc.*, Vol. 63, 300–305, 1980.

Munz, D., Himsolt, G., and Eschweiler, J., "Effect of Stable Crack Growth on Fracture Toughness Determination for Hot-Pressed Silicon Nitride at Elevated Temperatures," pp. 69–84 in Freiman, S. W., and Fuller, E. R., editors, *Fracture Mechanics Methods for Ceramics, Rocks, and Concrete*, ASTM Special Technical Publication 745, American Society for Testing and Materials, Philadelphia, 1981.

Murakami, Y., *Stress Intensity Factors Handbook*, Pergamon Press, New York, 1986.

Nabarro, R. R. N., p. 75 in *Report of a Conference on Strength of Solids*, The Physical Society, London, 1948.

Nabarro, R. R. N., editor, *Dislocations in Solids*, Vols. 1–8, North-Holland, Amsterdam, 1979–1989.

Nemeth, N. N., and Gyekenyesi, J. P., "Probabilistic Design of Ceramic Components with the NASA CARES Computer Program," pp. 700–708 in Schneider, S. J., volume chairman, *Engineered Materials Handbook, Vol. 4, Ceramics and Glasses*, ASM International, Materials Park, OH, 1991.

Newman, J. C. Jr., and Raju, I. S., "An Empirical Stress-Intensity Factor Equation for the Surface Crack," *Eng. Fract. Mech.*, Vol. 15, 185–192, 1981.

Nieh, T. G., and Wadsworth, J., "Superplastic Behavior of a Fine-Grained Yttria-Stabilized, Tetragonal Zirconia Polycrystal," *Acta Metall. Mater.*, Vol. 38, 1121, 1990a.

Nieh, T. G., and Wadsworth, J., "Superplastic Ceramics," *Annu. Rev. Mater. Sci.*, Vol. 20, 117–140, 1990b.

Nielsen, L. F., "Elastic Properties of Two-Phase Materials," *Mater. Sci. Eng.*, Vol. 52, 39–62, 1982.

Nielsen, L. F., "Elasticity and Damping of Porous Materials and Impregnated Materials" *J. Am. Ceram. Soc.*, Vol. 67, 93–98, 1984.

Nixon, R. D., and Davis, R. F., "Diffusion-Accommodated Grain Boundary Sliding and Dislocation Glide in the Creep of Sintered Alpha Silicon Carbide," *J. Am. Ceram. Soc.*, Vol. 75, 1786–1795, 1992.

Nye, J. F., *Physical Properties of Crystals*, Oxford University Press, New York, 1957.

O'Brian, M. H., and Coon, D. N., "Mechanistic Analysis of Time-Dependent Failure of Oxynitride Glass-Joined Silicon Nitride below 1000 °C," *J. Am. Ceram. Soc.*, Vol. 74, 103–108, 1991.

Orowan, E., "Fracture and Strength of Solids," *Rep. Prog. Phys.*, Vol. 12, 185–232, 1949.

Padgett, G. C., "Review of the National and International Standardization of Advanced Ceramics," pp. 363–370 in de With, G., Terpstra, R. A., and Metselaar, R., editors, *Euro-Ceramics*, Vol. 3, Elsevier Applied Science, London, 1989.

Palaniswamy, A., and Knauss, W. G., "On the Problem of Crack Extension in Brittle Solids Under General Loading," *Mech. Today*, Vol. 4, 87–148, 1978.

Paluszny, A., "Proceedings of the Workshop on Ceramics for Advanced Heat Engines," Orlando, FL, January 24–26, 1977, pp. 231–246 in *Energy Research and Development Report*, ERDA *Rep. CONF-770110*, 1977.

Pasto, A. E., Van Schalkwyk, W. C., and Mahoney, F. M., "Creep Behavior of Yttria- and Alumina-Doped Silicon Nitride," pp. 776–785 in Tennery, V. J., editor, *Third International Symposium on Ceramic Components and Materials for Engines*, American Ceramic Society, Westerville, OH, 1989.

Petch, N. J. "The Cleavage Strength of Polycrystals," *J. Inst. Sci. Technol.*, Vol. 173, 25–28, 1953.

Petch, N. J., "Metallographic Aspects of Fracture," p. 351 in Liebowitz, H., editor, *Fracture*, Vol. 1, Academic Press, New York, 1968.

Phelps, G. W., and Wachtman, J. B., "Ceramics General Survey," *Uhlmann's Encyclopedia of Industrial Chemistry*, Vol. A6, 1–42, 1986.

Piroz, P., Lawler, B. F., Geiper, T., Bilde-Sørensen, J. B., Heuer, A. H., and Lagerlöf, K. P. D., "On Basal Slip and Basal Twinning in Sapphire (α-Al_2O_3): II. A New Model of Basal Twinning," *Acta Metall.*, in press, 1995.

Poirier, J.-P., *High Temperature Deformation Processes in Metallurgy, Ceramics, and Minerals*, Cambridge University Press, Cambridge, 1985.

Porter, D. L., and Heuer, A. H., "Microstructural Development in MgO–Partially Stabilized Zicronia (Mg-PSZ)," *J. Am. Ceram. Soc.*, Vol. 62, 298–305, 1979.

Prewo, K. M., "Fiber-Reinforced Ceramics: New Opportunities for Composite Materials," *Ceram. Bull.*, Vol. 68, 395–400, 1989.

Prewo, K. M., and Brennan, J. J., "High Strength Silicon Carbide Fibre Reinforced Glass Matrix Composites," *J. Mater. Sci.*, Vol. 15, 463, 1980.

Pyzik, A. J., and Carroll, D. F., "Technology of Self-Reinforced Silicon Nitride," *Annu. Rev. Mater. Sci.*, Vol. 24, 189–214, 1994.

Quinn, G. D., "Review of Static Fatigue in Silicon Nitride and Silicon Carbide," *Ceram. Eng. Sci. Proc.*, Vol. 3, 77–98, 1982.

Quinn, G. D., *Static Fatigue in High-Performance Ceramics*, ASTM Special Technical Testing Publication 844, 1984a. Report of a conference in San Francisco in 1982.

Quinn, G. D., "Fracture Mechanism Map for Hot-Pressed Silicon Nitride," *Ceram. Eng. Sci. Proc.*, Vol. 5, 596–602, 1984b.

Quinn, G. D., "Fracture Mechanism Maps for Silicon Nitride," pp. 931–939 in Bunk, W., and Hausner, H., editors, *Ceramic Materials and Components for Engines—Proceedings of the Second International Symposium*, Verlag Deutsche Keramische Gesellschaft, 1986.

Quinn, G. D., "Fracture Mechanism Maps for Advanced Structural Ceramics," *J. Mater. Sci.*, Vol. 25, 4361–4376, 1990.

Quinn, G. D., and Braue, W. R., "Fracture Mechanism Maps for Advanced Structural Ceramics, Part 2, Sintered Silicon Nitride," *J. Mater. Sci.*, Vol. 25, 4377–4392, 1990.

Quinn, G. D., and Gettings, R. J., "Fracture Toughness by the Surface Crack in Flexure (SCF) Method: Results of the Vamas Round Robin," *Ceram. Eng. Sci. Proc.*, Vol. 15, 846–855, 1994.

Quinn, G. D., and Morrell, R., "Design Data for Engineering Ceramics: A Review of the Flexure Test," *J. Am. Ceram. Soc.*, Vol. 74, 2037–2066, 1991.

Quinn, G. D., Swab, J. J., and Slavin, M., *Fractography and Characterization of Fracture Origins in Advanced Structural Ceramics*, Military Handbook 790, U.S. Government Printing Office, Washington, DC, 1 July 1992.

Raj, R., "Separation of Cavitation-Strain and Creep-Strain During Deformation," *J. Am. Ceram. Soc.*, Vol. 63, C-46, 1982.

Raj, R., and Dang, C. H., "De-adhesion by Growth of Penny-Shaped Bubbles in an Adhesive Layer," *Philos. Mag.*, Vol. 32, 909–922, 1975.

Ramakrishnan, N., and Arunachalam, V. S., "Effective Elastic Moduli of Porous Ceramic Materials," *J. Am. Ceram. Soc.*, Vol. 76, 2745–2752, 1993.

Rao, S. S., Tsakalakos, T., and Cannon, W. R., "Stress Distributions in Ceramic Composites Containing Faceted Inclusions," *J. Am. Ceram. Soc.*, Vol. 75, 1807–1817, 1992.

Rawson, H., *Glasses and Their Applications*, Instituite of Metals, London, 1991.

Read, W. T. Jr., *Dislocations in Crystals*, McGraw-Hill, New York, 1953.

Reece, M., and Guiu, F., "Indentation Fatigue of High-Purity Alumina in Fluid Environments," *J. Am. Ceram. Soc.*, Vol. 74, 148–154, 1991.

Rice, J. R., "A Path-Independent Integral and the Approximate Analysis of Strain Concentration by Notches and Cracks," *J. Appl. Mech.*, Vol. 35, XX–XX, 1968a.

Rice, J. R., "Mathematical Analysis in the Mechanics of Fracture," pp. 192–308 in Liebowitz, H., editor, *Fracture, An Advanced Treatise*, Vol. 2, Academic Press, New York, 1968b.

Rice, J. R., "Elastic Fracture Mechanics Concepts for Interfacial Cracks," *J. Appl. Mech.*, (*Trans. ASME*), Vol. 55, 98–103, 1988.

Rice, R. W., "Microstructural Dependence of Mechanical Properties," p. 199 in McCrone, R. K., editor, *Treatise on Materials Science and Technology, Vol. 11, Properties and Microstructure*, Academic Press, New York, 1977.

Rice, R. W., "Mechanisms of Toughening in Ceramic Matrix Composites," *Ceram. Eng. Sci. Proc.*, Vol. 2, 661–701, 1981.

Rice, R. W., "Mechanically Reliable Ceramics," *J. Phys. Chem. Solids*, Vol. 45, 1033–1050, 1984.

Rice, R. W., "Ceramic Matrix Composite Toughening Mechanisms: An Update," *Ceram. Eng. Sci. Proc.*, Vol. 6, 589–607, 1985.

Rice, R. W., "Evaluating Porosity Parameters for Porosity-Property Relations," *J. Am. Ceram. Soc.*, Vol. 76, 1801–1808, 1993.

Rice, R. W., "Effects of Environment and Temperature on Ceramic Tensile Strength-Grain Size Relations," *J. Mater. Sci. Technol.*, 1996 (submitted).

Richerson, D. W., "Failure Analysis of Ceramics," pp. 755–757 in *Metals Handbook*, Vol. 11, 9th ed., ASM International, Materials Park, OH, 1986.

Richerson, D. W., *Modern Ceramic Engineering*, 2nd ed., Marcel Dekker, New York, 1992.

Riley, F. L., *Nitrogen Ceramics*, Nordhoff International Publishing, Leyden, 1977.

Ripperger, A., and Davids, N., "Critical Stresses in a Circular Ring," *Trans. Am. Soc. Civ. Eng.*, Vol. 112, 619–628, 1947.

Ritchie, R. O., and Dauskardt, R. H., "Cyclic Fatigue of Ceramics: A Fracture Mechanics Approach to Subcritical Crack Growth and Life Prediciton," *J. Ceram. Soc. Japan, Int. Ed.*, Vol. 99, 1009–1023, 1991.

Ritter, J. E. Jr., "Engineering Design and Fatigue Failure of Brittle Materials," pp. 667–686 in Bradt, R. C., Hasselman, D. P. H., and Lange, F. F., editors, *Fracture Mechanics of Ceramics, Volume 4 Crack Growth and Microstructure*, Plenum Press, New York, 1978.

Ritter, J. E. Jr., Bandyopadhyay, N., and Jakus, K., "Statistical Reproducibility of the Crack Propagation Parameter *N* in Dynamic Fatigue Tests," *J. Am. Ceram. Soc.*, Vol. 62, 542–543, 1979.

Ritter, J. E., Jr., Bandyopadhyay, N., and Jakus, K., "Statistical Reproducibility of the Dynamic and Static Fatigue Experiments," Bulletin of the American Ceramic Society, Vol. 60, 7989–806, 1981.

Ritter, J. E. Jr., Jakus, El, Batakis, A., and Bandyopadhyay, N., "Appraisal of Biaxial Strength Testing," J. Non-Crystalline Solids, Vol. 38–39, 419–424, 1980.

Rodel, Jurgen, "Interaction Between Crack Deflection and Crack Bridging," *J. European Cer. Soc.*, Vol 10, 143–150, 1992.

Ruhle, M., Claussen, N., and Heuer, A. H., "Transformation and Microcrack Toughening as Complementary Processes in ZrO_2 Toughened Al_2O_3," *J. Am. Ceram. Soc.*, Vol. 69, 195, 1986.

Russo, Charles, J., Harmer, Martin P., Chan, Helen M., and Miller, Gary A., "Design of a Laminated Ceramic Composite for Improved Strength and Toughness," *J. Am. Ceram. Soc.*, Vol. 75, 3396–3400, 1992.

Sakai, M., and Bradt, R. C., "Fracture Toughness Testing of Brittle Materials," International Materials Reviews, Vol. 38, 53–78, 1993.

Sakai, Mototsugu, and Kouzou, Yamasaki, "Numerical Fracture Analysis of Chevron-Notched Specimens: I, Shear Correction Factor, k," *J. Am. Ceram. Soc.*, Vol. 66, No. 5, 371–375, 1983. Numerical Fracture Analysis of Chevron-Notched specimens II, Stability Condition for Crack Growth", *J. Am. Ceram. Soc.*, Vol. 66, No. 5, 376–379, 1983.

Sambel, R. A., Bowen, D., and Phillips, D. C., "Carbon Fiber Composites with Ceramic and Glass Matrices, Part I, Discontinuous Fibers" *J. Mater. Sci.*, Vol. 7, 663–675, 1976.

Sbaizero, O., and Evans, A. G., "Tensile and Shear Properties of Laminated Ceramic Matrix Composites," *J. Am. Ceram. Soc.*, Vol. 69, 481–486, 1986.

Schaeffer, H. A., "Thermal and Chemical Strengthening of Glass - Review and Outlook," pp. 469–483 in Kurkjian, C. R., *Strength of Inorganic Glass*, NATO Conference Series, VI, Materials Science, Vol. 11, Plenum Press, New York, 1983.

Scherer, G. W., "Editorial Comments on a Paper by Gordon S. Fulcher," *J. Am. Ceram. Soc.*, Vol. 75, 1060–1062, 1992.

Schneider, S. J., "Advanced Ceramics—What's in a Name," *ASTM Standardization News*, pp. 28–30, 1989.

Schneider, S. J., volume chairman, *Engineered Materials Handbook, Vol. 4, Ceramics and Glasses*, ASM International, Materials Park, OH, 1991.

Schonhorn, H., Kurkjian, C. R., Jaeger, R. E., Vazirani, H. N., Albarino, R. V., and DiMarcello, F. V., "Epoxy-Acrylate-Coated Fused Silica Fibers with Tensile Strengths 500 ksi (3.5 GN/m^2) in 1-km Gauge Lengths," *Appl. Phys. Lett.*, Vol. 29, 712–714, 1976.

Scott, W. D., and Orr, K. K., "Rhombohedral Twinning in Alumina," *J. Am. Ceram. Soc.*, Vol. 66, 27–32, 1983.

Service, T. H., Ritter, J. E. Jr., Jakus, K., and Sonderman, D., "Bimodal Strength Populations," *J. Am. Ceram. Soc.*, Vol. 64, 1276–1280, 1985.

Shaffer, P. T. B., "Engineering Properties of Carbides," pp. 804–811 in Schneider, D. J., volume chairman, *Engineered Materials Handbook, Vol. 4, Ceramics and Glasses*, ASM International, Materials Park, OH, 1991.

Shah, S. P., and Ouyang, C. O., "Mechanical Behavior of Fiber-Reinforced Cement-Based Composites," *J. Am. Ceram. Soc.*, Vol. 74, 2727–2738 and 2947–2953, 1991.

Shalek, P. D., Petrovic, J. J., Hurley, G. F., and Gac, F. D., "Hot-Pressed SiC Whisker/Si_3N_4 Matrix Composites," *Am. Ceram. Soc. Bull.*, Vol. 65, 351–356, 1986.

Shetty, D. K., Rosenfield, A. R., Duckworth, W. H., and Held, P. R., "A Biaxial-Flexure Test for Evaluating Ceramaic Strengths," *J. Am. Ceram. Soc.*, Vol. 66, 36–42, 1983.

Shetty, D. K., and Wang, J.-S., "Crack Stability and Strength Distribution of Ceramics that Exhibit Rising Crack-Growth-Resistance (R-Curve) Behavior," *J. Am. Ceram. Soc.*, Vol. 72, 1158–1162, 1989.

Sigl, L. S., and Evans, A. G., "Effects of Residual Stress and Frictional Sliding on Cracking and Pull-Out in Brittle Matrix Composites," *Mech. Mater.*, Vol. 8, 1–9, 1989.

Simmons, J. H., and Simmons, C. J., "Nonlinear Viscous Flow in Glass Forming," *Ceram. Bull.*, Vol. 68, 1949–1955, 1989.

Sines, G., *Elasticity and Strength*, Allyn & Bacon, Boston, 1969.

Singpurwalla, N. D., Wilson, S. P., and Fuller, E. R. Jr., "Statistical Aspects of Failure Processes in Ceramics," in Bernardeo, J. M., Berger, J. O., David, A. P., and Smith, A. F. M., editors, *Bayesian Statistics*, Oxford University Press, New York, 1995.

Smith, S. M., and Scattergood, R. O., "Crack-Shape Effects for Indentation Fracture Toughness Measurements," *J. Am. Ceram. Soc.*, Vol. 75, 305–315, 1992.

Sneddon, I. N., and Lowengrub, M., *Crack Problems in the Classical Theory of Elasticity*, Wiley, New York, 1969.

Snow, J. D., and Heuer, A. H., "Slip Systems in Al_2O_3," *J. Am. Ceram. Soc.*, Vol. 56, 153–157, 1973.

Somiya, S., Yamamoto, N., Yanagida, H., *Science and Technology of Zirconia III*, Advances in Ceramics Vol. 24B, American Ceramic Society, Westerville, OH, 1986.

Sonderman, D., Jakus, K., Ritter, J. E. Jr., Yuhaski, S. Jr., and Service, T. H., "Maximum Likelihood Estimation Techniques for Concurrent Flaw Subpopulations," *J. Mater. Sci.*, Vol. 20, 207–212, 1985.

Sprackling, M. T., *The Plastic Deformation of Simple Ionic Crystals*, Academic Press, London, 1976.

Spriggs, R. M., "Expression for Effect of Porosity on Elastic Modulus of Polycrystalline Refrctory Materials, Particularly Aluminum Oxide," *J. Am. Ceram. Soc.*, Vol. 45, 94, 1961.

Spriggs, R. M., and Brissette, L. A., "Expressions for Shear Modulus and Poisson's Ratio of Porous Refractory Oxides," *J. Am. Ceram. Soc.*, Vol. 45, 198, 1962.

Spriggs, R. M., Brissette, L. A., and Vasilos, T., "Effect of Porosity on Elastic and Shear Moduli of Polycrystalline Magnesium Oxide," *J. Am. Ceram. Soc.*, Vol. 45, 400, 1962.

Srawley, J. E., and Brown, W. F. Jr., pp. 133–198 in *Fracture Toughness Testing Methods*, ASTM Special Technical Publication 381, American Society for Testing and Materials, Philadelphia, 1965.

Sridhar, N., Yang, W., Srolovitz, D. J., and Fuller, E. R. Jr., "Microstructural Mechanics Model of Anisotropic-Thermal-Expansion-Induced Microcracking," *J. Am. Ceram. Soc.*, Vol. 77, 1123–1138, 1994.

Srinivasan, M., "The Silicon Carbide Family of Structural Ceramics," pp. 99–159 in Wachtman, J. B. Jr., editor, *Structural Ceramics*, Vol. 29 of *Treatise on Materials Science and Technology*, edited by H. Herman, Academic Press, Orlando, 1989.

Srinivasan, M., and Seshadri, S. G., "Application of Single Edge Notched Beam and Indentation Techniques to Determine Fracture Toughness of Alpha Silicon Carbide," pp. 46–68 in Freiman, S. W., and Fuller, E. R., editors, *Fracture Mechanics Methods for Ceramics, Rocks, and Concrete*, ASTM Special Technical Publication 745, American Society for Testing and Materials, Philadelphia, 1981.

Stein, D. F., and Low, J. R. Jr., "Mobility of Edge Dislocations in Silicon-Iron Crystals," *J. Appl. Phys.*, Vol. 31, 362–369, 1960.

Steinbrech, R. W., "Toughening Mechanisms for Ceramic Materials," *J. Eur. Ceram. Soc.*, Vol. 10, 131–142, 1992.

Steinbrech, R. W., and Schmenkel, O., "Crack-Resistance Curves of Surface Cracks in Alumina," *J. Am. Ceram. Soc.*, Vol. 71, C271–C272, 1988.

Stevens, R., "Engineering Properties of Zirconia," pp. 775–786 in Schneider, S. J., volume chairman, *Engineered Materials Handbook, Vol. 4, Ceramics and Glasses*, ASM International, Materials Park, OH, 1991.

Stokes, R. J., "Dislocation Sources and the Strength of Magnesium Oxide Single Crystals," *Trans. Metall. Soc. AIME*, Vol. 222, 1227–1237, 1962.

Stokes, R. J., "Thermal-Mechanical History and the Strength of Magnesium Oxide Single Crystals: I, Mechanical Tests," *J. Am. Ceram. Soc.*, Vol. 48, 60–67, 1965.

Stokes, R. J., "Thermal-Mechanical History and the Strength of Magnesium Oxide Single Crystals: II, Etch Pit and Electron Transmission Studies," *J. Am. Ceram. Soc.*, Vol. 49, 39–46, 1966a.

Stokes, R. J., "The Role of Defects on the Mechanical Properties of Nonfissionable Ceramics," p. 3 in *Nuclear Applications of Non-Fissionable Ceramics*, American Nuclear Society, Hinsdale, Ill., 1966b.

Stokes, R. J., "Mechanical Behavior of Polycrystalline Ceramics," p. 379 in Fulrath, J. M. and Pask, J. A., editors, *Ceramic Microstructures*, Wiley, New York, 1968.

Stokes, R. J., "Microscopic Aspects of Fracture in Ceramics," p. 157 in *Fracture*, Vol. VII, Academic Press, New York, 1972.

Stokes, R. J., Johnston, T. L., and Li, C. H., "Environmental Effects of the Mechanical Properties of Ionic Solids with Particular Reference to the Joffe Effect," *Trans. Metall. Soc. AIME*, Vol. 218, 655–662, 1960.

Stokes, R. J., and Li, C. H., "The Sodium Chloride–Silver Chloride Alloy System," *Acta Metall.*, Vol. 10, 535–542, 1962.

Stokes, R. J., and Li, C. H., "The Anisotropic Extension of Microcracks by Plastic Flow in Semi-Brittle Solids," p. 289 in Drucker, D. C., and Gilman, J. J., editors, *Fracture of Solids*, Wiley-Interscience, New York, 1963.

Stokes, R. J., and Li, C. H., "Dislocation Configurations and the Initiation of Yielding in Magnesium Oxide," *Discuss. Faraday Soc.*, No. 38, 233–242, Butterworth, London, 1964.

Stump, D. M., "Toughening and Strengthening of Ceramics Reinforced by Dilatant Transformations and Ductile Particles," *Int. J. Solids*, Vol. 28, 669–689, 1991.

Suresh, S., *Fatigue of Materials*, Cambridge University Press, New York, 1991.

Suresh, S., Shih, C. F., Morrone, A., and O'Dowd, N. P., "Mixed Mode Fracture Toughness of Ceramic Materials," *J. Am. Ceram. Soc.*, Vol. 73, 1257–1267, 1990.

Suresh, S., and Tschegg, E. K., "Combined Mode I–Mode III Fracture of Fatigue-Precracked Alumina," *J. Am. Ceram. Soc.*, Vol. 70, 726–733, 1987.

Swab, J. J., and Quinn, G. D., "Fractography of Advanced Structural Ceramics: Results from Topic #2 of the VAMAS Round Robin Exercise," *Ceram. Eng. Sci. Proc.*, Vol. 16, 929–938, 1995.

Swain, M. V., "R-Curve Behavior and Thermal Shock Resistance of Ceramics," *J. Am. Ceram. Soc.*, Vol. 73, 621–28, 1990.

Swain, M. V., and Hannink, R. H. J., "R-Curve Behavior in Zirconia Ceramics," pp. 225–239 in Claussen, N., Ruhle, M., and Heuer, A., editors, *Science and Technology of Zirconia II*, Advances in Ceramics Vol. 12, American Ceramic Society, Westerville, OH, 1983.

Swain, M. V., Hannink, R. H. J., and Garvie, R. C., "The Influence of Precipitate Size and Temperature on the Fracture Toughness of Calcia– and Magnesia–Partially Stabilized Zirconia," pp. 339–354 in Bradt, R. C., Evans, A. G., Hasselman, D. P. H., and Lange, F. F., editors, *Fracture Mechanics of Ceramics, Volume 6 Concepts, Measurements, Transformations, and High-Temperature Fracture*, Plenum Press, New York, 1983.

Swanson, P. L., Fairbanks, C., J., Lawn, B. R., Mai, Y.-W., and Hockey B. J., "Crack-Interface Grain Bridging as a Fracture Resistance Mechanism in Ceramics: I, Experimental Study on Alumina," *J. Am. Ceram. Soc.*, Vol. 70, 279–289, 1987.

Tada, H., Paris, P., and Irwin, G., *The Stress Analysis Handbook*, Del Research Corporation, St. Louis, MO, 1985.

Tandon, R., Green, D. J., and Cook, R. F., "Strength Variability in Brittle Materials with Stabilizing and Destabilizing Resistance Fields," *Acta Metall. Mater.*, Vol. 41, 399–408, 1993.

Tennery, V. J., editor, *Third International Symposium on Ceramic Materials and Components for Engines*, American Ceramic Society, Westerville, OH, 1989.

Thiemeier, T., and Bruckner-Foit, A., Influence of the Fracture Criterion on the Failure Prediction of Ceramics Loaded in Biaxial Flexure," *J. Am. Ceram. Soc.*, Vol. 74, 48–52, 1991.

Thoman, D. R., Bain, L. J., and Antle, C. E., "Inferences on the Parameters of the Weibull Distribution," *Technometrics*, Vol. 11, 445–460, 1969.

Thouless, M. D., and Evans, A. G., "Effects of Pull-Out on the Mechanical Properties of Ceramic-Matrix Composites," *Acta Metall.*, Vol. 36, 517–522, 1988.

Thouless, M. D., Sbaizero, O., Sigl, L. S., and Evans, A. G., "Effect of Interface Mechanical Properties in a SiC-Fiber Reinforced Lithium Aluminum Silicate Glass-Ceramic," *J. Am. Ceram. Soc.*, Vol. 72, 525–532, 1989

Timoshenko, S., and Goodier, J. N., *Theory of Elasticity*, 2nd edition, McGraw-Hill, New York, 1951.

Tomozawa, M., "Fracture of Glasses", Annual Reviews of Materials Science", Vol. 26, in press.

Tressler, R. E., and Barber, D. J., "Yielding and Flow of c-Axis Sapphire Filaments," *J. Am. Ceram. Soc.*, Vol. 57, 13–19, 1974.

Tsai, R. L., and Raj, R., "Creep Fracture in Ceramics Containing Small Amounts of a Liquid Phase," *Acta Metall.*, Vol. 30, 1043–1058, 1982.

Tsukuma, K., Kubota, Y., and Tsukidate, T., "Thermal and Mechanical Properties of Y_2O_3-Stabilized Tetragonal Zirconia Polycrystals," pp. 382–390 in Claussen, N., Ruhle, M., and Heuer, A., editors, *Science and Technology of Zirconia II*, Advances in Ceramics Vol. 12, American Ceramic Society, Westerville, OH, 1983.

Tsukuma, K., and Ueda, K., "Strength and Fracture Toughness of Isostatically Hot-Pressed Composites of Al_2O_3 and Y_2O_3–Partially-Stabilized ZrO_2," *J. Am. Ceram. Soc.*, Vol. 68, C-4–C-5, 1985.

Tsukuma, K., Ueda, K., Matsushita, K., and Shimada, M., "High-Temperature Strength and Fracture Toughness of Y_2O_3–Partially Stabilized ZrO_2/Al_2O_3 Composites," *J. Am. Ceram. Soc.*, Vol. 68, C-56, 1985.

Tucker, W. T., and Johnson, C. A., "The Multiaxial Equivalent of Stressed Volume," pp. 265–279 in Brinkman, C. R., and Duffy, S. F., editors, *Life Prediction Methodologies and Data for Ceramic Materials*, ASTM STP 1201, American Society for Testing and Materials, Philadelphia, 1993.

Varner, J. R., "Descriptive Fractography," pp. 635–644 in Schneider, S. J., volume chairman, *Engineered Materials Handbook, Vol. 4, Ceramics and Glasses*, ASM International, Materials Park, OH, 1991.

Varshneya, A. K., *Fundamentals of Inorganic Glass*, Academic Press, San Diego, CA, 1994.

Veknis, G., Ashby, M. F., and Beaumont, P. W. R., "R-Curve Behavior of Al_2O_3 Ceramics," *Acta Metall. Mater.*, Vol. 38, 1151–1162, 1990.

Virher, A. V., and Matsumoto, R. L. K., "Ferroelastic Domain Switching as a Toughening Mechanism in Tetragonal Zirconia," *J. Am. Ceram. Soc.*, Vol. 69, C224, 1986.

Wachtman, J. B. Jr., "Elastic Deformation of Ceramics and Other Refractory Materials," pp. 139–168 in Wachtman, J. B. Jr., editor, *Mechanical and Thermal Properties of Ceramics*, NBS Special Publication 303, U.S. Government Printing Office, Washington, DC, 1969.

Wachtman, J. B. Jr., "Highlights of Progress in the Science of Fracture of Ceramics and Glass," *J. Am. Ceram. Soc.*, Vol. 57, 509–519, 1974.

Wachtman, J. B. Jr., editor, *Structural Ceramics*, Vol. 29 of *Treatise on Materials Science and Technology*, edited by H. Herman, Academic Press, Orlando, 1989.

Wachtman, J. B. Jr., Capps, W., and Mandel, J., "Biaxial Flexure Tests of Ceramic Substrates," *J. Mater.*, Vol. 7, 188–194, 1972.

Wachtman, J. B. Jr., and Lam, D. G. Jr. "Young's Modulus of Various Refractory Materials as a Function of Temperature," *J. Am. Ceram. Soc.*, Vol. 42, 254–260, 1959.

Wachtman, J. B. Jr., and Maxwell, L. H., "Plastic Deformation of Ceramic-Oxide Single Crystals," *J. Am. Ceram. Soc.*, Vol. 37, 291–299, 1954.

Wachtman, J. B. Jr., and Maxwell, L. H., "Plastic Deformation of Ceramic-Oxide Single Crystals, II," *J. Am. Ceram. Soc.*, Vol. 40, 377–385, 1957.

Wachtman, J. B. Jr., and Maxwell, L. H., "Strength of Synthetic Single Crystal Sapphire and Ruby as a Function of Temperature and Orientation," *J. Am. Ceram. Soc.*, Vol. 42, 432–433, 1959.

Wachtman, J. B. Jr., Tefft, W. E., Lam, D. G. Jr., and Apstein, C. S., "Elastic Constants of Synthetic Single Crystal Corundum at Room Temperature," *J. Res. Natl. Bur. Stand.*, Vol. 64A, 213–228, 1960.

Wachtman, J. B. Jr., Tefft, W. E., Lam, D. G. Jr., and Apstein, C. S., "Exponential Temperature Dependence of Young's Modulus for Several Oxides," *Phys. Rev.*, Vol. 122, 1754, 1961.

Walsh, J. B., "Effect of Cracks on the Compressibility of Rock," *J. Geophys. Res.*, Vol. 70, 382–389, 1965.

Weertman, J., "Steady-State Creep of Crystals," *J. Appl. Phys.*, Vol. 28, 1185–1189, 1957.

Weertman, J., and Weertman, J. R., "Mechanical Properties, Strongly Temperature Dependent," pp. 1039–1340 in Kahn, R. W., editor, *Physical Metallurgy*, North-Holland, Third Edition Amsterdam, 1983.

Weertman, J., and Weertman, J. R., *Elementary Dislocation Theory*, Oxford University Press, New York, 1992.

Weibull, W., "A Statistical Theory of the Strength of Materials," *Ingenioersvetenskapsakad. Handl.*, Vol. 151, 1–45, 1939.

Weiss, J., "Silicon Nitride Ceramics: Composition, Fabrication Parameters, and Properties," *Annu. Rev. Mater. Sci.*, Vol. 11, 381–389, 1981.

Westergaard, H. M., "Bearing Pressures and Cracks," *J. Appl. Mech.*, Vol. 6, A49–53, 1939.

White, K. W., and Guazzone, "Elevated-Temperature Toughening Mechanisms in a SiC_w/Al_2O_3 Composite," *J. Am. Ceram. Soc.*, Vol. 74, 2280–2285, 1991.

Wiederhorn, S. M., "Influence of Water Vapor on Crack Propagation in Soda-Lime Glass," *J. Am. Ceram. Soc.*, Vol. 50, 407–414, 1967.

Wiederhorn, S. M., "Subcritical Crack Growth in Ceramics," pp. 613–646 in Bradt, R. C., Hasselman, D. P. H. and Lange, F. F., editors, *Fracture Mechanics of Ceramics*, Plenum Press, New York, 1974.

Wiederhorn, S. M., "Brittle Fracture and Toughening Mechanisms in Ceramics," *Annu. Rev. Mater. Sci.*, Vol. 14, 374–403, 1984.

Wiederhorn, S. M. and Bolz, L. H., "Stress Corrosion and Static Fatigue of Glass," *J. Am. Ceram. Soc.*, Vol. 53, 543–548, 1970.

Wiederhorn, S. M., and Fuller, E. R. Jr., "Structural Reliability of Ceramic Materials," *Mater. Sci. Eng.*, Vol. 71, 169–186, 1985.

Wiederhorn, S. M., and Hockey, B. J., "High Temperature Degradation of Structural Composites," pp. 365–380 in Vincenzini, P., editor, *Advanced Structural Ceramics*, Elsevier Science Publishers, New York, 1991.

Wiederhorn, S. M., Hockey, B. J., and Chuang, T.-J., "Creep and Creep Rupture of Structural Ceramics," pp. 555–575 in Shah, S. P., editor, *Toughening Mechanisms in Quasi-Brittle Materials*, Kluwer Academic Publishers, Norwell, MA, 1991.

Wiederhorn, S. M., and Johnson, H., "Effect of Electrolyte pH on Crack Propagation in Glass," *J. Am. Ceram. Soc.*, Vol. 56, 192–197, 1973.

Wiederhorn, S. M., Johnson, H., Diness, A. M., and Heuer, A. H., "Fracture of Glass in Vacuum," *J. Am. Ceram. Soc.*, Vol. 57, 8, 1974.

Wiederhorn, S. M., Quinn, G. D., and Krause, R., "Fracture Mechanism Maps: Their Applicability to Silicon Carbide," pp. 36–61 in Brinkman, C. R., and Duffy, S. f., editors, *Life Prediction Methodologies and Data for Ceramic Materials*, ASTM STP 1201, American Society for Testing and Materials, Philadelphia, 1994.

Wiederhorn, S. M., Roberts, D. E., Chuang, T.-J., and Chuck, L., "Damage-Enhanced Creep in a Siliconized Silicon Carbide: Phenomenology," *J. Am. Ceram. Soc.*, Vol. 71, 602–608, 1988.

Wilkinson, D. S., "Creep Damage Mechanisms in Structural Ceramics," pp. 349–366 in Bradt, R. C., Hasselman, D. P. H., Munz, D., Sakai, M., and Shevchenko, V. Y., editors, *Fracture Mechanics of Ceramics, Volume 10 Fracture Fundamentals, High-Temperature Deformation, Damage, and Design*, Plenum Press, New York, 1992.

Woodilla, D., Buonomo, M., Bar-On, I., Katz, R. N., and Whalen, T., "Elevated-Temperature Behavior of High-Strength Silicon Carbide," *J. Am. Ceram. Soc.*, Vol. 76, 249–252, 1993.

Xu, A., and Solomon, A. A., "Diffusional Creep and Cavitational Strains in High-Purity Alumina Under Tension and Subsequent Hydrostatic Compression," *J. Am. Ceram. Soc.*, Vol. 75, 985–995, 1992.

Yaruz, B. D., and Tressler, R. E., "Threshold Stress Intensity for Crack Growth in Silicon Carbide Ceramics," *J. Am. Ceram. Soc.*, Vol. 76, 1017–1024, 1993.

Yajima, S., Hayashi, J., Omori, M., and Okamura, K., "Development of a Silicon Carbon Fibre with High Tensile Strength," *Nature (London)*, Vol. 261, 683–685, 1976.

Yang, S. F., and Knowles, K. M., "The One-Dimensional Car Parking Problem and Its Application to the Distribution of Spacings Between Matrix Cracks in Unidirectional Fiber-Reinforced Brittle Materials," *J. Am. Ceram. Soc.*, Vol. 75, 141–147, 1992.

Young, W. C., *Roark's Formulas for Stress and Strain*, 6th ed., McGraw-Hill, New York, 1989.

Zeng, J., Tanaka, I., Miyamoto, Y., Yamada, O. and Niihara, K., "High-Temperature Strength and Cavitation Threshold of Silicon Nitride-Silica Ceramics," *J. Am. Ceram. Soc.*, Vol. 75, 195–200, 1992.

Zok, F. W., Evans, A. G., and Mackin, T. J., "The Mechanical Properties of Fiber-Reinforced Ceramic-Matrix Composites," in *Handbook on Continuous Fiber Reinforced Ceramic Matrix Composites*, R. L. Lehman, technical editor, S. K. El-Rahaiby, managing editor, and J. B. Wachtman, American Ceramic Society editor, jointly published by Purdue University and the American Ceramic Society, 1996.

INDEX